Study Guide for Basic Fluid Mechanics

Second Edition

by

Christopher P. Landry

and

David C. Wilcox

Study Guide for Basic Fluid Mechanics
Second edition copyright © 1999 by DCW Industries, Inc. All rights reserved.

First Printing: May, 1999

First edition copyright © 1998 by DCW Industries, Inc.

No part of this book may be reproduced or transmitted in any form or by any means, electronic or mechanical, including photocopying, recording, or any information storage and retrieval system, without permission in writing from DCW Industries, Inc.

> DCW Industries, Inc.
> 5354 Palm Drive, La Cañada, California 91011
> 818/790-3844 (FAX) 818/952-1272
> e-mail: dcwilcox@ix.netcom.com
> World Wide Web: http://www.dcwindustries.com

This book was prepared with LaTeX as implemented by Personal TeX, Inc. of Mill Valley, California. It was printed and bound in the United States of America by KNI, Inc., Anaheim, California.

Library of Congress Cataloging in Publication Data

Landry, Christopher P. and Wilcox, David C.
   Study Guide for Basic Fluid Mechanics/C. P. Landry and D. C. Wilcox—2nd ed.
   1. Fluid Mechanics.
   2. Computational Fluid Dynamics.
Catalog Card Number 99-72858

ISBN   0-9636051-9-4

Dedicated to our Moms
>LILA and HELEN

and our Dads
>PAUL and ROBERT

About the Authors

Christopher P. Landry was born in Mineola, New York. He received his Bachelor of Science degree in 1992 and Master of Science degree in 1996, both in Aerospace Engineering from the University of Southern California. From 1993 to 1998, he worked for the Northrop Grumman Corporation in Palmdale, California. In 1998 he began and is currently working at the Jet Propulsion Laboratory in Pasadena, California. He has served as a consultant and associate editor for DCW Industries, Inc. since 1994. He is a member of Phi Kappa Phi, Tau Beta Pi and Sigma Gamma Tau honor societies.

Dr. David C. Wilcox, was born in Wilmington, Delaware. He did his undergraduate studies from 1963 to 1966 at the Massachusetts Institute of Technology, graduating with a Bachelor of Science degree in Aeronautics and Astronautics. From 1967 to 1970, he attended the California Institute of Technology, graduating with a Ph.D. in Aeronautics. Between 1966 and 1973 he worked for several Southern California aerospace companies. In 1973, he founded DCW Industries, Inc., a La Cañada, California firm engaged in engineering research, software development and publishing, for which he is currently the President. He has taught several fluid mechanics and applied mathematics courses at the University of Southern California and at the University of California, Los Angeles. His publications include an undergraduate-level text *Basic Fluid Mechanics* and two graduate-level texts, *Turbulence Modeling for CFD* and *Perturbation Methods in the Computer Age*. He is an Associate Fellow of the American Institute of Aeronautics and Astronautics (AIAA) and has served as an Associate Editor for the AIAA Journal.

Contents

Preface ix

1 Introduction 1
    1.1   The Continuum Limit 2
    1.2   The Perfect-Gas Law 2
    1.3   Surface Tension 2
    1.4   Couette Flow in Steady-State Motion I 3
    1.5   Couette Flow in Steady-State Motion II 4
    1.6   Couette Flow in Transient Motion 5
    1.7   Pipe Flow 7

2 Dimensional Analysis 9
    2.1   Dimensionless Groupings 10
    2.2   Dimensional Homogeneity 10
    2.3   Indicial Method 11
    2.4   E. S. Taylor's Method 12
    2.5   Dynamic Similitude I 14
    2.6   Dynamic Similitude II 15

3 Effects of Gravity on Pressure 17
    3.1   Hydrostatic Relation 18
    3.2   U. S. Standard Atmosphere 19
    3.3   Manometer 20
    3.4   Vertical Gate 21
    3.5   Tilted Gate—Tilted Coordinate Frame 22
    3.6   Tilted Gate—Normal Coordinate Frame 24
    3.7   Curved Gate 28
    3.8   Superposition 30
    3.9   Buoyancy in a Uniform Fluid 32
    3.10   Buoyancy in a Dual-Fluid Tank 34

4 Kinematics 35
    4.1   Eulerian Description 36
    4.2   Lagrangian Description 37
    4.3   Vorticity—Rotational Flow 38
    4.4   Vorticity—Irrotational Flow 39
    4.5   Circulation 40
    4.6   Streamlines—Cartesian Coordinates 42

	4.7	Streamlines—Cylindrical Coordinates	43
	4.8	Surface Fluxes	44
	4.9	Reynolds Transport Theorem I	45
	4.10	Reynolds Transport Theorem II	46

5 Conservation of Mass and Momentum 49

5.1	Irrotationality and Incompressibility I	50
5.2	Irrotationality and Incompressibility II	51
5.3	Irrotationality and Incompressibility III	52
5.4	Continuity Equation	52
5.5	Bernoulli's Equation	53
5.6	Bernoulli's Equation—Quasi-Steady Flow	54
5.7	Euler's Equation	55
5.8	Euler's Equation—Rotating Tank	58
5.9	Pitot-Static Tube	59
5.10	Galilean Invariance	60

6 Control Volume Method 63

6.1	Basic Control Volume Concepts	65
6.2	Mass—Stationary Control Volume I	66
6.3	Mass—Stationary Control Volume II	68
6.4	Mass—Translating Control Volume	69
6.5	Mass—Deforming Control Volume I	71
6.6	Mass—Deforming Control Volume II	72
6.7	Mass—Nonuniform Profiles	74
6.8	Mass and Momentum—Stationary Control Volume I	75
6.9	Mass and Momentum—Stationary Control Volume II	77
6.10	Mass and Momentum—Translating Control Volume	80
6.11	Mass and Momentum—Rotational Flow	82
6.12	Mass and Momentum—Irrotational Flow	85
6.13	Mass and Momentum—Nonuniform Profiles	88
6.14	Mass and Momentum—Accelerating Control Volume	90
6.15	Mass and Momentum—Three-Dimensional Flow	93

7 Conservation of Energy 97

7.1	Thermodynamics	98
7.2	Energy—Integral Form	100
7.3	Energy—Differential Form	102
7.4	Moody Diagram	104
7.5	Flow Through a Turbine	105
7.6	Minor Losses I	106
7.7	Minor Losses II	108
7.8	Pipe System	109
7.9	Free-Surface Waves	111
7.10	Open-Channel Flow I	111
7.11	Open-Channel Flow II	113
7.12	Hydraulic Jump	115

8 One-Dimensional Compressible Flow — 117
- 8.1 Speed of Sound — 118
- 8.2 Isentropic Flow I — 119
- 8.3 Isentropic Flow II — 119
- 8.4 Normal Shock I — 121
- 8.5 Normal Shock II — 122
- 8.6 Laval Nozzle I — 123
- 8.7 Laval Nozzle II — 124

9 Turbomachinery — 127
- 9.1 Pump Analysis — 128
- 9.2 Performance Curves — 129
- 9.3 Power — 131
- 9.4 Turbine Analysis — 133
- 9.5 Pelton Wheel — 134

10 Vorticity and Viscosity — 137
- 10.1 The Vorticity Equation — 138
- 10.2 Circulation — 139
- 10.3 Vorticity in a Boundary Layer — 140
- 10.4 Drag — 142
- 10.5 Lift — 143

11 Potential Flow — 145
- 11.1 Basic Concepts—Cartesian Coordinates — 146
- 11.2 Basic Concepts—Cylindrical Coordinates — 147
- 11.3 Computing ψ from a Given ϕ — 148
- 11.4 Computing ϕ from a Given ψ — 149
- 11.5 Bernoulli's Equation — 150
- 11.6 Superposition of Fundamental Solutions I — 151
- 11.7 Superposition of Fundamental Solutions II — 152
- 11.8 Flow Past a Rotating Cylinder I — 153
- 11.9 Flow Past a Rotating Cylinder II — 155
- 11.10 Unsteady Potential Flow — 156
- 11.11 Linearized Airfoil Theory — 156
- 11.12 CFD—Richardson Extrapolation — 159

12 Viscous Effects — 161
- 12.1 Kinematics — 162
- 12.2 Strain-Rate Tensor — 163
- 12.3 Viscous-Stress Tensor — 164
- 12.4 Navier-Stokes Equation — 166
- 12.5 CFD—Stability Analysis — 168

13 Navier-Stokes Solutions — 169
- 13.1 Couette Flow — 170
- 13.2 Rotating Channel Flow — 171
- 13.3 Trailing-Vortex Flow — 173
- 13.4 Stokes' Flow with Suction — 174
- 13.5 CFD—Thomas' Algorithm — 175

14 Boundary Layers — 179
- 14.1 Boundary-Layer Equations I 180
- 14.2 Boundary-Layer Equations II 181
- 14.3 Momentum Integral Equation I 182
- 14.4 Momentum Integral Equation II 184
- 14.5 Blasius Solution 186
- 14.6 Transitional Boundary Layer 187
- 14.7 Turbulent Boundary Layers 189
- 14.8 Law of the Wall I 190
- 14.9 Law of the Wall II 191
- 14.10 CFD—Truncation Error 195

15 Viscous and 2-D Compressible Flow — 197
- 15.1 Fanno Flow I 198
- 15.2 Fanno Flow II 199
- 15.3 Rayleigh Flow I 202
- 15.4 Rayleigh Flow II 203
- 15.5 Oblique Shock Wave 205
- 15.6 Reflection of an Oblique Shock Wave 207
- 15.7 Prandtl-Meyer Expansion I 209
- 15.8 Prandtl-Meyer Expansion II 210
- 15.9 Compressible Law of the Wall 211
- 15.10 CFD—Numerical Dissipation and Dispersion 213

A Fluid Properties — 215
- A.1 Perfect-Gas Properties 215
- A.2 Pressure 215
- A.3 Density 216
- A.4 Compressibility and Speed of Sound 216
- A.5 Viscosity 217
- A.6 Surface Tension 218

B Hydraulics Properties — 219
- B.1 Fluid Statics 219
- B.2 Open-Channel Flow 220
- B.3 Pipe Flow 222

C Compressible-Flow Properties — 227
- C.1 Isentropic Flow and Normal-Shock Relations 227
- C.2 Prandtl-Meyer Function and Mach Angle 235
- C.3 Fanno Flow 237
- C.4 Rayleigh Flow 240

D Drag Data — 243

E Conversion Factors — 245

Preface

Study Guide for Basic Fluid Mechanics is a collection of solved problems in fluid mechanics. The types of problems considered are those normally presented in introductory and intermediate fluids courses. This study guide is self contained with tables and graphs of important fluid flow properties in its many appendices. While it can be used in conjunction with most undergraduate fluid mechanics texts, it is especially effective when used as a companion text for the introductory fluid mechanics book entitled *Basic Fluid Mechanics* by David C. Wilcox. While *Basic Fluid Mechanics* includes examples for all key concepts, the number of worked problems has been held to a minimum to eliminate disruption of a continuous train of thought in presenting basic concepts. Preserving the focus in this manner is a key problem every author must address. This study guide contains 129 examples worked in complete detail to help students benefit from step-by-step solutions, while maintaining the integrity and readability of *Basic Fluid Mechanics*.

Study Guide for Basic Fluid Mechanics is an illustrative-example book, while a text such as *Basic Fluid Mechanics* focuses on the theoretical development and explanation of fluid-mechanical principles. The study guide explains how to solve typical problems in fluid mechanics based on theoretical concepts. It presents complete and explicit solutions to common types of fluid-flow problems. Each chapter begins with an overview of the types of problems included and a discussion of some of the most important details of general solution methodology. Often, there is a recurrent theme running through all of the solved problems, and the introductory comments identify such themes. On the one hand, as with a teacher's lecture notes, the study guide amplifies the most important concepts. On the other hand, unlike a teacher's lecture notes, theoretical explanations are not covered here. Consequently, maximum value will be gained by first studying the material as presented in class and in the main text, and then examining the solved problems in the study guide.

Another significant problem faced by an author in preparing a textbook for use in a classroom setting is deciding on the level of complexity of the examples embodied in the narrative. Usually, to permit complete solution of a problem in a typical class lecture, only the simplest geometries are considered. *Basic Fluid Mechanics*, for example, even "rigs" answers for many problems to be rational numbers, with an eye on emphasizing concepts rather than stressing operations on a hand-held calculator. While convenient for the classroom, real engineering problems usually involve much more complex examples than standard textbook problems. Even though *Basic Fluid Mechanics* has numerous complicated examples in the problems sections, none are explicitly worked start to finish with comments indicating why certain steps are taken. This study guide has been created to fill this void.

We are confident that conscientious use of *Basic Fluid Mechanics* in conjunction with *Study Guide for Basic Fluid Mechanics* will give the reader a solid foundation in the theoretical foundations of fluid mechanics from the text and a mastery of the consequent methodology used to solve typical fluid-flow problems from the study guide. Combined use of these two books also benefits the instructor by providing detailed examples that can be studied independently by students outside of the classroom, thus permitting more time for in-class discussion of basic concepts.

Comments About the Second Edition

In creating this Study Guide, we chose to focus most of the emphasis on the first eleven chapters of *Basic Fluid Mechanics*. This part of the book is normally used for a first course in fluid mechanics, and worked examples are especially helpful for students at this level. By contrast, the final four chapters are normally covered in an intermediate- or a graduate-level course, where the audience consists of more-advanced students.

The response to the *Study Guide* has surpassed our expectations, and we have had considerable feedback already. It has proven to be nearly as popular with students using it for an intermediate fluid mechanics course as for students having their first exposure to the subject. Overwhelmingly, professors and students have expressed a desire for more worked examples in the final four chapters. In response, we have added numerous problems to the final four chapters to satisfy this desire.

Christopher P. Landry
and David C. Wilcox

NOTE: We have taken great pains to assure the accuracy of this manuscript. However, if you find errors, please report them to DCW Industries' Home Page on the Worldwide Web at **http://www.dcwindustries.com**. As long as this book remains in print, we will maintain an updated list of known typographical errors.

Chapter 1

Introduction

Chapter 1 is intended to provide a overview of and introduction to fundamental concepts in fluid mechanics. Since the chapter serves mainly as an introduction to the field of fluid mechanics, including a discussion of mechanical and thermodynamic properties of fluids, the problems involve application of relatively simple equations. Perhaps the most important thing to observe in the solutions discussed in this chapter is the careful handling of units. To help avoid confusion, you should treat units the same as you would variables in an algebraic equation. The chapter includes the following problems.

Section 1.1 The Continuum Limit: A computation of the number of molecules in a small droplet.

Section 1.2 The Perfect-Gas Law: A simple problem using the perfect-gas law.

Section 1.3 Surface Tension: A surface-tension problem with a pressure difference supported by liquid-air interfaces.

Section 1.4 Couette Flow in Steady-State Motion I: The first of two steady-state Couette-flow problems—a simple tension calculation with a thin gap.

Section 1.5 Couette Flow in Steady-State Motion II: The second of two steady-state Couette-flow problems—another simple tension calculation.

Section 1.6 Couette Flow in Transient Motion: A Couette-flow application in which the boundary accelerates. The solution involves setting up the equation of motion for a two-mass system sliding on two oil films with different thicknesses.

Section 1.7 Pipe Flow: A Hagen-Poiseuille pipe-flow problem. Using known formulas, the solution involves determining the fluid viscosity from tables and then computing the maximum velocity and Reynolds number in a circular pipe.

1.1 The Continuum Limit

Statement of the Problem: *The diameter of a droplet ejected in an ink-jet printer can be as small as 0.01 mm. Observing that the U. S. national debt has increased by roughly $1.75 trillion in the 1990s, if the money had been used to purchase an ink-jet droplet, determine the cost of each molecule. You may assume ink contains the same number of molecules as water.*

Solution: We know from the text that the number density of water, n, is

$$n = 3.34 \cdot 10^{19} \text{ molecules/mm}^3 \tag{1.1}$$

The volume, V, of the droplet is

$$V = \frac{1}{6}\pi d^3 = \frac{1}{6}\pi(.01 \text{ mm})^3 = 5.236 \cdot 10^{-7} \text{ mm}^3 \tag{1.2}$$

Therefore, the number of molecules in the droplet, N, is

$$N = nV = (3.34 \cdot 10^{19} \text{ molecules/mm}^3) \cdot (5.236 \cdot 10^{-7} \text{ mm}^3) = 1.75 \cdot 10^{13} \text{ molecules} \tag{1.3}$$

Finally, since 1.75 trillion dollars = $1.75 \cdot 10^{12}$ dollars, each molecule would cost 10 cents.

1.2 The Perfect-Gas Law

Statement of the Problem: *We wish to find the density of steam at the boiling point of water. Assume steam is a perfect gas with a perfect-gas constant of R = 2768 ft·lb/(slug·°R). At the boiling point, the measured pressure is 300 psi.*

Solution: When we use the perfect-gas law, temperature must be expressed in terms of the absolute scale. Since the problem is stated in USCS units, we must use the Rankine scale. The temperature at the boiling point of water is 212° F. Converting to absolute temperature, we find

$$T = 212° \text{ F} + 459.67° \text{ R} = 671.67° \text{ R} \tag{1.4}$$

From the perfect-gas law, the density is given by

$$\rho = \frac{p}{RT} = \frac{\left(300 \frac{\text{lb}}{\text{in}^2}\right)\left(144 \frac{\text{in}^2}{\text{ft}^2}\right)}{\left(2768 \frac{\text{ft}\cdot\text{lb}}{\text{slug}\cdot°\text{R}}\right)(671.67°\text{R})} = 0.023 \frac{\text{slug}}{\text{ft}^3} \tag{1.5}$$

1.3 Surface Tension

Statement of the Problem: *Sitting by the ocean enjoying your summer vacation, you see a jet skier spray water on the table near you. Having a keen interest in fluid mechanics, you wonder how the pressure difference between the interior and exterior of a water drop, Δp_{H_2O}, might compare to the corresponding difference for a drop of mercury from a broken thermometer, Δp_{Hg}. The water drop is 1/3 cm in diameter and the drop of mercury is 1/2 cm in diameter. Assume the drops are spherical and that the air temperature is 20° C. Compare Δp_{H_2O} and Δp_{Hg}.*

Solution: The pressure difference for a liquid drop of general shape is given by the *Landau formula*, viz.,

$$\Delta p = \frac{\sigma}{\mathcal{R}_1} + \frac{\sigma}{\mathcal{R}_2} \qquad (1.6)$$

where $\mathcal{R}_1$ and $\mathcal{R}_2$ are the principal radii of curvature and σ is surface tension. For a sphere of radius R, we know that $\mathcal{R}_1 = \mathcal{R}_2 = R$. Therefore, the pressure difference for a spherical drop is

$$\Delta p = \frac{2\sigma}{R} \qquad (1.7)$$

From Table A.7, the values of surface tension for water and mercury at 20° C are

$$\sigma_{H_2O} = 0.073 \text{ N/m}, \qquad \sigma_{Hg} = 0.466 \text{ N/m} \qquad (1.8)$$

Also, from what is given, the radii of the drops are

$$R_{H_2O} = \frac{1}{2}\left(\frac{1}{3}\text{ cm}\right)\left(\frac{1}{100}\frac{\text{m}}{\text{cm}}\right) = \frac{1}{600}\text{ m}, \qquad R_{Hg} = \frac{1}{2}\left(\frac{1}{2}\text{ cm}\right)\left(\frac{1}{100}\frac{\text{m}}{\text{cm}}\right) = \frac{1}{400}\text{ m} \qquad (1.9)$$

Thus, the pressure differences supported by the surface tension acting on the two drops are

$$\left.\begin{array}{rclclcl}
\Delta p_{H_2O} &=& \dfrac{2\sigma_{H_2O}}{R_{H_2O}} &=& \dfrac{2\,(0.073\text{ N/m})}{1/600\text{ m}} &=& 87.6\,\dfrac{\text{N}}{\text{m}^2} = 0.0876\text{ kPa} \\[2ex]
\Delta p_{Hg} &=& \dfrac{2\sigma_{Hg}}{R_{Hg}} &=& \dfrac{2\,(0.466\text{ N/m})}{1/400\text{ m}} &=& 372.8\,\dfrac{\text{N}}{\text{m}^2} = 0.3728\text{ kPa}
\end{array}\right\} \qquad (1.10)$$

Consequently, the pressure difference between the interior and exterior of the mercury drop is more than four times that of the water drop.

1.4 Couette Flow in Steady-State Motion I

Statement of the Problem: *Insulating paint is applied to a wire by pulling it through a cylindrical orifice of radius R. The wire, of radius r, is centered in the orifice whose length is ℓ. The viscosity of the paint is μ. What force, F, is required to pull the wire at a constant velocity, U? You may assume $(R - r) \ll R$ and $(R - r) \ll \ell$.*

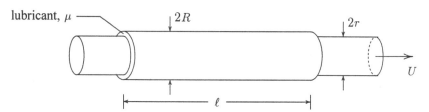

Figure 1.1: *Geometry for wire-insulation process.*

Solution: The forces on the wire are the pulling force, F, and the friction force on the surface of the wire. In steady state, these forces balance so that

$$F = 2\pi\tau_w r\ell \qquad (1.11)$$

where τ_w is the shear stress on the surface of the wire. Since we are given that the gap width, $h = (R - r)$, is sufficiently narrow to have $h \ll R$, the flow looks locally two dimensional. Also, since $(R - r) \ll \ell$, we can use the Couette-flow solution. Thus, the shear stress in the gap (and at the surface of the wire) is

$$\tau_w = \frac{U\mu}{h} = \frac{\mu U}{R - r} \tag{1.12}$$

So, the pulling force is given by

$$F = 2\pi \frac{\mu U r \ell}{R - r} \tag{1.13}$$

which can be rewritten in terms of R/r as

$$F = 2\pi \frac{\mu U \ell}{R/r - 1} \tag{1.14}$$

1.5 Couette Flow in Steady-State Motion II

Statement of the Problem: *You were bored waiting for commercials to scan by while watching your favorite documentary on fluid mechanics. You decided to fix your VCR motor, increasing the scanning speed for a 6-hour tape. In Super Long Play (SLP) mode, 3 minutes of commercials now scan in 15 seconds instead of 50 seconds. Compute the tape speed, U, before and after the alteration. By what factor did you increase the tension in the tape as it passes by the head? You may assume the Couette-flow solution holds in the small gap between the tape and the head. A standard 6-hour tape is 807 feet long.*

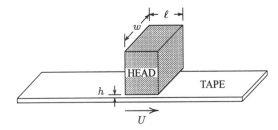

Figure 1.2: *VCR tape and play-back head.*

Solution: Since this is a thin-gap problem, we will be using the Couette-flow solution in which the shear stress on the tape is directly proportional to the tape velocity. Hence, we must determine the tape velocities before and after the modification. We have sufficient information to do this from what is given. That is, we know the total length of the tape, total recording time and the true time of the commercials. From this, we can determine the length of tape needed to record the commercials. Dividing this length by the scanning times before and after the modification yields the required velocities.

First, note that if the 807 foot tape contains 6 hours of video, then 3 minutes of commercials occupy a length L of tape, where

$$L = \frac{3 \text{ min}}{(6 \text{ hr}) \cdot (60 \text{ min/hr})} (807 \text{ ft}) = 6.725 \text{ ft} \tag{1.15}$$

1.6. COUETTE FLOW IN TRANSIENT MOTION

Before the alteration, the commercials scan in $T_{\text{before}} = 50$ sec. The velocity, U_{before}, is given by

$$U_{\text{before}} = \frac{L}{T_{\text{before}}} = \frac{6.725 \text{ ft}}{50 \text{ sec}} = 0.1345 \frac{\text{ft}}{\text{sec}} \quad (1.16)$$

After the alteration, the commercials scan in $T_{\text{after}} = 15$ sec. The velocity, U_{after}, is

$$U_{\text{after}} = \frac{L}{T_{\text{after}}} = \frac{6.725 \text{ ft}}{15 \text{ sec}} = 0.4483 \frac{\text{ft}}{\text{sec}} \quad (1.17)$$

In general, the tension on the tape, T, is the product of the shear stress, τ, and the area of the head, $A = w\ell$, i.e.,

$$T = \tau w \ell \quad (1.18)$$

where w and ℓ are the width and length of the head, respectively. Assuming the Couette-flow solution holds, the velocity varies linearly with distance from the tape, so that $u = Uy/h$. Hence, the shear stress is

$$\tau = \mu \frac{du}{dy} = \frac{\mu U}{h} \quad (1.19)$$

where U is the tape velocity, μ is the viscosity of the air in the gap and h is the gap width. Combining Equations (1.18) and (1.19) yields

$$T = \frac{\mu \ell w}{h} U \quad (1.20)$$

Finally, since μ, ℓ, w and h are unchanged, the ratio of the tension on the tape after the alteration to the value before is

$$\frac{T_{\text{after}}}{T_{\text{before}}} = \frac{U_{\text{after}}}{U_{\text{before}}} = \frac{0.4483 \text{ ft/sec}}{0.1345 \text{ ft/sec}} = 3.33 \quad (1.21)$$

Therefore, the tape tension increases by a factor of 3.33.

1.6 Couette Flow in Transient Motion

Statement of the Problem: *Two cubes are arranged on an incline with a pulley as indicated in Figure 1.3. The objects are free to slide on thin oil films of different thicknesses with the same viscosity, μ. Using the notation indicated in the figure, set up the differential equation of motion for the system. Assume that $h_1 \ll s_1$ and $h_2 \ll s_2$.*

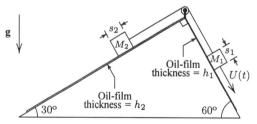

Figure 1.3: *Dual mass/pulley system.*

Solution: To solve, the first step is to draw free-body diagrams for each cube, including all pertinent forces. Figure 1.4 shows the free-body diagram for Cube 1. There are 4 forces acting

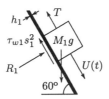

Figure 1.4: *Free-body diagram for Cube 1.*

on the cube, viz., the gravitational force, the tension in the cable connecting the two cubes, the friction force at the inclined surface and a reaction force normal to the incline. The reaction force, R_1, balances the component of the gravitational force normal to the incline, and needn't be considered further. Since the area of the cube face exposed to the lubricating film is s_1^2, the friction force is $\tau_{w1} s_1^2$, where τ_{w1} is the shear stress acting on the cube. Balancing forces in the direction tangent to the incline gives

$$M_1 \frac{dU}{dt} = M_1 g \sin 60° - T - \tau_{w1} s_1^2 \qquad (1.22)$$

where $U(t)$ is the instantaneous velocity of the cubes and T is the tension in the cable. Since we are given $h_1 \ll s_1$, we can assume that the Couette-flow solution is valid in the oil film. For the sake of rigor, note that we are using the Couette-flow solution in what is termed a *quasi-steady* sense for this problem. That is, we are using the Couette-flow solution, which is strictly valid only for constant velocity, with the instantaneous flow velocity, $U(t)$. This assumption is valid provided the acceleration is not too large. This means the velocity in the film is a function of both distance, y, and time, t, given by $u(y,t) = U(t) y / h_1$. Thus, the shear stress is

$$\tau_{w1} = \mu \frac{\partial u}{\partial y} = \frac{\mu U(t)}{h_1} \qquad (1.23)$$

Combining Equations (1.22) and (1.23), and observing that $\sin 60° = \sqrt{3}/2$, we find

$$M_1 \frac{dU}{dt} = \frac{\sqrt{3}}{2} M_1 g - T - \frac{\mu s_1^2}{h_1} U \qquad (1.24)$$

Focusing now on Cube 2, Figure 1.5 shows the free-body diagram. Again, there are 4 forces acting on the cube, viz., the gravitational force, the tension in the cable connecting the two cubes, the friction force at the inclined surface and a reaction force normal to the incline. As with Cube 1, the reaction force, R_2, balances the component of the gravitational force normal to the incline, and has no effect on motion tangent to the incline. Since the area of the cube face exposed to the lubricating film is s_2^2, the friction force is $\tau_{w2} s_2^2$. Balancing forces

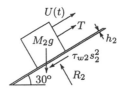

Figure 1.5: *Free-body diagram for Cube 2.*

1.7. PIPE FLOW

in the direction tangent to the incline gives

$$M_2 \frac{dU}{dt} = T - M_2 g \sin 30° - \tau_{w2} s_2^2 \tag{1.25}$$

Since we are given $h_2 \ll s_2$, we can again assume that the Couette-flow solution is valid in the oil film. Hence, the shear stress is

$$\tau_{w2} = \mu \frac{\partial u}{\partial y} = \frac{\mu U(t)}{h_2} \tag{1.26}$$

Combining Equations (1.25) and (1.26), and using the fact that $\sin 30° = 1/2$, we obtain

$$M_2 \frac{dU}{dt} = T - \frac{1}{2} M_2 g - \frac{\mu s_2^2}{h_2} U \tag{1.27}$$

We can now add Equations (1.24) and (1.27) to eliminate the tension, T. This yields the following equation for the combined motion of the two cubes:

$$(M_1 + M_2) \frac{dU}{dt} = \frac{1}{2} \left(\sqrt{3} M_1 - M_2 \right) g - \mu \left(\frac{s_1^2}{h_1} + \frac{s_2^2}{h_2} \right) U \tag{1.28}$$

1.7 Pipe Flow

Statement of the Problem: *Water is flowing very slowly through a drain pipe of diameter $D = 5$ cm and length $L = 120$ m. The pressure difference between inlet and outlet is $(p_1 - p_2) = 0.04$ kPa, and the temperature is $20°$ C. Compute the maximum velocity and the Reynolds number of the water flowing in the pipe.*

Solution: From the laminar pipe-flow solution, the maximum velocity, u_m, and Reynolds number, Re, are given by

$$u_m = \frac{(p_1 - p_2)}{4\mu L} R^2 \quad \text{and} \quad Re = \frac{\rho u_m R}{\mu} \tag{1.29}$$

where R is pipe radius, μ is fluid viscosity and ρ is fluid density. For this problem, we are given $(p_1 - p_2) = 0.04$ kPa, $R = 2.5$ cm and $L = 120$ m. Also, since the fluid is water at $20°$ C, reference to Tables A.3 and A.6 shows that the density is $\rho = 998$ kg/m^3 and the kinematic viscosity is $\nu = 1.00 \cdot 10^{-6}$ m^2/sec. Therefore, the molecular viscosity, μ, is

$$\mu = \rho \nu = \left(998 \text{ kg/m}^3\right) \left(1.00 \cdot 10^{-6} \text{ m}^2/\text{sec}\right) = 9.98 \cdot 10^{-4} \text{ kg/(m} \cdot \text{sec)} \tag{1.30}$$

To simplify the calculations, it is worthwhile to express the pressure difference in terms of the basic units, viz., 1 Pa = 1 N/m^2 = 1 (kg·m/sec^2)/m^2 = 1 kg/(m·sec^2). Thus, the pressure difference and the radius are

$$p_1 - p_2 = 40 \text{ kg/} \left(\text{m} \cdot \text{sec}^2 \right) \quad \text{and} \quad R = 0.025 \text{ m} \tag{1.31}$$

Substituting into the first of Equations (1.29), the maximum velocity is

$$u_m = \frac{\left[40 \text{ kg/} \left(\text{m} \cdot \text{sec}^2\right)\right] (0.025 \text{ m})^2}{4 \left[9.98 \cdot 10^{-4} \text{ kg/(m} \cdot \text{sec)}\right] (120 \text{ m})} = 0.052 \text{ m/sec} \tag{1.32}$$

Also, substituting into the second of Equations (1.29), the Reynolds number is

$$Re = \frac{\left(998 \text{ kg/m}^3\right) (0.052 \text{ m/sec}) (0.025 \text{ m})}{9.98 \cdot 10^{-4} \text{ kg/(m} \cdot \text{sec)}} = 1300 \tag{1.33}$$

Chapter 2

Dimensional Analysis

Chapter 2 focuses on dimensional analysis. Of greatest interest are the issues of deriving dimensionless groupings, dynamic similitude and dimensional homogeneity. The astute reader will observe that, in Chapter 1, we handle *units* the same as variables in an algebraic equation. In this chapter, the solutions treat *dimensions* the same as variables in algebraic equations. They are, in fact, explicity named as algebraic variables, thus leaving no choice about how they should be handled. This is the fundamental difference between units (e.g., kilograms, meters, seconds) and dimensions (e.g., M, L and T), and underscores the rationale of dealing with units as algebraic variables in Chapter 1. The problems in this chapter address elements of dimensional analysis as follows.

Section 2.1 Dimensionless Groupings: A straightforward problem verifying that the Weber number is dimensionless through standard manipulation of dimensional quantities.

Section 2.2 Dimensional Homogeneity: A problem that demonstrates how all terms in a physically-based equation have the same dimensions, and uses this fact to deduce the dimensions of a particular quantity appearing in the equation.

Section 2.3 Indicial Method: A standard dimensional-analysis problem—the solution makes use of the Buckingham Π Theorem and the classical indicial method.

Section 2.4 E. S. Taylor's Method: The same dimensional-analysis problem as in Section 2.3—this time, the solution makes use of E. S. Taylor's method.

Section 2.5 Dynamic Similitude I: The first of two similitude problems for which the appropriate dimensionless groupings are known.

Section 2.6 Dynamic Similitude II: Another similitude problem for which the appropriate dimensionless groupings are known.

2.1 Dimensionless Groupings

Statement of the Problem: *Show that the Weber number, defined by $We = \rho U^2 R/\sigma$ is dimensionless, where ρ is density, U is velocity, R is length and σ is surface tension.*

Solution: The first thing we must do is make a list of the dimensions of ρ, U, R and σ. On the one hand, the dimensions of the first three quantities are obvious, and are

$$[\rho] = \frac{M}{L^3}, \quad [U] = \frac{L}{T}, \quad [R] = L \tag{2.1}$$

where M, L and T denote dimensions of mass, length and time, respectively. The dimensions of surface tension are more subtle and require a bit more care. Since its dimensions are the ratio of force, a *secondary dimension*, to length, an *independent dimension*, we have

$$[\sigma] = \frac{\text{Force}}{L} = \frac{ML/T^2}{L} = \frac{M}{T^2} \tag{2.2}$$

We have used the fact that force has the dimensions of mass (M) times acceleration (L/T^2) to arrive at the desired expression for surface tension in which only independent dimensions appear. Substituting into the Weber number, We, we have

$$[We] = \left[\frac{\rho U^2 R}{\sigma}\right] = \frac{\frac{M}{L^3}\frac{L^2}{T^2}L}{\frac{M}{T^2}} = 1 \tag{2.3}$$

Therefore, the Weber number is dimensionless.

2.2 Dimensional Homogeneity

Statement of the Problem: *One equation of motion for a rocket burning fuel is*

$$\frac{dm}{dt} + \rho_e w_e A_e = 0$$

where m is the mass of the rocket (including fuel), t is time, ρ_e is fuel density at the rocket-nozzle exit plane, w_e is the exit velocity and A_e is the nozzle exit area. Assuming this equation is dimensionally homogeneous, verify that the velocity has dimensions length per unit time.

Solution: First, we must identify the dimensions of all quantities involved, with the exception of the exit velocity, w_e (whose dimensions we must solve for). Clearly, from the statement of the problem,

$$[m] = M, \quad [t] = T, \quad [\rho_e] = \frac{M}{L^3}, \quad [A_e] = L^2 \tag{2.4}$$

where M, L and T denote dimensions of mass, length and time, respectively.

Next, we determine the dimensions of each term in the equation of motion as follows.

$$\left.\begin{aligned}\left[\frac{dm}{dt}\right] &= \frac{M}{T} \\ [\rho_e w_e A_e] &= \frac{M}{L^3}[w_e]L^2 = \frac{M}{L}[w_e]\end{aligned}\right\} \tag{2.5}$$

2.3. INDICIAL METHOD

Finally, assuming the equation is dimensionally homogeneous, necessarily both terms have the same dimensions. Thus, we must have

$$\frac{M}{T} = \frac{M}{L}[w_e] \implies [w_e] = \frac{L}{T} \tag{2.6}$$

Therefore, the exit velocity, w_e, has dimensions of length per unit time.

2.3 Indicial Method

Statement of the Problem: *We are given a force, F, that is a function of velocity, U, fluid viscosity, μ, characteristic length, ℓ, fluid density, ρ, and gravitational acceleration, g. Using the Buckingham Π Theorem, determine the number of dimensionless groupings. Solve for the groupings with the indicial method.*

Solution: We are given a force and five dimensional quantities upon which it depends. The first step is to list the dimensions of these six quantities in terms of independent dimensions, mass (M), length (L) and time (T). The dimensional quantities and their dimensions are

$$[F] = [\text{Force}] = \frac{ML}{T^2} \tag{2.7}$$

$$[U] = \frac{L}{T}, \quad [\mu] = \frac{M}{LT}, \quad [\ell] = L, \quad [\rho] = \frac{M}{L^3}, \quad [g] = \frac{L}{T^2} \tag{2.8}$$

There are 6 dimensional quantities and 3 independent dimensions (M, L, T), so that, according to the Buckingham Π Theorem, there are 3 dimensionless groupings.

The appropriate dimensional equation is

$$[F] = [U]^{a_1}[\mu]^{a_2}[\ell]^{a_3}[\rho]^{a_4}[g]^{a_5} \tag{2.9}$$

Substituting the dimensions for each quantity yields

$$\begin{aligned} MLT^{-2} &= L^{a_1}T^{-a_1}M^{a_2}L^{-a_2}T^{-a_2}L^{a_3}M^{a_4}L^{-3a_4}L^{a_5}T^{-2a_5} \\ &= M^{a_2+a_4}L^{a_1-a_2+a_3-3a_4+a_5}T^{-a_1-a_2-2a_5} \end{aligned} \tag{2.10}$$

Thus, equating exponents, we arrive at the following three (indicial) equations:

$$\left. \begin{aligned} 1 &= a_2 + a_4 \\ 1 &= a_1 - a_2 + a_3 - 3a_4 + a_5 \\ -2 &= -a_1 - a_2 - 2a_5 \end{aligned} \right\} \tag{2.11}$$

We have 3 equations for the 5 unknown exponents, a_1, a_2, a_3, a_4 and a_5. Our goal is to solve for 3 of the exponents in terms of the other two. We select a_1, a_2 and a_3 as the three exponents for which we will solve as functions of a_4 and a_5.

We can solve immediately for a_2 from the first of Equations (2.11), viz.,

$$a_2 = 1 - a_4 \tag{2.12}$$

Substituting this result into the third of Equations (2.11) yields

$$-2 = -a_1 - (1 - a_4) - 2a_5 \implies a_1 = 1 + a_4 - 2a_5 \tag{2.13}$$

Substituting Equations (2.12) and (2.13) into the second of Equations (2.11), we have

$$1 = (1 + a_4 - 2a_5) - (1 - a_4) + a_3 - 3a_4 + a_5 \quad \Longrightarrow \quad a_3 = 1 + a_4 + a_5 \quad (2.14)$$

Now, substituting the solutions for a_1, a_2 and a_3 into the dimensional equation, Equation (2.9), gives

$$[F] = [U]^{1+a_4-2a_5}[\mu]^{1-a_4}[\ell]^{1+a_4+a_5}[\rho]^{a_4}[g]^{a_5} = [\mu U \ell]\left[\frac{\rho U \ell}{\mu}\right]^{a_4}\left[\frac{g\ell}{U^2}\right]^{a_5} \quad (2.15)$$

This equation is of the form

$$[F] = [\text{Quantity with dimensions of force}][P_1]^{a_4}[P_2]^{a_5} \quad (2.16)$$

where P_1 and P_2 are dimensionless. Therefore, the 3 dimensionless groupings are:

$$\frac{F}{\mu U \ell}, \quad \frac{\rho U \ell}{\mu}, \quad \frac{g\ell}{U^2} \quad (2.17)$$

2.4 E. S. Taylor's Method

Statement of the Problem: *We are given a force, F, that is a function of velocity, U, fluid viscosity, μ, characteristic length, ℓ, fluid density, ρ, and gravitational acceleration, g. Using the Buckingham Π Theorem, determine the number of dimensionless groupings. Solve for the groupings with E. S. Taylor's method.*

Solution: We are given a force and five dimensional quantities upon which the force depends. The first step is to list the dimensions of these six quantities in terms of independent dimensions, mass (M), length (L) and time (T). The dimensional quantities and their dimensions are

$$[F] = [\text{Force}] = \frac{ML}{T^2} \quad (2.18)$$

$$[U] = \frac{L}{T}, \quad [\mu] = \frac{M}{LT}, \quad [\ell] = L, \quad [\rho] = \frac{M}{L^3}, \quad [g] = \frac{L}{T^2} \quad (2.19)$$

There are 6 dimensional quantities and 3 independent dimensions (M, L, T), so that, according to the Buckingham Π Theorem, there are 3 dimensionless groupings.

To solve using E. S. Taylor's method, we proceed as follows. First, we set up a matrix with the dimensional quantities on the rows and dimensions along the columns. To establish the elements of the matrix, we list the exponents of each dimension. For example, a quantity of dimension length has $M = 0$, $L = 1$ and $T = 0$. For the problem at hand, the matrix is as follows.

	M	L	T
F	1	1	-2
U	0	1	-1
μ	1	-1	-1
ℓ	0	1	0
ρ	1	-3	0
g	0	1	-2

2.4. E. S. TAYLOR'S METHOD

Next, we choose the simplest column with the aim of leaving a single row with a quantity having that dimension. In this context, "simple" means the minimum number of nonzero entries. For the matrix above, the M column is the simplest. We can accomplish the desired end by subtracting the third (μ) row from the first (F) and fifth (ρ) rows. These operations correspond to dividing F by μ and dividing ρ by μ, respectively.

The matrix becomes:

	M	L	T
F/μ	0	2	-1
U	0	1	-1
μ	1	-1	-1
ℓ	0	1	0
ρ/μ	0	-2	1
g	0	1	-2

$\rightarrow$

	L	T
F/μ	2	-1
U	1	-1
ℓ	1	0
ρ/μ	-2	1
g	1	-2

We have eliminated the μ row as it is the only one left containing a quantity whose dimensions involve mass.

Now, we select the next easiest column, i.e., the T column. We subtract the second (U) row from the first (F/μ) row to arrive at $F/(\mu U)$. Then, we add the second row to the fourth (ρ/μ) row to give $\rho U/\mu$. Finally, we subtract two times the second row from the fifth (g) row to yield g/U^2. What remains of the matrix after these operations is the following.

	L	T
$F/(\mu U)$	1	0
U	1	-1
ℓ	1	0
$\rho U/\mu$	-1	0
g/U^2	-1	0

$\rightarrow$

	L
$F/(\mu U)$	1
ℓ	1
$\rho U/\mu$	-1
g/U^2	-1

We have eliminated the U row, which is the only one left containing a quantity whose dimensions involve length.

Finally, we subtract the second (ℓ) row from the first ($F/(\mu U)$) row and add it to the third ($\rho U/\mu$) and fourth (g/U^2) rows. The matrix simplifies to:

	L
$F/(\mu U \ell)$	0
ℓ	1
$\rho U \ell/\mu$	0
$g\ell/U^2$	0

Only the second row involves a dimensional quantity, namely ℓ on row 2. By contrast, the quantities on rows 1, 3 and 4 are all dimensionless, and thus correspond to the three dimensionless groupings for this problem. That is, we have shown that the dimensionless groupings are

$$\frac{F}{\mu U \ell}, \quad \frac{\rho U \ell}{\mu}, \quad \frac{g\ell}{U^2} \qquad (2.20)$$

2.5 Dynamic Similitude I

Statement of the Problem: *A researcher is conducting experiments in a wind tunnel to simulate operating conditions for a sonic buoy. The prototype buoy is $\ell_p = 6$ ft long and will travel at a speed of $U_p = 8$ ft/sec. The wind tunnel test section can accommodate a model buoy that is $\ell_m = 3$ ft long. Assuming flow is incompressible for prototype and model, at what speed must the air in the wind tunnel be moving to insure dynamic similitude? If the fluid temperature for prototype and model are 60° F, compare the Mach numbers.*

Comment on the suitability of the simulation, noting that a flow is considered to be incompressible if the Mach number is less than 0.3. Assume the kinematic viscosity of water for typical operating conditions is $\nu_p = 1.08 \cdot 10^{-5}$ ft^2/sec, and wind tunnel conditions are such that the kinematic viscosity in the test section is $\nu_m = 1.62 \cdot 10^{-4}$ ft^2/sec.

Solution: In general, for aerodynamic applications and for hydrodynamic applications with objects submerged far below the surface, dynamic similitude will be achieved if the Reynolds and Mach numbers are the same. For the present problem, we assume a priori that the flow for both prototype and model is incompressible, so that the Mach number can be ignored. Thus, we will have dynamic similitude provided the Reynolds numbers are the same, viz., provided we design the experiments so that

$$\frac{U_m \ell_m}{\nu_m} = \frac{U_p \ell_p}{\nu_p} \tag{2.21}$$

Solving for the model velocity, U_m, we have

$$U_m = \frac{\nu_m}{\nu_p} \frac{\ell_p}{\ell_m} U_p \tag{2.22}$$

For the given conditions, the value of the model velocity is

$$U_m = \left(\frac{1.62 \cdot 10^{-4} \text{ ft}^2/\text{sec}}{1.08 \cdot 10^{-5} \text{ ft}^2/\text{sec}}\right) \left(\frac{6 \text{ ft}}{3 \text{ ft}}\right) \left(8 \frac{\text{ft}}{\text{sec}}\right) = 240 \frac{\text{ft}}{\text{sec}} \tag{2.23}$$

From Table A.4, the speed of sound in water is $a_p = 4859$ ft/sec and the speed of sound in air is $a_m = 1119$ ft/sec. Thus, the prototype Mach number, M_p, and the model Mach number, M_m, are as follows.

$$M_p = \frac{U_p}{a_p} = \frac{8 \text{ ft/sec}}{4859 \text{ ft/sec}} = 0.0016 \tag{2.24}$$

$$M_m = \frac{U_m}{a_m} = \frac{240 \text{ ft/sec}}{1119 \text{ ft/sec}} = 0.21 \tag{2.25}$$

Although the Mach numbers are not matched, both are less than 0.3, so that the flow is incompressible for both the prototype and the model. Hence, matching Reynolds numbers is sufficient to achieve dynamic similitude for these experiments.

2.6 Dynamic Similitude II

Statement of the Problem: *A man named George is teaching his wife how to swing on a vine in the jungle. George's wife—who took a fluid mechanics course in college—has used dimensional analysis to show that the velocity required to topple a tree as a result of a collision is a function, $\mathcal{F}$, of m/M and ℓ/L, i.e.,*

$$U = \sqrt{g\ell}\; \mathcal{F}\left(\frac{m}{M}, \frac{\ell}{L}\right)$$

where g is gravitational acceleration, ℓ is the height above the ground at which the swinger collides with the tree, m is the mass of the swinger, M is the mass of the tree and L is the height of the tree. George and his wife weigh 100 kg and 50 kg, respectively. After a heavy rainfall, George can topple a 1000 kg tree with a collision height of $\ell = 5$ m.

(a) How massive a tree can George's wife topple after the rainfall?

(b) Assuming $M \propto L^3$, at what height above the ground should her collision occur?

Solution: From the problem statement, we know that the dimensionless groupings are

$$P_0 = \frac{U}{\sqrt{g\ell}}, \quad P_1 = \frac{m}{M}, \quad P_2 = \frac{\ell}{L} \tag{2.26}$$

2.6(a): In order to achieve dynamic similitude, all three dimensionless groupings of Equation (2.26) must be identical. In particular, P_1, which is the only parameter involving mass must be matched. Thus, letting subscript g denote George and subscript w his wife, then

$$\frac{m_w}{M_w} = \frac{m_g}{M_g} \quad \Longrightarrow \quad M_w = \frac{m_w}{m_g} M_g \tag{2.27}$$

Therefore, from the given data, the mass of the tree George's wife can topple is

$$M_w = \frac{50 \text{ kg}}{100 \text{ kg}}(1000 \text{ kg}) = 500 \text{ kg} \tag{2.28}$$

2.6(b): Achieving dynamic similitude also requires having P_2 constant, wherefore (using the additional fact that $M \propto L^3$):

$$\frac{\ell_w}{L_w} = \frac{\ell_g}{L_g} \quad \Longrightarrow \quad \ell_w = \frac{L_w}{L_g}\ell_g = \left(\frac{M_w}{M_g}\right)^{1/3} \ell_g \tag{2.29}$$

Therefore, the distance above the ground at which George's wife must collide with the tree is

$$\ell_w = \left(\frac{500 \text{ kg}}{1000 \text{ kg}}\right)^{1/3}(5 \text{ m}) = 3.97 \text{ m} \tag{2.30}$$

Chapter 3

Effects of Gravity on Pressure

Chapter 3 focuses on the manner in which gravity affects pressure in a non-moving fluid. We begin by using the hydrostatic relation to calculate pressure in tanks and manometers. Next, we turn to the calculation of forces on planar and curved surfaces. Usually these are gates that either hold or separate fluid. Finally, we focus on the buoyancy force, another manifestation of the gravitational field. To underscore how the gravitational field directly controls the forces in a static fluid, Section 3.9 illustrates how the forces change when the gravitational acceleration changes.

In reading the solutions in this chapter, observe that we always develop an algebraic result prior to computing numerical results. This approach helps eliminate errors by permitting checks for dimensional consistency. Also, an algebraic result permits us to reach general conclusions that sometimes explain subtle, and sometimes unexpected, answers. Section 3.9 is a quintessential example of this approach and its ramifications. The chapter includes problems in all of these areas as follows.

Section 3.1 Hydrostatic Relation: A problem that uses the hydrostatic relation to calculate pressure at various depths in a tank. The medium is three layers of unmixed fluids of different densities.

Section 3.2 U. S. Standard Atmosphere: A simple problem of calculating pressure and temperature in the U. S. Standard Atmosphere.

Section 3.3 Manometer: This problem involves calculating pressure in a slanted manometer. As in Section 3.1 the solution requires use of the hydrostatic relation to calculate pressure differences. The only subtle aspect of the solution is recognizing that distances are measured along the vertical direction.

Section 3.4 Vertical Gate: This is a problem of calculating forces on a vertical gate that has fluid on both sides, and with differing depths. It requires first calculating the center of pressure and the horizontal force on the gate. Balancing moments completes the solution.

Section 3.5 Tilted Gate—Tilted Coordinate Frame: This is another problem involving forces on a planar gate. However, this time the gate is tilted at an angle to the vertical. The solution uses a tilted coordinate system aligned with the gate, including computation of the center of pressure and the force normal to the gate. Balancing moments yields the minimum force required to keep the gate closed.

Section 3.6 Tilted Gate—Normal Coordinate Frame: This is the same application considered in Section 3.5. However, the solution uses the standard xz coordinate system. The reason for this is to illustrate how much more work is added to the problem by choosing an inconvenient coordinate system. The solution requires calculating the x and z components of the center of pressure. Also, the force has x and z components, and the moment computation is more complicated. Note the final answer is the same as in Section 3.5.

Section 3.7 Curved Gate: The problem in this section is similar to the preceding three. It requires computing forces on a curved surface. The problem also involves determining both the x and z components of the center of pressure.

Section 3.8 Superposition: A simple example of how to use the superposition principle. It illustrates how we replace a complicated problem with a number of simpler problems for which solutions are easy to obtain. Then, we simply add or subtract the results (depending on the defined superposition) to arrive at the final answer.

Section 3.9 Buoyancy in a Uniform Fluid: A simple problem in buoyancy. It requires calculating the depth of submergence of an object on Earth and on the Moon.

Section 3.10 Buoyancy in a Dual-Fluid Tank: A more difficult problem in buoyancy. It requires calculating the depth of submergence of an object in a dual-fluid tank and then determining the unknown density of one of the fluids.

3.1 Hydrostatic Relation

Statement of the Problem: *Consider the tank shown with three layers of unmixed fluids. Also, the tank is open to the atmosphere at the top. Determine the pressure, p, as a function of depth, z, and make a graph of $p - p_o$, where p_o is atmospheric pressure.*

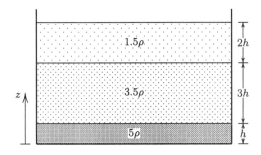

Figure 3.1: *Tank with three layers of unmixed fluids.*

Solution: The key point of this problem is that the hydrostatic relation applies in each layer. Because the layers have different densities, we must use the relation separately in each layer. The pressure at the free surface is

$$p(6h) = p_o \tag{3.1}$$

The pressure at the interface between the two fluids at $z = 4h$ is

$$p(4h) = p(6h) + 1.5\rho g(2h) = p_o + 3\rho g h \tag{3.2}$$

3.2. U. S. STANDARD ATMOSPHERE

The pressure at the interface between the two fluids at $z = h$ is

$$p(h) = p(4h) + 3.5\rho g(3h) = p(4h) + 10.5\rho gh = p_o + 13.5\rho gh \tag{3.3}$$

Finally, the pressure at the bottom of the tank, $p(0)$, is

$$p(0) = p(h) + 5\rho gh = p_o + 13.5\rho gh + 5\rho gh = p_o + 18.5\rho gh \tag{3.4}$$

Since the pressure varies linearly in each layer, the variation of $p - p_o$ with z is as shown in Figure 3.2.

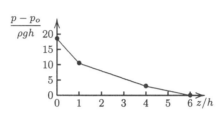

Figure 3.2: *Computed pressure variation.*

3.2 U. S. Standard Atmosphere

Statement of the Problem: *An airplane loses an engine at an altitude of $z_1 = 34,000$ ft and falls to an altitude of $z_2 = 21,000$ ft before regaining control. Assuming pressure and temperature are those of the U. S. Standard Atmosphere, compute the temperature and pressure differences at these two altitudes.*

Solution: The first thing we need to determine is the part of the atmosphere in which the airplane is flying. Before and after losing the engine, the altitudes are

$$z_1 = \frac{34000 \text{ ft}}{5280 \text{ ft/mi}} = 6.439 \text{ mi}, \qquad z_2 = \frac{21000 \text{ ft}}{5280 \text{ ft/mi}} = 3.977 \text{ mi} \tag{3.5}$$

Both altitudes lie within the *troposphere*, which, according to the U. S. Standard Atmosphere, extends from sea level, $z = 0$, up to $z = 6.84$ miles.

The temperature varies linearly with altitude in the troposphere according to

$$T = T_o - \alpha z \tag{3.6}$$

where $T_o = 518.4°$ R is the surface temperature and $\alpha = 18.85°$ R/mi is the *adiabatic lapse rate*. Also, the pressure is given by

$$p = p_o \left[1 - \frac{\alpha z}{T_o}\right]^{g/(\alpha R)} \tag{3.7}$$

where $p_o = 14.7$ psi is the pressure at the surface and $R = 1716$ ft·lb/(slug·°R) is the perfect-gas constant for air. The exponent, $g/(\alpha R)$, appearing in Equation (3.7) is 5.26. Hence, the change in temperature, ΔT, is

$$\Delta T = -\alpha \Delta z = -(18.85° \text{ R/mi})(3.977 \text{ mi} - 6.439 \text{ mi}) = 46.41° \text{ R} \tag{3.8}$$

Turning to pressure, the change, Δp, is

$$\begin{aligned}
\Delta p &= p_o \left[1 - \frac{\alpha z_2}{T_o} \right]^{g/(\alpha R)} - p_o \left[1 - \frac{\alpha z_1}{T_o} \right]^{g/(\alpha R)} \\
&= (14.7 \text{ psi}) \left[1 - \frac{(18.85^\circ \text{ R/mi})(3.977 \text{ mi})}{518.4^\circ \text{ R}} \right]^{5.26} \\
&\quad - (14.7 \text{ psi}) \left[1 - \frac{(18.85^\circ \text{ R/mi})(6.439 \text{ mi})}{518.4^\circ \text{ R}} \right]^{5.26} \\
&= 6.46 \text{ psi} - 3.61 \text{ psi} = 2.85 \text{ psi}
\end{aligned} \quad (3.9)$$

Therefore, as the airplane falls from 34,000 ft to 21,000 ft, the temperature increases by 46.41° R and the pressure increases by 2.85 psi.

3.3 Manometer

Statement of the Problem: *The manometer shown in Figure 3.3 contains fluids of density ρ_1 and ρ_2. It is open to the atmosphere at Point C. Determine the pressure at Point A as a function of atmospheric pressure, p_o, gravitational acceleration, g, the length h and the densities ρ_1 and ρ_2.*

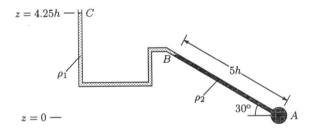

Figure 3.3: *Manometer with fluids of two densities.*

Solution: There are two key points amplified by this problem. First, as in Section 3.1, the hydrostatic relation applies in each fluid. Because the layers have different densities, we must use the relation separately in each fluid. Second, the hydrostatic relation tells us the pressure depends only upon vertical distance, z. Now, the vertical distance from Point A to Point B is

$$z_B = 5h \sin 30^\circ = 2.5h \quad (3.10)$$

In general, if $z = 0$ at Point A and z increases upward, the hydrostatic relation tells us that, in Fluid 1,

$$p(z) = p_o - \rho_1 g(z - 4.25h) \quad (3.11)$$

which yields $p = p_o$ at the free surface, $z = 4.25h$. The pressure at Point B, which lies at $z_B = 2.5h$, is

$$p(2.5h) = p_o - \rho_1 g(2.5h - 4.25h) = p_o + 1.75 \rho_1 g h \quad (3.12)$$

Similarly, in Fluid 2, the hydrostatic relation becomes

$$p(z) = p(2.5h) - \rho_2 g(z - 2.5h) \quad (3.13)$$

3.4. VERTICAL GATE

Thus, at Point A, we have

$$p(0) = p(2.5h) - \rho_2 g(0 - 2.5h) = p(2.5h) + 2.5\rho_2 gh \qquad (3.14)$$

Finally, combining Equations (3.12) and (3.14), the pressure at Point A is

$$p(0) = p_o + (1.75\rho_1 + 2.5\rho_2)gh \qquad (3.15)$$

3.4 Vertical Gate

Statement of the Problem: *A square gate of side $4h$ separates fluids of density ρ and $\lambda\rho$ as shown, where λ is a constant to be determined. For what value of λ will there be zero force on the stop? Also, compute the reaction force on the hinge.*

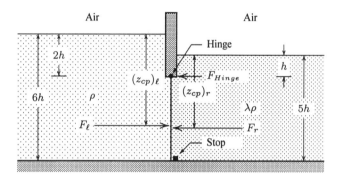

Figure 3.4: *Gate separating fluids of different densities.*

Solution: There are five unknown quantities in this problem, viz., λ and four forces acting on the gate. The forces are from the hydrostatic pressure on each side and reaction forces at the stop and the hinge. The solution strategy is as follows. First, we use standard formulas to compute the hydrostatic forces (two equations). Then, observing that the force at the stop is zero (one equation), we take moments about the hinge (one equation). Using these four equations, we have sufficient information to solve for λ. Finally, balancing forces on the gate yields a fifth equation from which the force on the hinge can be determined.

The area of the gate is $A = (4h)^2 = 16h^2$. Also, the centroids for the fluid to the left, $\bar{z}_\ell$, and right, $\bar{z}_r$, (see Figure B.1) are

$$\bar{z}_\ell = 4h \quad \text{and} \quad \bar{z}_r = 3h \qquad (3.16)$$

Hence, the hydrostatic forces from the left, F_ℓ, and right, F_r, are

$$F_\ell = \rho g A \bar{z}_\ell = \rho g \left(16h^2\right)(4h) = 64\rho g h^3 \qquad (3.17)$$
$$F_r = \lambda \rho g A \bar{z}_r = \lambda \rho g \left(16h^2\right)(3h) = 48\lambda \rho g h^3 \qquad (3.18)$$

Taking moments about the hinge, the moment arm for the force to the left is $(z_{cp})_\ell - 2h$. Also, the moment arm for the force to the right is $(z_{cp})_r - h$. If there is no force on the stop, we must have

$$\left[(z_{cp})_\ell - 2h\right] F_\ell = \left[(z_{cp})_r - h\right] F_r \qquad (3.19)$$

Now, reference to Figure B.1 shows that the moment of inertia of the gate is

$$I = \frac{1}{12}(4h)(4h)^3 = \frac{64}{3}h^4 \qquad (3.20)$$

So, the center of pressure for the hydrostatic force from the left is

$$(z_{cp})_\ell = \bar{z}_\ell + \frac{I}{\bar{z}_\ell A} = 4h + \frac{\frac{64}{3}h^4}{(3h)(16h^2)} = 4h + \frac{1}{3}h = \frac{13}{3}h \qquad (3.21)$$

Similarly, the center of pressure for the hydrostatic force from the right is

$$(z_{cp})_r = \bar{z}_r + \frac{I}{\bar{z}_r A} = 3h + \frac{\frac{64}{3}h^4}{(4h)(16h^2)} = 3h + \frac{4}{9}h = \frac{31}{9}h \qquad (3.22)$$

Using these values for $(z_{cp})_\ell$ and $(z_{cp})_r$ in the moment equation, Equation (3.19), gives

$$\left[\frac{13}{3}h - 2h\right][64\rho g h^3] = \left[\frac{31}{9}h - h\right][48\lambda \rho g h^3] \implies \frac{7 \cdot 64}{3}\rho g h^4 = \frac{22 \cdot 48}{9}\lambda \rho g h^4 \qquad (3.23)$$

Thus, solving for λ, we find

$$\lambda = \frac{14}{11} \qquad (3.24)$$

Finally, balancing forces on the gate yields

$$F_{Hinge} + F_r = F_\ell \qquad (3.25)$$

Substituting Equations (3.17), (3.18) and (3.24) into Equation (3.25), the force on the hinge is given by

$$F_{Hinge} = F_\ell - F_r = 64\rho g h^3 - 48\lambda \rho g h^3 = 64\rho g h^3 - 48\frac{14}{11}\rho g h^3 = \frac{32}{11}\rho g h^3 \qquad (3.26)$$

3.5 Tilted Gate—Tilted Coordinate Frame

Statement of the Problem: *Consider the tilted gate shown in Figure 3.5. Find the force, F, necessary to keep the gate hinged at Point A closed. Do your computations using a tilted coordinate system aligned with the gate, which is rectangular with width $2L$. You may ignore the weight of the gate and assume that F acts normal to the gate.*

Solution: As shown in Figure 3.6, there are three forces acting on the gate, viz., the hydrostatic force from below, F_N, the force needed to hold the gate closed, F, and a reaction force at the hinge, F_A. Hence, we need three equations to solve the problem.

The first equation follows from balancing the forces acting on the gate, i.e.,

$$F_N = F + F_A \qquad (3.27)$$

The second equation follows from computing the hydrostatic force in terms of the geometry of the gate, i.e.,

$$F_N = \rho g \bar{\zeta} A \sin 30° \qquad (3.28)$$

3.5. TILTED GATE—TILTED COORDINATE FRAME

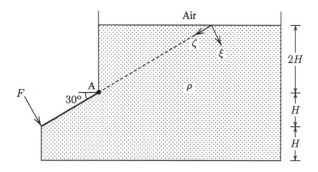

Figure 3.5: *Tilted gate—tilted coordinate system.*

where ρ is the fluid density, g is gravitational acceleration, $\bar{\zeta}$ is the centroid measured along the ζ axis and A is the area of the gate. Reference to Figure 3.6 shows that $\bar{\zeta}$ and A are

$$\bar{\zeta} = 4H + H = 5H \quad \text{and} \quad A = (2L)(2H) = 4LH \tag{3.29}$$

Thus, since $\sin 30° = \frac{1}{2}$, the hydrostatic force is

$$F_N = \rho g(5H)(4LH)(1/2) = 10\rho g LH^2 \tag{3.30}$$

The third equation follows from balancing moments. For convenience, we can take moments about the hinge. Again referring to Figure 3.6, the lever arm for the hydrostatic force relative to the hinge is $(\zeta_{cp} - 4H)$, where ζ_{cp} is the center of pressure. The lever arm for the force F is $2H$. Because we are taking moments about its point of application, the lever arm for the force at the hinge is 0. Thus, we have

$$F_N(\zeta_{cp} - 4H) - F(2H) - F_A(0) = 0 \tag{3.31}$$

Solving for F and substituting for F_N from Equation (3.30), we find

$$F = 10\left(\frac{\zeta_{cp}}{2H} - 2\right)\rho g LH^2 \tag{3.32}$$

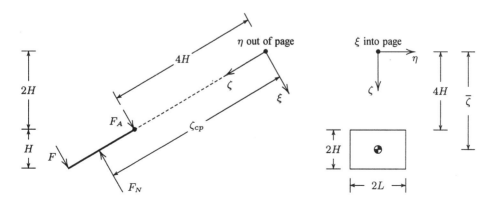

Figure 3.6: *Forces and moment arms.*

The center of pressure is

$$\zeta_{cp} = \bar{\zeta} + \frac{I}{\bar{\zeta} A} \tag{3.33}$$

where I is the moment of inertia. Referring to Figure B.1, the moment of inertia is

$$I = \frac{1}{12}(2L)(2H)^3 = \frac{4}{3}LH^3 \tag{3.34}$$

Therefore, we find

$$\zeta_{cp} = 5H + \frac{\frac{4}{3}LH^3}{(5H)(4LH)} = 5H + \frac{1}{15}H = \frac{76}{15}H \tag{3.35}$$

Substituting Equation (3.35) into Equation (3.32) gives the force F.

$$F = 10\left(\frac{38}{15} - 2\right)\rho g LH^2 = \frac{16}{3}\rho g LH^2 \tag{3.36}$$

As a final comment, note that we have not used Equation (3.27). It is superfluous because we took moments about the hinge, and thus only two equations are necessary.

3.6 Tilted Gate—Normal Coordinate Frame

Statement of the Problem: *Find the force, F, necessary to keep the gate hinged at Point A closed. Do your computations using the xz coordinate system. The gate is rectangular with width $2L$. You may ignore the weight of the gate and assume that F acts normal to the gate.*

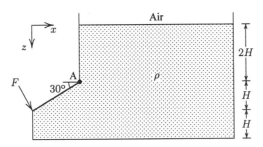

Figure 3.7: *Tilted gate—standard coordinate system.*

Solution: The simplest approach to this problem is to first recognize that, with a change of signs, this problem is equivalent to the same gate holding back a fluid of density ρ and the same depth from the left. As illustrated in Figure 3.8, since pressure depends only upon depth, z, the pressure acting at all points of the gate is identical in magnitude, regardless of which side of the gate the fluid is on. As shown in Figure 3.9 there are three forces acting on the gate, viz., the hydrostatic force, $\mathbf{F}_N$, the force needed to hold the gate closed, $\mathbf{F}_R$, and a reaction force at the hinge, $\mathbf{F}_A$. As in Section 3.5, we can circumvent the need to determine $\mathbf{F}_A$ if we take moments about the hinge. Thus, we have four unknowns to solve for, viz., the x and z components of both $\mathbf{F}_R$ and $\mathbf{F}_N$, and we need four equations to solve.

3.6. TILTED GATE—NORMAL COORDINATE FRAME

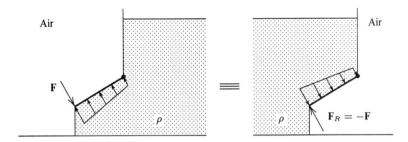

Figure 3.8: *Equivalent problems.*

The first equation follows from the statement of the problem, which tells us the force acts normal to the gate. Therefore, for the problem with the fluid to the left of the gate,

$$\mathbf{F}_R = -F \sin 30° \mathbf{i} - F \cos 30° \mathbf{k} = -\frac{1}{2} F \mathbf{i} - \frac{\sqrt{3}}{2} F \mathbf{k} \tag{3.37}$$

where $F = |\mathbf{F}_R|$. Therefore, we conclude that

$$F_{Rx} = -\frac{1}{2} F \quad \text{and} \quad F_{Rz} = -\frac{\sqrt{3}}{2} F \tag{3.38}$$

The second equation follows from computing the force on the projection of the gate on the vertical (yz) plane. That is, the horizontal component of the hydrostatic force, F_{Nx}, is

$$F_{Nx} = \rho g \bar{z} A \tag{3.39}$$

where $\bar{z}$ is the centroid measured along the z axis and A is the area of the gate's projection on the yz plane. Reference to Figure 3.9 shows that

$$\bar{z} = 2H + \frac{1}{2}H = \frac{5}{2}H \quad \text{and} \quad A = 2LH \tag{3.40}$$

Thus, the horizontal component of $\mathbf{F}_N$ is

$$F_{Nx} = \rho g \left(\frac{5}{2}H\right)(2LH) = 5\rho g L H^2 \tag{3.41}$$

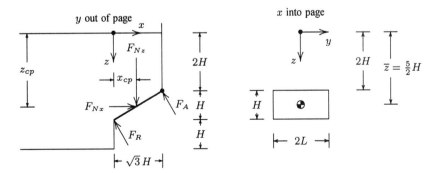

Figure 3.9: *Forces, moment arms and projection of the gate cross section on the yz plane.*

The third equation follows from equating the vertical component of the hydrostatic force, F_{Nz}, to the weight of the column of fluid above the gate. The volume of the fluid above the gate, V, is the sum of the volume with triangular cross section just above the gate, V_ℓ, and the volume with rectangular cross section above the top of the gate, V_u. Reference to Figure 3.9 shows that

$$\left. \begin{aligned} V_\ell &= \tfrac{1}{2}(2L)(\sqrt{3}H)(H) = \sqrt{3}\,LH^2 \\ V_u &= (2L)(\sqrt{3}H)(2H) = 4\sqrt{3}\,LH^2 \end{aligned} \right\} \qquad (3.42)$$

Therefore, the normal component of the hydrostatic force is

$$F_{Nz} = \rho g\,(V_\ell + V_u) = 5\sqrt{3}\,\rho g L H^2 \qquad (3.43)$$

The fourth equation follows from balancing moments. As mentioned above, we will take moments about the hinge. Reference to Figure 3.9 reveals that the hinge is located at $x = \sqrt{3}\,H$ and $z = 2H$. Thus, the vector location of the hinge, $\mathbf{r}_h$, is given by

$$\mathbf{r}_h = x_h \mathbf{i} + z_h \mathbf{k} = \sqrt{3}\,H\,\mathbf{i} + 2H\,\mathbf{k} \qquad (3.44)$$

Hence, the lever arm for the hydrostatic force is $\mathbf{r}_{cp} - \mathbf{r}_h$, where

$$\mathbf{r}_{cp} = x_{cp}\mathbf{i} + z_{cp}\mathbf{k} \qquad (3.45)$$

Also, the lever arm for $\mathbf{F}_R$ is

$$\mathbf{r}_R = x_R \mathbf{i} + z_R \mathbf{k} = -\sqrt{3}\,H\,\mathbf{i} + H\,\mathbf{k} \qquad (3.46)$$

So, taking moments about the hinge, we have

$$\mathbf{r}_R \times \mathbf{F}_R + (\mathbf{r}_{cp} - \mathbf{r}_h) \times \mathbf{F}_N = 0 \qquad (3.47)$$

which can be expanded in terms of components symbolically as follows.

$$(x_R \mathbf{i} + z_R \mathbf{k}) \times (F_{Rx}\mathbf{i} + F_{Rz}\mathbf{k}) + [(x_{cp} - x_h)\mathbf{i} + (z_{cp} - z_h)\mathbf{k}] \times (F_{Nx}\mathbf{i} + F_{Nz}\mathbf{k}) = 0 \qquad (3.48)$$

Then, noting that $\mathbf{i} \times \mathbf{k} = -\mathbf{j}$ and $\mathbf{k} \times \mathbf{i} = \mathbf{j}$, we have

$$z_R F_{Rx} - x_R F_{Rz} + (z_{cp} - z_h) F_{Nx} - (x_{cp} - x_h) F_{Nz} = 0 \qquad (3.49)$$

Combining Equations (3.44), (3.45), (3.46) and (3.49), the moment equation simplifies to:

$$H F_{Rx} + \sqrt{3}\,H F_{Rz} + (z_{cp} - 2H) F_{Nx} + \left(\sqrt{3}\,H - x_{cp}\right) F_{Nz} = 0 \qquad (3.50)$$

Substituting the force components from Equations (3.38), (3.41) and (3.43) into Equation (3.50) gives

$$H\left(-\tfrac{1}{2}F\right) + \sqrt{3}\,H\left(\frac{\sqrt{3}}{2}F\right) + (z_{cp} - 2H)\left(5\rho g L H^2\right) \\ + \left(\sqrt{3}\,H - x_{cp}\right)\left(5\sqrt{3}\,\rho g L H^2\right) = 0 \qquad (3.51)$$

which can be expanded to yield the following.

$$-2HF + 5\rho g L H^2 \left[z_{cp} - 2H + 3H - \sqrt{3}\,x_{cp}\right] = 0 \qquad (3.52)$$

3.6. TILTED GATE—NORMAL COORDINATE FRAME

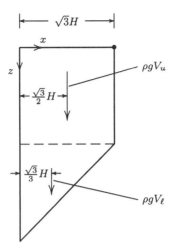

Figure 3.10: *Moment arms for computing x_{cp}.*

Solving for F, we arrive at

$$F = \frac{5}{2}\rho g L H \left[z_{cp} - \sqrt{3}\, x_{cp} + H \right] \tag{3.53}$$

In order to complete the solution, we must determine the components of the center of pressure. To determine z_{cp}, we know that

$$z_{cp} = \bar{z} + \frac{I}{\bar{z}A} \tag{3.54}$$

where the moment of inertia, I, for the projection of the gate on the yz plane is (see Figure B.1):

$$I = \frac{1}{12}(2L)H^3 = \frac{1}{6}LH^3 \tag{3.55}$$

Thus, combining Equations (3.40), (3.54) and (3.55), we have

$$z_{cp} = \frac{5}{2}H + \frac{\frac{1}{6}LH^3}{\left(\frac{5}{2}H\right)(2LH)} = \frac{5}{2}H + \frac{1}{30}H = \frac{38}{15}H \tag{3.56}$$

To determine x_{cp}, the most convenient method is to take moments about the z axis. Referring to Figure 3.10, the moment for the complete column of fluid above the gate is equal to the sum of the moments from volumes V_ℓ and V_u by

$$x_{cp} F_{Nz} = x_{cp}^\ell \rho g V_\ell + x_{cp}^u \rho g V_u \tag{3.57}$$

Inspection of the figure shows that

$$x_{cp}^\ell = \frac{\sqrt{3}}{3}H \quad \text{and} \quad x_{cp}^u = \frac{\sqrt{3}}{2}H \tag{3.58}$$

Substituting for F_{Nz} from Equation (3.43) and for V_ℓ and V_u from Equation (3.42), we have

$$x_{cp}\left(5\sqrt{3}\,\rho g L H^2\right) = \frac{\sqrt{3}}{3}H\,(\rho g)\left(\sqrt{3}\,LH^2\right) + \frac{\sqrt{3}}{2}H\,(\rho g)\left(4\sqrt{3}\,LH^2\right) \tag{3.59}$$

or,

$$5\sqrt{3}\rho g L H^2 x_{cp} = \rho g L H^3 + 6\rho g L H^3 \quad \Longrightarrow \quad x_{cp} = \frac{7\sqrt{3}}{15}H \tag{3.60}$$

Finally, substituting Equations (3.56) and (3.60) into Equation (3.53) gives

$$F = \frac{5}{2}\rho g L H \left[\frac{38}{15}H - \sqrt{3}\frac{7\sqrt{3}}{15}H + H\right] = \frac{16}{3}\rho g L H^2 \tag{3.61}$$

This result agrees with the solution of Section 3.5, as it must.

Comparison of this solution and the solution in Section 3.5 shows that using a tilted coordinated system aligned with the gate is much simpler. On the one hand, a clever choice of coordinates simplifies the analysis of a given problem, and this is an excellent example. On the other hand, the methodology in this problem is completely general, and can be used for much more complex geometries. By contrast, the tilted coordinate system is advantageous only for planar geometries.

3.7 Curved Gate

Statement of the Problem: *The gate shown is a quarter circle of radius $2H$, and is $3W$ wide (out of the page). Determine the applied horizontal force, F, required to hold the gate closed. You may ignore the weight of the gate.*

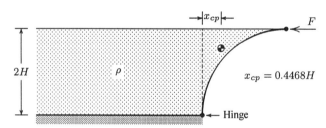

Figure 3.11: *Quarter-circle gate.*

Solution: As shown in Figure 3.12, there are three forces acting on the gate, viz., the hydrostatic force, $\mathbf{F}_N$, the force needed to hold the gate closed, $\mathbf{F}$, and a reaction force at the hinge, $\mathbf{F}_h$. We can circumvent the need to determine $\mathbf{F}_h$ if we take moments about the hinge. Thus, since $\mathbf{F}$ has only a horizontal component, we have three unknowns to solve for, viz., the x and z components of $\mathbf{F}_N$ and $F = |\mathbf{F}|$.

The first equation follows from computing the force on the projection of the gate on the vertical (yz) plane. That is, the horizontal component of the hydrostatic force, F_{Nx}, is

$$F_{Nx} = \rho g \bar{z} A \tag{3.62}$$

The quantity $\bar{z}$ is the centroid measured along the z axis and A is the area of the gate's projection on the yz plane. The projection of the gate on the yz plane is a rectangle of height $2H$ and width $3W$. Reference to Figure B.1 tells us that

$$\bar{z} = H \quad \text{and} \quad A = (3W)(2H) = 6WH \tag{3.63}$$

3.7. CURVED GATE

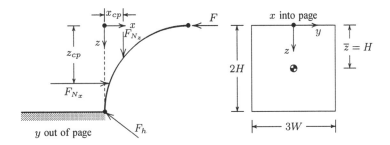

Figure 3.12: *Forces and center of pressure components—the gate cross section shown is its projection on the yz plane.*

Thus, the horizontal component of $\mathbf{F}_N$ is

$$F_{Nx} = \rho g H (6WH) = 6\rho g W H^2 \tag{3.64}$$

The second equation follows from equating the vertical component of the hydrostatic force, F_{Nz}, and the weight of the column of fluid above the gate. That is,

$$F_{Nz} = \rho g V \tag{3.65}$$

The volume of the fluid above the gate, V, is $3W$ multiplied by the difference between the area of a square of side $2H$ and a quarter circle of radius $2H$. Thus,

$$V = 3W \left[(2H)^2 - \frac{\pi}{4}(2H)^2 \right] = 3(4-\pi)WH^2 \tag{3.66}$$

Therefore, the normal component of the hydrostatic force is

$$F_{Nz} = 3(4-\pi)\rho g W H^2 \tag{3.67}$$

The third equation follows from balancing moments. Inspection of Figure 3.12 shows that, relative to the hinge, the lever arms for F, F_{Nz} and F_{Nx} are $2H$, x_{cp} and $(2H - z_{cp})$, respectively. Therefore, we have

$$2HF - x_{cp}F_{Nz} - (2H - z_{cp})F_{Nx} = 0 \tag{3.68}$$

Combining Equations (3.64), (3.67) and (3.68), we have

$$F = \frac{1}{2}\frac{x_{cp}}{H}\left[3(4-\pi)\rho g W H^2\right] + \left(1 - \frac{1}{2}\frac{z_{cp}}{H}\right)\left[6\rho g W H^2\right] \tag{3.69}$$

To complete the solution, we must determine the components of the center of pressure. We are given

$$x_{cp} = 0.4468H \tag{3.70}$$

The vertical component of the center of pressure, z_{cp}, is

$$z_{cp} = \bar{z} + \frac{I}{\bar{z}A} \tag{3.71}$$

where $\bar{z}$, I and A are the centroid, moment of inertia and area of the projection of the gate on the yz plane. The quantities $\bar{z}$ and A are given in Equation (3.63), while reference to Figure B.1 shows that

$$I = \frac{1}{12}(3W)(2H)^3 = 2WH^3 \tag{3.72}$$

Hence, z_{cp} is

$$z_{cp} = H + \frac{2WH^3}{H(6WH)} = H + \frac{1}{3}H = \frac{4}{3}H \qquad (3.73)$$

Thus, combining Equations (3.69), (3.70) and (3.73), the force required to hold the gate in place is

$$F = 0.2234\left[3(4-\pi)\rho g W H^2\right] + \left(1-\frac{2}{3}\right)\left[6\rho g W H^2\right] = 2.575\rho g W H^2 \qquad (3.74)$$

3.8 Superposition

Statement of the Problem: *Using superposition, determine the hydrostatic force on gate AB. The gate is rectangular with width 4W (out of the page).*

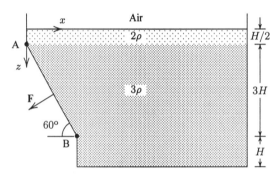

Figure 3.13: *Gate submerged in a two-layer fluid.*

Solution: To use superposition to solve any fluid-statics problem, we must establish equivalent problems whose sum or difference yields the same overall pressure in the fluid. Figure 3.14 shows the most obvious way of approaching the problem at hand. To expand on this point, let's determine the pressure in the original problem and in the sum of the two replacement problems. For the original problem, the pressure in the upper fluid, relative to atmospheric pressure, is

$$p(z) = 2\rho g z \qquad (3.75)$$

In the lower fluid, we have

$$p(z) = p(H/2) + 3\rho g(z - H/2) = \rho g H + 3\rho g z - \frac{3}{2}\rho g H = -\frac{1}{2}\rho g H + 3\rho g z \qquad (3.76)$$

For the replacement problems, we have

$$p_1(z) = 2\rho g z \quad \text{and} \quad p_2 = \rho g(z - H/2) \qquad (3.77)$$

Hence, the sum of the pressures in the two replacement problems is

$$p_1(z) + p_2(z) = \begin{cases} 2\rho g z, & z \leq \dfrac{H}{2} \\ -\dfrac{1}{2}\rho g H + 3\rho g z, & z \geq \dfrac{H}{2} \end{cases} \qquad (3.78)$$

3.8. SUPERPOSITION

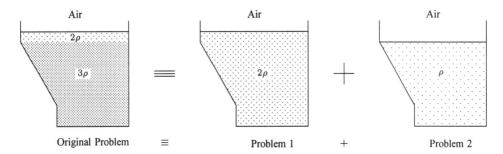

Figure 3.14: *Equivalent problems.*

This is identical to the pressure for the original problem, so the proposed superposition is valid.

The horizontal force can be computed by working with the projection of the gate on the vertical (yz) plane. Thus, the horizontal component of the hydrostatic forces for Problems 1 and 2, F_{x1} and F_{x2}, are

$$F_{x1} = 2\rho g \bar{z}_1 A \quad \text{and} \quad F_{x2} = \rho g \bar{z}_2 A \tag{3.79}$$

As shown in Figure 3.15, the projection is a rectangle of height $3H$ and width $4W$. Thus, we have

$$\bar{z}_1 = 2H, \quad \bar{z}_2 = \frac{3}{2}H, \quad A = (4W)(3H) = 12WH \tag{3.80}$$

Therefore, the horizontal forces for the two problems are

$$F_{x1} = 2\rho g(2H)(12WH) = 48\rho g W H^2 \tag{3.81}$$

$$F_{x2} = \rho g \left(\frac{3}{2}H\right)(12WH) = 18\rho g W H^2 \tag{3.82}$$

so that the total horizontal component of the hydrostatic force is

$$F_x = F_{x1} + F_{x2} = 66\rho g W H^2 \tag{3.83}$$

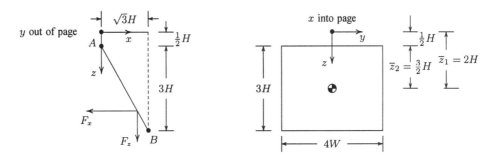

Figure 3.15: *Closeup of gate geometry—the gate cross section shown is its projection on the yz plane.*

The vertical force is the weight of the column of fluid above the gate. Thus, the vertical component of the hydrostatic forces for Problems 1 and 2, F_{z1} and F_{z2}, are

$$F_{z1} = 2\rho g V_1 \quad \text{and} \quad F_{z2} = \rho g V_2 \tag{3.84}$$

From the geometry indicated in Figure 3.15, the volume for Problem 1 is

$$V_1 = \underbrace{\frac{H}{2}\left(\sqrt{3}\,H\right)(4W)}_{upper\ rectangle} + \underbrace{\frac{1}{2}(3H)\left(\sqrt{3}\,H\right)(4W)}_{lower\ triangle} = 8\sqrt{3}\,WH^2 \tag{3.85}$$

Also, the volume for Problem 2 is

$$V_2 = \frac{1}{2}(3H)\left(\sqrt{3}\,H\right)(4W) = 6\sqrt{3}\,WH^2 \tag{3.86}$$

Therefore, the vertical forces for the two problems are

$$F_{z1} = 2\rho g\left(8\sqrt{3}\,WH^2\right) = 16\sqrt{3}\,\rho g WH^2 \tag{3.87}$$

$$F_{z2} = \rho g\left(6\sqrt{3}\,WH^2\right) = 6\sqrt{3}\,\rho g WH^2 \tag{3.88}$$

so that the total vertical component of the hydrostatic force is

$$F_z = F_{z1} + F_{z2} = 22\sqrt{3}\,\rho g WH^2 \tag{3.89}$$

Therefore, in vector form, the hydrostatic force on the gate is

$$\mathbf{F} = 66\rho g WH^2\left(-\mathbf{i} + \frac{\sqrt{3}}{3}\mathbf{k}\right) \tag{3.90}$$

We can check the validity of our answer by observing that, because the pressure is everywhere normal to the gate, the total force must also be normal to the gate. Noting that $\cos 60° = \sqrt{3}/2$ and $\sin 60° = 1/2$, we can rewrite the final answer as follows.

$$\mathbf{F} = 44\sqrt{3}\,\rho g WH^2\left(-\sin 60°\mathbf{i} + \cos 60°\mathbf{k}\right) \tag{3.91}$$

The unit vector in parentheses is indeed normal to the gate, which is inclined at 60° to the horizontal.

3.9 Buoyancy in a Uniform Fluid

Statement of the Problem: *A floating object of volume V has 1/2 of its volume submerged in water, with density* $\rho_w = 1000\ kg/m^3$.

(a) *What fraction of the object's volume will be submerged if the fluid is carbon tetrachloride, with density* $\rho_c = 1590\ kg/m^3$?

(b) *What fraction of its volume will be submerged in water on the moon where gravitational acceleration is one-sixth that on Earth?*

3.9. BUOYANCY IN A UNIFORM FLUID

Solution: First, note that the only forces acting on the body are its weight, mg, and the buoyancy force, F_{buoy}. Denoting the density of the body by ρ_o, we have

$$mg = \rho_o g V \tag{3.92}$$

We appeal to *Archimedes Principle*, which tells us the buoyancy force is equal to the weight of the displaced fluid. In terms of the density of water, ρ_w, the buoyancy force is

$$F_{buoy} = \rho_w g \left(\frac{1}{2}V\right) = \frac{1}{2}\rho_w g V \tag{3.93}$$

Balancing these forces yields

$$\rho_o g V = \frac{1}{2}\rho_w g V \implies \rho_o = \frac{1}{2}\rho_w \tag{3.94}$$

3.9(a): Since the density of carbon tetrachloride is ρ_c, the gravitational and buoyancy forces are

$$mg = \rho_o g V = \frac{1}{2}\rho_w g V \quad \text{and} \quad F_{buoy} = \rho_c g (\lambda V) = \lambda \rho_c g V \tag{3.95}$$

where λ is the fraction of the volume that is submerged. Thus,

$$F_{buoy} = mg \implies \lambda \rho_c g V = \frac{1}{2}\rho_w g V \tag{3.96}$$

Therefore, solving for λ yields

$$\lambda = \frac{\rho_w}{2\rho_c} = \frac{1000 \text{ kg/m}^3}{2\left(1590 \text{ kg/m}^3\right)} = 0.314 \tag{3.97}$$

Hence, the object will be 31.4% submerged in carbon tetrachloride.

3.9(b): Let g_m denote gravitational acceleration on the moon. The weight of the object will be

$$mg_m = \rho_o g_m V = \frac{1}{2}\rho_w g_m V \tag{3.98}$$

Assuming a fraction ξ of the volume is submerged, the buoyancy force will be

$$F_{buoy} = \rho_w g_m (\xi V) = \xi \rho_w g_m V \tag{3.99}$$

Balancing the gravitational and buoyancy forces gives

$$\xi \rho_w g_m V = \frac{1}{2}\rho_w g_m V \implies \xi = \frac{1}{2} \tag{3.100}$$

Therefore, half of the object's volume is submerged. This is the same as on Earth as both forces are proportional to the gravitational acceleration.

3.10 Buoyancy in a Dual-Fluid Tank

Statement of the Problem: *A cube is placed in a tank containing two unmixed layers of fluid as shown in Figure 3.16. The bottom of the cube sinks to an unknown distance, D, below the surface. Assuming the cube floats, determine D as a function of H and λ. Then, compute the value of λ for which the cube just becomes totally submerged.*

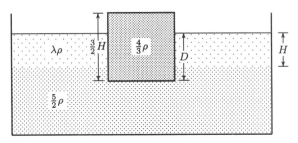

Figure 3.16: *Cube immersed in a dual-fluid tank.*

Solution: The weight of the cube, W, is

$$W = \frac{4}{3}\rho g V = \frac{4}{3}\rho g \left(\frac{3}{2}H\right)^3 = \frac{9}{2}\rho g H^3 \qquad (3.101)$$

We appeal to *Archimedes Principle*, which tells us the buoyancy force is equal to the weight of the displaced fluid. The volume of fluid displaced in the upper layer of fluid of density $\lambda\rho$ is

$$V_u = H\left(\frac{3}{2}H\right)^2 = \frac{9}{4}H^3 \qquad (3.102)$$

The volume displaced in the lower layer of density $\frac{5}{2}\rho$ is

$$V_\ell = (D-H)\left(\frac{3}{2}H\right)^2 = \frac{9}{4}(D-H)H^2 \qquad (3.103)$$

So, the buoyancy force, F_{buoy}, is

$$\begin{aligned} F_{buoy} &= \lambda\rho g V_u + \frac{5}{2}\rho g V_\ell = \frac{9}{4}\lambda\rho g H^3 + \frac{45}{8}\rho g(D-H)H^2 \\ &= \frac{9}{8}\rho g H^2[2\lambda H + 5(D-H)] = \frac{9}{8}\rho g H^2[(2\lambda-5)H + 5D] \end{aligned} \qquad (3.104)$$

The buoyancy force is balanced by the weight of the cube, wherefore

$$\frac{9}{2}\rho g H^3 = \frac{9}{8}\rho g H^2[(2\lambda-5)H + 5D] \implies 4H = (2\lambda-5)H + 5D \qquad (3.105)$$

Thus, solving for D, we find

$$D = \frac{(9-2\lambda)}{5}H \qquad (3.106)$$

Finally, the cube just becomes totally submerged when $D = \frac{3}{2}H$. This will occur when

$$\frac{3}{2}H = \frac{(9-2\lambda)}{5}H \implies \lambda = \frac{3}{4} \qquad (3.107)$$

Chapter 4

Kinematics

This chapter deals with the mathematical formulation needed to describe fluid motion. First, it focuses on how we define the position, velocity and acceleration in both the Eulerian and Lagrangian descriptions. The Lagrangian description follows a fluid particle as it moves, while the Eulerian description is fixed at the same point in space for all time. Next, the chapter deals with vorticity and circulation, important fluid-flow properties that play a key role in describing and quantifying fluid motion. The presence or absence of vorticity determines whether a given flow is rotational or irrotational, respectively. Then, the chapter turns to the calculation of the streamlines from a given velocity vector. This is followed by problems involving surface fluxes, i.e., the rate at which mass or momentum carried by the fluid moves through a given area. The chapter concludes with two problems illustrating the Reynolds Transport Theorem. An understanding of kinematics as presented in the preceding problems is needed to fully comprehend the meaning and significance of this theorem. It permits us to compute the rate of change of properties in a finite volume that contains different fluid particles from one instant to the next in terms of properties of the fixed volume. In this sense, it bridges the gap between the Lagrangian and Eulerian descriptions, and lays the foundation for derivation of the conservation laws. The chapter includes the following problems.

Section 4.1 Eulerian Description: This is a simple problem introducing the concept of the Eulerian derivative and how to calculate acceleration. We compute unsteady and convective accelerations, and then sum them to arrive at the total rate of change following a fluid particle, i.e., the Eulerian derivative of the velocity.

Section 4.2 Lagrangian Description: This problem illustrates how we compute the position of a fluid particle in the Lagrangian description when the velocity field is given.

Section 4.3 Vorticity—Rotational Flow: A straightforward calculation of vorticity, including determination of whether or not the given flow is irrotational.

Section 4.4 Vorticity—Irrotational Flow: This is a calculation in which the given velocity vector includes unknown constants. Their values are determined by setting each component of the vorticity equal to zero.

Section 4.5 Circulation: This is a calculation of the circulation attending a given velocity field over a specified contour. Note that the integrals are always done in the positive sense. The sign of each part of the contour integral is determined by the vector differential arclength, $d\mathbf{s}$. The solution also verifies that the circulation can be computed as the integral of the vorticity vector over the area bounded by the contour.

Section 4.6 Streamlines—Cartesian Coordinates: Calculation of the streamlines for a given velocity field in Cartesian coordinates.

Section 4.7 Streamlines—Cylindrical Coordinates: Calculation of the streamlines for a given velocity field in cylindrical coordinates.

Section 4.8 Surface Fluxes: This problem illustrates computation of the mass flux and the momentum flux across a given area. The area is a square with unit normal parallel to the x axis. The solution method is sufficiently general to apply to complex geometries and velocities.

Section 4.9 Reynolds Transport Theorem I: This is the first of two problems that use the Reynolds Transport Theorem. The problem specifies the rate of change of students in a room, and asks for the speed of a line of students leaving the room after a fluid-mechanics exam.

Section 4.10 Reynolds Transport Theorem II: This is the second problem using the Reynolds Transport Theorem, involving development of an equation for the movement of two lines. It also is more subtle since some people are leaving or entering a line at a constant rate, which adds another nonvanishing term to the governing equation.

4.1 Eulerian Description

Statement of the Problem: *Consider a flow for which the fluid velocity is*

$$\mathbf{u} = ax^2 t\,\mathbf{i} - ay^3 t^2\,\mathbf{j} + e^{2z^2}\,\mathbf{k}$$

where a is a constant. Also, x, y and z are spatial coordinates, while t is time. All quantities are dimensionless. Compute the unsteady, convective and total acceleration for this flow.

Solution: The total acceleration, $\mathbf{a}$ is given by the Eulerian derivative of the velocity vector, i.e.,

$$\mathbf{a} = \underbrace{\frac{d\mathbf{u}}{dt}}_{Total} = \underbrace{\frac{\partial \mathbf{u}}{\partial t}}_{Unsteady} + \underbrace{(\mathbf{u}\cdot\nabla)\mathbf{u}}_{Convective} \qquad (4.1)$$

First, consider the unsteady acceleration. We have

$$\mathbf{a}_{unsteady} = \frac{\partial \mathbf{u}}{\partial t} = \frac{\partial}{\partial t}\left(ax^2 t\,\mathbf{i} - ay^3 t^2\,\mathbf{j} + e^{2z^2}\,\mathbf{k}\right) = ax^2\,\mathbf{i} - 2ay^3 t\,\mathbf{j} \qquad (4.2)$$

Turning to the convective acceleration, there follows

$$\begin{aligned}\mathbf{a}_{convective} &= (\mathbf{u}\cdot\nabla)\mathbf{u} = \left(u\frac{\partial}{\partial x} + v\frac{\partial}{\partial y} + w\frac{\partial}{\partial z}\right)(u\mathbf{i} + v\mathbf{j} + w\mathbf{k}) \\ &= \left(u\frac{\partial u}{\partial x} + v\frac{\partial u}{\partial y} + w\frac{\partial u}{\partial z}\right)\mathbf{i} + \left(u\frac{\partial v}{\partial x} + v\frac{\partial v}{\partial y} + w\frac{\partial v}{\partial z}\right)\mathbf{j} \\ &\quad + \left(u\frac{\partial w}{\partial x} + v\frac{\partial w}{\partial y} + w\frac{\partial w}{\partial z}\right)\mathbf{k}\end{aligned} \qquad (4.3)$$

Substituting the given velocity components yields

$$\mathbf{a}_{convective} = \left[ax^2 t\frac{\partial}{\partial x}\left(ax^2 t\right)\right]\mathbf{i} + \left[-ay^3 t^2\frac{\partial}{\partial y}\left(-ay^3 t^2\right)\right]\mathbf{j} + \left[e^{2z^2}\frac{\partial}{\partial z}\left(e^{2z^2}\right)\right]\mathbf{k} \qquad (4.4)$$

4.2. LAGRANGIAN DESCRIPTION

So, performing the indicated differentiation,

$$\begin{aligned}\mathbf{a}_{convective} &= \left(ax^2t\right)(2axt)\mathbf{i} + \left(ay^3t^2\right)\left(3ay^2t^2\right)\mathbf{j} + \left(e^{2z^2}\right)\left(4ze^{2z^2}\right)\mathbf{k} \\ &= 2a^2x^3t^2\mathbf{i} + 3a^2y^5t^4\mathbf{j} + 4ze^{4z^2}\mathbf{k}\end{aligned} \quad (4.5)$$

Therefore, the total acceleration is

$$\begin{aligned}\mathbf{a} &= \mathbf{a}_{unsteady} + \mathbf{a}_{convective} \\ &= \left(ax^2\mathbf{i} - 2ay^3t\,\mathbf{j}\right) + \left(2a^2x^3t^2\mathbf{i} + 3a^2y^5t^4\mathbf{j} + 4ze^{4z^2}\mathbf{k}\right) \\ &= ax^2\left(1 + 2axt^2\right)\mathbf{i} + ay^3t\left(3ay^2t^3 - 2\right)\mathbf{j} + 4ze^{4z^2}\mathbf{k}\end{aligned} \quad (4.6)$$

4.2 Lagrangian Description

Statement of the Problem: *We are given the following velocity vector:*

$$\mathbf{u} = Ax\,\mathbf{i} + By^2\mathbf{j} + Cz^2e^{Dt}\mathbf{k}$$

where A and D are constants of dimensions 1/time. Also, B and C are constants of dimensions 1/(length·time). Find the Lagrangian coordinates of the position vector, $\mathbf{r} = x\,\mathbf{i} + y\,\mathbf{j} + z\,\mathbf{k}$, *in terms of A, B, C and D and the initial (t = 0) coordinates* x_o, y_o *and* z_o.

Solution: By definition, the Lagrangian coordinates are given by

$$u = \left(\frac{\partial x}{\partial t}\right)_{x_o, y_o, z_o} = Ax \quad \Longrightarrow \quad \frac{dx}{x} = A\,dt \quad (4.7)$$

$$v = \left(\frac{\partial y}{\partial t}\right)_{x_o, y_o, z_o} = By^2 \quad \Longrightarrow \quad \frac{dy}{y^2} = B\,dt \quad (4.8)$$

$$w = \left(\frac{\partial z}{\partial t}\right)_{x_o, y_o, z_o} = Cz^2e^{Dt} \quad \Longrightarrow \quad \frac{dz}{z^2} = Ce^{Dt}\,dt \quad (4.9)$$

Before proceeding, note that we can integrate each equation independently because of the special form of the given velocity. That is, for example, the given u is a function only of x and t, with no dependence upon y or z. Similarly, v depends only upon y and t, while w is a function only of z and t. Consequently, there is no coupling amongst the three equations for the Lagrangian coordinates. For more general velocities where each velocity component is a function of x, y, z and t, the three equations are coupled, and the solution is far more complicated.

Solving for x, we have

$$\ell n\, x = At + \ell n\, x_o \quad (4.10)$$

where we use the fact that $x = x_o$ at $t = 0$. Exponentiating both sides of this equation gives

$$e^{\ell n\, x} = e^{At + \ell n\, x_o} \quad \Longrightarrow \quad x = x_o e^{At} \quad (4.11)$$

Solving for y, there follows

$$-\frac{1}{y} = Bt - \frac{1}{y_o} \quad \Longrightarrow \quad \frac{1}{y} = \frac{1 - By_o t}{y_o} \quad (4.12)$$

where we use the initial condition that tells us $y = y_o$ at $t = 0$. Thus, taking the reciprocal of both sides of this equation, the solution for y is

$$y = \frac{y_o}{1 - By_o t} \tag{4.13}$$

Solving for z, straightforward integration yields

$$-\frac{1}{z} = \frac{C}{D}e^{Dt} - \frac{1}{z_o} \implies \frac{1}{z} = \frac{D - Cz_o e^{Dt}}{Dz_o} \tag{4.14}$$

where we note that $z = z_o$ at $t = 0$. Taking the reciprocal of both sides of this equation, the solution for z is

$$z = \frac{Dz_o}{D - Cz_o e^{Dt}} \tag{4.15}$$

Therefore, the complete Lagrangian description of the position vector is

$$\mathbf{r} = x_o e^{At} \mathbf{i} + \frac{y_o}{1 - By_o t} \mathbf{j} + \frac{Dz_o}{D - Cz_o e^{Dt}} \mathbf{k} \tag{4.16}$$

4.3 Vorticity—Rotational Flow

Statement of the Problem: *A flowfield has the following velocity vector*

$$\mathbf{u} = y^2 z^2 \mathbf{i} + 2xyz^2 \mathbf{j} + 3x^3 z \mathbf{k}$$

where all quantities are dimensionless.

(a) Determine whether or not this flow is irrotational.

(b) If we drop the z component of the velocity, is the flow irrotational?

Solution: The flow is irrotational if the vorticity vector is zero. Thus, we must compute the vorticity, which is $\boldsymbol{\omega} = \nabla \times \mathbf{u}$.

4.3(a): The vorticity is given by the following determinant.

$$\begin{aligned}
\boldsymbol{\omega} &= \begin{vmatrix} \mathbf{i} & \mathbf{j} & \mathbf{k} \\ \frac{\partial}{\partial x} & \frac{\partial}{\partial y} & \frac{\partial}{\partial z} \\ y^2 z^2 & 2xyz^2 & 3x^3 z \end{vmatrix} = \mathbf{i}\left[\frac{\partial}{\partial y}\left(3x^3 z\right) - \frac{\partial}{\partial z}\left(2xyz^2\right)\right] \\
&\quad + \mathbf{j}\left[\frac{\partial}{\partial z}\left(y^2 z^2\right) - \frac{\partial}{\partial x}\left(3x^3 z\right)\right] + \mathbf{k}\left[\frac{\partial}{\partial x}\left(2xyz^2\right) - \frac{\partial}{\partial y}\left(y^2 z^2\right)\right] \\
&= \mathbf{i}\left(0 - 4xyz\right) + \mathbf{j}\left(2y^2 z - 9x^2 z\right) + \mathbf{k}\left(2yz^2 - 2yz^2\right) \\
&= -4xyz\,\mathbf{i} + \left(2y^2 - 9x^2\right) z\,\mathbf{j} \tag{4.17}
\end{aligned}$$

Thus, since the vorticity is nonzero, this velocity field is **rotational**.

4.4. VORTICITY—IRROTATIONAL FLOW

4.3(b): Dropping the z component of the velocity, the vorticity is now given by

$$\omega = \begin{vmatrix} \mathbf{i} & \mathbf{j} & \mathbf{k} \\ \dfrac{\partial}{\partial x} & \dfrac{\partial}{\partial y} & \dfrac{\partial}{\partial z} \\ y^2 z^2 & 2xyz^2 & 0 \end{vmatrix} = \mathbf{i}\left[\dfrac{\partial}{\partial y}(0) - \dfrac{\partial}{\partial z}(2xyz^2)\right]$$

$$+ \mathbf{j}\left[\dfrac{\partial}{\partial z}(y^2 z^2) - \dfrac{\partial}{\partial x}(0)\right] + \mathbf{k}\left[\dfrac{\partial}{\partial x}(2xyz^2) - \dfrac{\partial}{\partial y}(y^2 z^2)\right]$$

$$= \mathbf{i}(0 - 4xyz) + \mathbf{j}(2y^2 z - 0) + \mathbf{k}(2yz^2 - 2yz^2)$$

$$= -4xyz\,\mathbf{i} + 2y^2 z\,\mathbf{j} \tag{4.18}$$

Thus, since the vorticity is still nonzero, this velocity field is also **rotational**.

4.4 Vorticity—Irrotational Flow

Statement of the Problem: *Consider the following velocity vector.*

$$\mathbf{u} = \left(3Ay^2 z^2 - 5y^2 z^2\right)\mathbf{i} - 4Bxyz^2\mathbf{j} + 5xy^2 z\,\mathbf{k}$$

All quantities, including the unknown constants A and B, are dimensionless. Determine the values of A and B that yield an irrotational flow.

Solution: The flow is irrotational if the vorticity vector is zero. Thus, we must compute the vorticity, which is $\omega = \nabla \times \mathbf{u}$. The vorticity is given by the following determinant.

$$\omega = \begin{vmatrix} \mathbf{i} & \mathbf{j} & \mathbf{k} \\ \dfrac{\partial}{\partial x} & \dfrac{\partial}{\partial y} & \dfrac{\partial}{\partial z} \\ \left(3Ay^2 z^2 - 5y^2 z^2\right) & -4Bxyz^2 & 5xy^2 z \end{vmatrix}$$

$$= \mathbf{i}\left[\dfrac{\partial}{\partial y}(5xy^2 z) - \dfrac{\partial}{\partial z}(-4Bxyz^2)\right]$$

$$+ \mathbf{j}\left[\dfrac{\partial}{\partial z}(3Ay^2 z^2 - 5y^2 z^2) - \dfrac{\partial}{\partial x}(5xy^2 z)\right]$$

$$+ \mathbf{k}\left[\dfrac{\partial}{\partial x}(-4Bxyz^2) - \dfrac{\partial}{\partial y}(3Ay^2 z^2 - 5y^2 z^2)\right] \tag{4.19}$$

So, performing the indicated differentiation, the vorticity is

$$\omega = \mathbf{i}(10xyz + 8Bxyz) + \mathbf{j}(6Ay^2 z - 10y^2 z - 5y^2 z)$$
$$+ \mathbf{k}(-4Byz^2 - 6Ayz^2 + 10yz^2)$$
$$= (10 + 8B)xyz\,\mathbf{i} + (6A - 15)\,y^2 z\,\mathbf{j} + (10 - 4B - 6A)yz^2\mathbf{k} \tag{4.20}$$

Clearly, the x and y components of the vorticity are zero provided

$$A = \dfrac{5}{2} \quad \text{and} \quad B = -\dfrac{5}{4} \tag{4.21}$$

Also, for this pair of values, we have

$$10 - 4B - 6A = 10 - 4\left(-\frac{5}{4}\right) - 6\left(\frac{5}{2}\right) = 10 + 5 - 15 = 0 \qquad (4.22)$$

Therefore, the z component of the vorticity is also zero, wherefore the vorticity vector is zero. Thus, the flow is **irrotational** when A and B are given by Equation (4.21).

4.5 Circulation

Statement of the Problem: *Consider the two-dimensional velocity field given by*

$$\mathbf{u} = U\frac{xy^2}{H^3}\mathbf{i} - U\frac{x^2 y}{H^3}\mathbf{j}$$

where U and H are constants of dimensions length/time and length, respectively. Compute the circulation, Γ, for the contour C shown in Figure 4.1. Check your answer by integrating the vorticity over the area bounded by the contour.

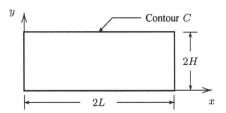

Figure 4.1: *Closed contour for computing circulation.*

Solution: By definition, the circulation is

$$\Gamma = \oint_C \mathbf{u} \cdot d\mathbf{s} \qquad (4.23)$$

where we use the standard mathematical convention that integration is positive in the counter-clockwise direction. Note that aerodynamicists generally prefer to integrate in the clockwise direction when computing circulation. This yields the *aerodynamic circulation*, Γ_a, which is given by $\Gamma_a = -\Gamma$.

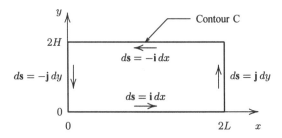

Figure 4.2: *Rectangular contour with differential direction vectors.*

4.5. CIRCULATION

In general, to evaluate a line integral, we treat the integral as the sum of conventional integrals on each segment of the contour, with the sign determined by the differential distance vector, $d\mathbf{s}$. Hence, referring to Figure 4.2, we see that

$$\Gamma = \int_0^{2L} \mathbf{u}(x,0) \cdot (\mathbf{i}\,dx) + \int_0^{2H} \mathbf{u}(2L,y) \cdot (\mathbf{j}\,dy)$$

$$+ \int_0^{2L} \mathbf{u}(x,2H) \cdot (-\mathbf{i}\,dx) + \int_0^{2H} \mathbf{u}(0,y) \cdot (-\mathbf{j}\,dy)$$

$$= \int_0^{2L} [u(x,0) - u(x,2H)]\,dx + \int_0^{2H} [v(2L,y) - v(0,y)]\,dy \qquad (4.24)$$

Thus, for the given velocity vector, we have

$$\left.\begin{array}{rcl} \mathbf{u}(x,0) &=& 0 \\[4pt] \mathbf{u}(x,2H) &=& U\dfrac{x(2H)^2}{H^3}\mathbf{i} - U\dfrac{x^2(2H)}{H^3}\mathbf{j} = \dfrac{4Ux}{H}\mathbf{i} - \dfrac{2Ux^2}{H^2}\mathbf{j} \\[8pt] \mathbf{u}(2L,0) &=& U\dfrac{2Ly^2}{H^3}\mathbf{i} - U\dfrac{(2L)^2 y}{H^3}\mathbf{j} = \dfrac{2ULy^2}{H^3}\mathbf{i} - \dfrac{4UL^2 y}{H^3}\mathbf{j} \\[8pt] \mathbf{u}(0,y) &=& 0 \end{array}\right\} \qquad (4.25)$$

This tells us that the relevant velocity differences are

$$\left.\begin{array}{rcl} u(x,0) - u(x,2H) &=& 0 - \dfrac{4Ux}{H} = -\dfrac{4Ux}{H} \\[8pt] v(2L,y) - v(0,y) &=& -\dfrac{4UL^2 y}{H^3} - 0 = -\dfrac{4UL^2 y}{H^3} \end{array}\right\} \qquad (4.26)$$

Therefore, the circulation is

$$\Gamma = \int_0^{2L}\left[-\frac{4Ux}{H}\right]dx + \int_0^{2H}\left[-\frac{4UL^2 y}{H^3}\right]dy$$

$$= -\frac{2Ux^2}{H}\bigg|_{x=0}^{x=2L} - \frac{2UL^2 y^2}{H^3}\bigg|_{y=0}^{y=2H} = -\frac{8UL^2}{H} - \frac{8UL^2}{H} = -\frac{16UL^2}{H} \qquad (4.27)$$

To check for consistency, note that the vorticity is

$$\boldsymbol{\omega} = \left(\frac{\partial v}{\partial x} - \frac{\partial u}{\partial y}\right)\mathbf{k} = \left(-\frac{2Uxy}{H^3} - \frac{2Uxy}{H^3}\right)\mathbf{k} = -\frac{4Uxy}{H^3}\mathbf{k} \qquad (4.28)$$

So, the integral of $\boldsymbol{\omega} \cdot \mathbf{k}$ over the area bounded by the contour C is

$$\iint_A \boldsymbol{\omega} \cdot \mathbf{k}\,dA = \int_0^{2H}\int_0^{2L}\left[-\frac{4Uxy}{H^3}\right]dx\,dy = \int_0^{2H}\left[-\frac{2Ux^2 y}{H^3}\right]_{x=0}^{x=2L}dy$$

$$= \int_0^{2H}\left[-\frac{8UL^2 y}{H^3}\right]dy = -\frac{4UL^2 y^2}{H^3}\bigg|_{y=0}^{y=2H} = -\frac{16UL^2}{H} \qquad (4.29)$$

This is identical to the result obtained above.

4.6 Streamlines—Cartesian Coordinates

Statement of the Problem: *Consider the incompressible flow whose velocity field is*

$$\mathbf{u} = Ay^2 e^{x/y}\,\mathbf{i} + \frac{1}{x}e^{x/y}\,\mathbf{j}$$

where all quantities are dimensionless and A is a constant.

(a) Derive an equation defining the streamlines.

(b) Sketch the streamlines in the first quadrant for $A = 1/50$.

Solution: First, we note that since the flow is two dimensional, the differential equation defining the streamlines is

$$\frac{dy}{dx} = \frac{v}{u} \tag{4.30}$$

where u and v denote the x and y components of **u**, respectively.

4.6(a): Substituting the given velocity components, i.e., $u = Ay^2 e^{x/y}$ and $v = e^{x/y}/x$, the equation for the streamlines becomes

$$\frac{dy}{dx} = \frac{e^{x/y}/x}{Ay^2 e^{x/y}} = \frac{1}{Axy^2} \tag{4.31}$$

Hence, separating variables, we find

$$\frac{dx}{x} = Ay^2\,dy \tag{4.32}$$

so that the equation for the streamlines assumes the following form.

$$\int \frac{dx}{x} = \int Ay^2\,dy \tag{4.33}$$

Evaluating the integrals yields

$$\ln x = \frac{1}{3}Ay^3 + \tilde{C} \tag{4.34}$$

where $\tilde{C}$ is an integration constant. Exponentiating both sides of this equation shows that

$$x = e^{Ay^3/3}e^{\tilde{C}} \tag{4.35}$$

Finally, defining $C \equiv e^{\tilde{C}}$, the equation for the streamlines is as follows.

$$x = Ce^{Ay^3/3} \tag{4.36}$$

4.6(b): When $A = 1/50$, the streamlines in the first quadrant are as shown in Figure 4.3.

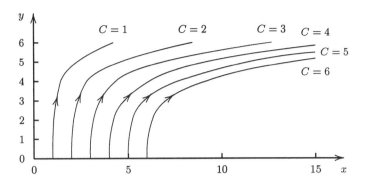

Figure 4.3: *Streamlines for $A = 1/50$.*

4.7 Streamlines—Cylindrical Coordinates

Statement of the Problem: *Derive an equation defining the streamlines for the incompressible flow whose velocity field is*

$$\mathbf{u} = \frac{2K \cos\theta \sin\theta}{r^2} \mathbf{e}_r + \frac{K \cos^2\theta}{r^2} \mathbf{e}_\theta$$

All quantities are dimensionless and K is a constant.

Solution: Note that since the flow is axisymmetric, the differential equation defining the streamlines is

$$\frac{dr}{u_r} = \frac{r d\theta}{u_\theta} \quad \Longrightarrow \quad \frac{dr}{d\theta} = \frac{r u_r}{u_\theta} \tag{4.37}$$

Substituting $u_r = 2K \cos\theta \sin\theta / r^2$ and $u_\theta = K \cos^2\theta / r^2$, the equation for the streamlines becomes

$$\frac{dr}{d\theta} = \frac{r \cdot 2K \cos\theta \sin\theta}{r^2} \frac{r^2}{K \cos^2\theta} = \frac{2r \sin\theta}{\cos\theta} \tag{4.38}$$

Hence, separating variables, we find

$$\frac{dr}{r} = \frac{2 \sin\theta \, d\theta}{\cos\theta} \tag{4.39}$$

Integration of this equation yields

$$\int \frac{dr}{r} = \int \frac{2 \sin\theta \, d\theta}{\cos\theta} \quad \Longrightarrow \quad \ln r = -2\ln(\cos\theta) + \tilde{C} \tag{4.40}$$

where $\tilde{C}$ is an integration constant. Exponentiating both sides of this equation shows that

$$r = e^{-2\ln(\cos\theta)} \cdot e^{\tilde{C}} = \frac{e^{\tilde{C}}}{\cos^2\theta} \tag{4.41}$$

Finally, defining $C \equiv e^{\tilde{C}}$, the equation for the streamlines is as follows.

$$r = \frac{C}{\cos^2\theta} \tag{4.42}$$

4.8 Surface Fluxes

Statement of the Problem: *An incompressible flow has a velocity vector given by*

$$\mathbf{u} = U\frac{x^2 y}{a^3}\mathbf{i} - U\frac{xy^2}{a^3}\mathbf{j}$$

where U and a are constant velocity and length scales, respectively. Compute the mass flux, $\dot{M}$, and momentum flux, $\dot{\mathbf{P}}$, across the plane shown in Figure 4.4.

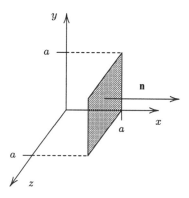

Figure 4.4: *Plane normal to the x axis.*

Solution: The unit normal to the plane is $\mathbf{n} = \mathbf{i}$, so that

$$\mathbf{u}\cdot\mathbf{n} = \left(U\frac{x^2 y}{a^3}\mathbf{i} - U\frac{xy^2}{a^3}\mathbf{j}\right)\cdot\mathbf{i} = U\frac{x^2 y}{a^3} \qquad (4.43)$$

Hence, the mass flux across the plane is

$$\begin{aligned}
\dot{M} &= \iint_A \rho\,(\mathbf{u}\cdot\mathbf{n})\,dA = \int_0^a\int_0^a \rho\left[U\frac{x^2 y}{a^3}\right]_{x=a}dy\,dz \\
&= \int_0^a\int_0^a \rho U\frac{y}{a}dy\,dz = \frac{\rho U}{a}\int_0^a\int_0^a y\,dy\,dz \\
&= \frac{\rho U}{a}\int_0^a\left[\frac{1}{2}y^2\right]_{y=0}^{y=a}dz = \frac{\rho U}{a}\int_0^a\frac{1}{2}a^2\,dz = \frac{1}{2}\rho U a\int_0^a dz \\
&= \frac{1}{2}\rho U a^2 \qquad (4.44)
\end{aligned}$$

Also, the momentum flux across the plane is

$$\begin{aligned}
\dot{\mathbf{P}} &= \iint_A \rho\mathbf{u}\,(\mathbf{u}\cdot\mathbf{n})\,dA = \int_0^a\int_0^a \rho\left[U\frac{x^2 y}{a^3}\mathbf{i} - U\frac{xy^2}{a^3}\mathbf{j}\right]_{x=a}\left[U\frac{x^2 y}{a^3}\right]_{x=a}dy\,dz \\
&= \int_0^a\int_0^a \frac{\rho U}{a^3}\left(a^2 y\,\mathbf{i} - ay^2\,\mathbf{j}\right)\frac{Ua^2 y}{a^3}dy\,dz = \frac{\rho U^2}{a^4}\int_0^a\int_0^a\left(a^2 y^2\,\mathbf{i} - ay^3\,\mathbf{j}\right)dy\,dz \\
&= \frac{\rho U^2}{a^4}\int_0^a\left[\frac{1}{3}a^2 y^3\,\mathbf{i} - \frac{1}{4}ay^4\,\mathbf{j}\right]_{y=0}^{y=a}dz = \frac{\rho U^2}{a^4}\int_0^a\left(\frac{1}{3}a^5\,\mathbf{i} - \frac{1}{4}a^5\,\mathbf{j}\right)dz \\
&= \rho U^2 a\left(\frac{1}{3}\mathbf{i} - \frac{1}{4}\mathbf{j}\right)\cdot a = \rho U^2 a^2\left(\frac{1}{3}\mathbf{i} - \frac{1}{4}\mathbf{j}\right) \qquad (4.45)
\end{aligned}$$

4.9 Reynolds Transport Theorem I

Statement of the Problem: *Students are running out of the classroom after their fluid mechanics exam at a rate of 4 students/sec. Including space between students, each occupies 40 ft³. Using the Reynolds Transport Theorem, with number density replacing mass density, determine the speed of the students as they leave the room. The door is $7\frac{1}{2}$ ft by $3\frac{1}{2}$ ft.*

Solution: We can apply the Reynolds Transport Theorem to this problem if we consider the students to be a system and the room a control volume. The Reynolds Transport Theorem tells us

$$\frac{dN}{dt} = \frac{d}{dt} \iiint_V \rho \, dV + \oiint_S \rho \left(\mathbf{u} \cdot \mathbf{n} \right) dS \qquad (4.46)$$

where N is the total number of students, and ρ is the number density of students, i.e., number of students per unit volume. We must have $dN/dt = 0$ as we are neither creating nor destroying students (it was not a "killer exam"). Also, by definition, the number of students in the room, N_R, is

$$N_R = \iiint_V \rho \, dV \qquad (4.47)$$

So, the Reynolds Transport Theorem simplifies to

$$\frac{dN_R}{dt} = - \oiint_S \rho \left(\mathbf{u} \cdot \mathbf{n} \right) dS \qquad (4.48)$$

In words, this equation says the rate of change of the number of students remaining in the room is minus the net flux of students out of the room.

Now, we know that the rate of decrease of the number of students in the control volume (the room) is 4 students/sec. Therefore, in terms of the notation above, we have

$$\frac{dN_R}{dt} = -4 \, \frac{\text{students}}{\text{sec}} \qquad (4.49)$$

Also, since the space surrounding each student is 40 ft³, the number density is

$$\rho = \frac{1}{40} \, \frac{\text{students}}{\text{ft}^3} \qquad (4.50)$$

Finally, the area of the only part of the control volume through which students pass, the door, is

$$A = (7.5 \text{ ft}) \times (3.5 \text{ ft}) = 26.25 \text{ ft}^2 \qquad (4.51)$$

Substituting into Equation (4.48), and denoting the speed of the students by $U = \mathbf{u} \cdot \mathbf{n}$, we have

$$-4 \, \frac{\text{students}}{\text{sec}} = -\rho U A = -\left(\frac{1}{40} \, \frac{\text{students}}{\text{ft}^3} \right) U \left(26.25 \text{ ft}^2 \right) \qquad (4.52)$$

Solving for U, we conclude that

$$U = 6.1 \, \frac{\text{ft}}{\text{sec}} \tag{4.53}$$

Moving this rapidly, we can reasonably conclude that they are racing home to boast about how well they did on the exam!

4.10 Reynolds Transport Theorem II

Statement of the Problem: *You have just arrived in Las Vegas at your favorite hotel. Once inside, you enter the general check-in line. The archway area leading to the front desk is A_g, and each person occupies a volume V, including space between people. The line moves at an average velocity U_g. While waiting in line, you observe that a new line opens up for those with casino cards. You also observe that someone has just hit the "megabucks" so that no new people are entering the check-in line (they're all watching the gleeful winner). The archway area leading to this line is A_c, and the line moves at an average velocity U_c. Using the Reynolds Transport Theorem, with number density replacing mass density, derive an equation for the rate of change of people in the two lines. Use two separate control volumes, one for each line.*

Solution: We can apply the Reynolds Transport Theorem to this problem if we first consider the total number of people in the original line to be a system and the room a control volume. The number of people in the system is decreasing because of the migration from the general check-in line to the casino-card check-in line. Now, the Reynolds Transport Theorem tells us

$$\frac{dN}{dt} = \frac{d}{dt} \iiint_V \rho \, dV + \oiint_S \rho (\mathbf{u} \cdot \mathbf{n}) \, dS \tag{4.54}$$

where N is the total number of people, and ρ is the number density of people, i.e., number of people per unit volume. Let $dN/dt = -\dot{\mathcal{N}}$ denote the rate at which people move from the general check-in line to the casino-card check-in line. Also, by definition, the number of people in the general check-in line, N_g, is

$$N_g = \iiint_V \rho \, dV \tag{4.55}$$

So, the Reynolds Transport Theorem simplifies to

$$-\dot{\mathcal{N}} = \frac{dN_g}{dt} + \oiint_S \rho (\mathbf{u} \cdot \mathbf{n}) \, dS \tag{4.56}$$

In words, this equation says the rate at which people move to the casino-card check-in line is equal to sum of the rate of change of the number of people in the original line and the rate at which they complete the check-in process and leave the line. From the given information, we know that

$$\oiint_S \rho (\mathbf{u} \cdot \mathbf{n}) \, dS = \frac{U_g A_g}{V} \tag{4.57}$$

Therefore, we conclude that

$$-\dot{\mathcal{N}} = \frac{dN_g}{dt} + \frac{U_g A_g}{V} \tag{4.58}$$

4.10. REYNOLDS TRANSPORT THEOREM II

Similarly, applying the Reynolds Transport Theorem to the casino-card check-in line, the system is gaining people at the same rate that the general check-in line is losing people due to the cross over. Thus, we have

$$\dot{\mathcal{N}} = \frac{dN_c}{dt} + \oiint_S \rho \left(\mathbf{u} \cdot \mathbf{n} \right) dS \qquad (4.59)$$

In words, this equation says the rate at which people move into the casino-card check-in line is equal to sum of the rate of change of the number of people in the casino-card line and the rate at which they complete the check-in process and leave the line. From the given information, we know that

$$\oiint_S \rho \left(\mathbf{u} \cdot \mathbf{n} \right) dS = \frac{U_c A_c}{V} \qquad (4.60)$$

Therefore, we conclude that

$$\dot{\mathcal{N}} = \frac{dN_c}{dt} + \frac{U_c A_c}{V} \qquad (4.61)$$

Finally, combining Equations (4.58) and (4.61), we have

$$-\frac{dN_g}{dt} - \frac{U_g A_g}{V} = \frac{dN_c}{dt} + \frac{U_c A_c}{V} \qquad (4.62)$$

Therefore, the desired equation is

$$\frac{dN_c}{dt} = -\frac{dN_g}{dt} - \frac{U_g A_g + U_c A_c}{V} \qquad (4.63)$$

Chapter 5

Conservation of Mass and Momentum

This chapter introduces the reader to elements of the basic conservation laws in fluid mechanics, most notably three fundamental differential equations. The first is the differential form of the mass-conservation principle, known as the *continuity equation*,. The second is the differential form of the momentum-conservation principle for an inviscid fluid, viz., *Euler's equation.* The third is *Bernoulli's equation*, which is a general solution to Euler's equation for steady, incompressible flows that are irrotational and subject only to conservative body forces. It can also be regarded as an equation for conservation of mechanical energy. The chapter concludes with problems involving velocity measurement techniques based on Bernoulli's equation, i.e., the *Pitot-static tube* and the important issue of *Galilean invariance.* The problems included in this chapter illustrate all of these concepts.

Section 5.1 Irrotationality and Incompressibility I: A simple problem that determines whether or not the given velocity field is irrotational and incompressible.

Section 5.2 Irrotationality and Incompressibility II: This problem is a little more involved than the one in Section 5.1. The given velocity vector includes a constant and we seek a value of the constant that corresponds to irrotational flow and/or incompressible flow.

Section 5.3 Irrotationality and Incompressibility III: A problem in which we solve for the y component of the velocity assuming the flow is incompressible and irrotational.

Section 5.4 Continuity Equation: This problem uses the continuity equation to solve for the density of a flowfield corresponding to a given one-dimensional velocity field.

Section 5.5 Bernoulli's Equation: This problem uses Bernoulli's equation to compute the pressure in terms of a given, three-dimensional velocity field.

Section 5.6 Bernoulli's Equation—Quasi-Steady Flow: This is another problem that uses Bernoulli's equation. Here the flow, although formally unsteady, varies so slowly that it can be considered quasi-steady, wherefore Bernoulli's equation can still be used.

Section 5.7 Euler's Equation: Given a three dimensional flow, this problem demonstrates that it is incompressible, irrotational and satisfies the continuity equation. After that, we use Euler's equation to solve for the pressure. Each component of Euler's equation is integrated separately to achieve the desired result.

Section 5.8 Euler's Equation—Rotating Tank: This problem also uses Euler's equation, but this time in a rotating coordinate frame. We analyze a rotating tank, calculating the shape of the free surface and the rotation rate based on the given conditions.

Section 5.9 Pitot-Static Tube: This problem focuses on the Pitot-static tube. The problem enunciates the relation between flow velocity and the measured values of static and stagnation pressure.

Section 5.10 Galilean Invariance: Here we use a Galilean transformation to transform a given unsteady flow to an equivalent steady flow. This permits us to use Bernoulli's equation to solve for the desired pressures and velocities in the flow.

5.1 Irrotationality and Incompressibility I

Statement of the Problem: *A flowfield has the following velocity vector*

$$\mathbf{u} = \frac{x^3 z}{y^2}\mathbf{i} - \frac{3x^2 z}{y}\mathbf{j} - \frac{3x^2 z^2}{y^2}\mathbf{k}$$

where all quantities are dimensionless.

(a) Determine whether or not this flow is irrotational.

(b) Determine whether or not this flow is incompressible.

Solution: The flow is irrotational if the vorticity vector is zero, and is incompressible if the divergence of the velocity is zero. Thus, in Part (a), we must compute and examine the vorticity, which is $\boldsymbol{\omega} = \nabla \times \mathbf{u}$, while we must compute and examine $\nabla \cdot \mathbf{u}$ in Part (b).

5.1(a): The vorticity vector for this flow, $\boldsymbol{\omega} = \nabla \times \mathbf{u}$, is obtained by evaluating the following determinant.

$$\begin{aligned}
\boldsymbol{\omega} &= \begin{vmatrix} \mathbf{i} & \mathbf{j} & \mathbf{k} \\ \dfrac{\partial}{\partial x} & \dfrac{\partial}{\partial y} & \dfrac{\partial}{\partial z} \\ \dfrac{x^3 z}{y^2} & \dfrac{-3x^2 z}{y} & \dfrac{-3x^2 z^2}{y^2} \end{vmatrix} \\
&= \left[\frac{\partial}{\partial y}\left(\frac{-3x^2 z^2}{y^2}\right) - \frac{\partial}{\partial z}\left(\frac{-3x^2 z}{y}\right)\right]\mathbf{i} \\
&\quad + \left[\frac{\partial}{\partial z}\left(\frac{x^3 z}{y^2}\right) - \frac{\partial}{\partial x}\left(\frac{-3x^2 z^2}{y^2}\right)\right]\mathbf{j} \\
&\quad + \left[\frac{\partial}{\partial x}\left(\frac{-3x^2 z}{y}\right) - \frac{\partial}{\partial y}\left(\frac{x^3 z}{y^2}\right)\right]\mathbf{k} \\
&= \left(\frac{6x^2 z^2}{y^3} + \frac{3x^2}{y}\right)\mathbf{i} + \left(\frac{x^3}{y^2} + \frac{6xz^2}{y^2}\right)\mathbf{j} + \left(\frac{-6xz}{y} + \frac{2x^3 z}{y^3}\right)\mathbf{k} \\
&= \frac{3x^2}{y}\left(1 + \frac{2z^2}{y^2}\right)\mathbf{i} + \frac{x}{y^2}\left(x^2 + 6z^2\right)\mathbf{j} + \frac{2xz}{y}\left(\frac{x^2}{y^2} - 3\right)\mathbf{k} \quad (5.1)
\end{aligned}$$

5.2. IRROTATIONALITY AND INCOMPRESSIBILITY II

Hence, since we have shown that the vorticity is nonzero, necessarily this velocity field is **rotational**.

5.1(b): The divergence of the velocity vector, $\nabla \cdot \mathbf{u}$, is

$$\begin{aligned}
\nabla \cdot \mathbf{u} &= \left(\mathbf{i}\frac{\partial}{\partial x} + \mathbf{j}\frac{\partial}{\partial y} + \mathbf{k}\frac{\partial}{\partial z}\right) \cdot \left(\frac{x^3 z}{y^2}\mathbf{i} - \frac{3x^2 z}{y}\mathbf{j} - \frac{3x^2 z^2}{y^2}\mathbf{k}\right) \\
&= \frac{\partial}{\partial x}\left(\frac{x^3 z}{y^2}\right) + \frac{\partial}{\partial y}\left(-\frac{3x^2 z}{y}\right) + \frac{\partial}{\partial z}\left(-\frac{3x^2 z^2}{y^2}\right) \\
&= \frac{3x^2 z}{y^2} + \frac{3x^2 z}{y^2} - \frac{6x^2 z}{y^2} = 0
\end{aligned} \tag{5.2}$$

Therefore, since we have demonstrated that the divergence of the velocity vector vanishes, the flow must be **incompressible**.

5.2 Irrotationality and Incompressibility II

Statement of the Problem: *A flowfield has the following velocity vector*

$$\mathbf{u} = Ar\cos\theta\, \mathbf{e}_r - r\sin\theta\, \mathbf{e}_\theta$$

where A is a constant and all quantities are dimensionless.

(a) Is there any value of A for which this flow is irrotational?

(b) Is there any value of A for which this flow is incompressible?

Solution: The flow is irrotational if the vorticity vector is zero, and is incompressible if the divergence of the velocity is zero. Thus, in Part (a), we must compute and examine the vorticity, which is $\boldsymbol{\omega} = \nabla \times \mathbf{u}$, while we must compute and examine $\nabla \cdot \mathbf{u}$ in Part (b).

5.2(a): For axisymmetric flow, the vorticity has only a z component and is as follows.

$$\begin{aligned}
\boldsymbol{\omega} &= \left[\frac{1}{r}\frac{\partial}{\partial r}(ru_\theta) - \frac{1}{r}\frac{\partial}{\partial \theta}(u_r)\right]\mathbf{k} \\
&= \left[\frac{1}{r}\frac{\partial}{\partial r}(-r^2\sin\theta) - \frac{1}{r}\frac{\partial}{\partial \theta}(Ar\cos\theta)\right]\mathbf{k} \\
&= \left[\frac{1}{r}(-2r\sin\theta) - \frac{1}{r}(-Ar\sin\theta)\right]\mathbf{k} \\
&= (A - 2)\sin\theta\, \mathbf{k}
\end{aligned} \tag{5.3}$$

Clearly, the flow is irrotational provided we choose $A = 2$.

5.2(b): The divergence of the velocity is

$$\begin{aligned}
\nabla \cdot \mathbf{u} &= \frac{1}{r}\frac{\partial}{\partial r}(ru_r) + \frac{1}{r}\frac{\partial u_\theta}{\partial \theta} = \frac{1}{r}\frac{\partial}{\partial r}(Ar^2\cos\theta) + \frac{1}{r}\frac{\partial}{\partial \theta}(-r\sin\theta) \\
&= \frac{1}{r}(2Ar\cos\theta) + \frac{1}{r}(-r\cos\theta) = (2A - 1)\cos\theta
\end{aligned} \tag{5.4}$$

Therefore, the flow is incompressible provided we select $A = 1/2$.

5.3 Irrotationality and Incompressibility III

Statement of the Problem: *The x component of the velocity vector for a two-dimensional, incompressible, irrotational flow is*

$$u(x,y) = U \left[1 - e^{-\lambda x} \cos \lambda y\right]$$

where U and λ are constant velocity and length scales, respectively. Determine the y component of the velocity vector, $v(x,y)$, assuming there is a stagnation point at $x = y = 0$.

Solution: Since the flow is two-dimensional and incompressible, we know that

$$\frac{\partial u}{\partial x} + \frac{\partial v}{\partial y} = 0 \tag{5.5}$$

So, differentiating the given $u(x,y)$,

$$\frac{\partial v}{\partial y} = -\frac{\partial u}{\partial x} = -\lambda U e^{-\lambda x} \cos \lambda y \tag{5.6}$$

Integrating over y yields

$$v(x,y) = -U e^{-\lambda x} \sin \lambda y + f(x) \tag{5.7}$$

where $f(x)$ is a function of integration. Also, since the flow is irrotational, necessarily the curl of the velocity vector is zero, i.e.,

$$\frac{\partial v}{\partial x} - \frac{\partial u}{\partial y} = 0 \implies \left[\lambda U e^{-\lambda x} \sin \lambda y + \frac{df}{dx}\right] - \left[\lambda U e^{-\lambda x} \sin \lambda y\right] = 0 \tag{5.8}$$

Canceling like terms, all that remains is

$$\frac{df}{dx} = 0 \implies f(x) = C \tag{5.9}$$

where C is a constant of integration. Finally, since there is a stagnation point at $x = y = 0$, we must have

$$v(0,0) = C = 0 \tag{5.10}$$

Therefore, the y component of the velocity vector is

$$v(x,y) = -U e^{-\lambda x} \sin \lambda y \tag{5.11}$$

5.4 Continuity Equation

Statement of the Problem: *For steady flow of a compressible fluid, the velocity vector is*

$$\mathbf{u} = u_o \left(\frac{x}{x_o}\right)^2 \mathbf{i}$$

where u_o and x_o are reference velocity and position, respectively. The fluid density is ρ_o at $x = x_o$. Determine the density, ρ, as a function of ρ_o, x and x_o.

5.5. BERNOULLI'S EQUATION

Solution: In general, the mass-conservation, or continuity, equation is

$$\frac{\partial \rho}{\partial t} + \nabla \cdot (\rho \mathbf{u}) = 0 \tag{5.12}$$

Because the flow is steady, $\partial \rho / \partial t = 0$. Also, since the flow is one-dimensional, the divergence simplifies to the ordinary derivative with respect to x, wherefore

$$\nabla \cdot (\rho \mathbf{u}) = \frac{d}{dx}(\rho u) = 0 \tag{5.13}$$

Integrating over x and using the fact that $\rho = \rho_o$ and $u = u_o$ at $x = x_o$, we find

$$\rho u = \rho_o u_o \implies \rho = \rho_o \frac{u_o}{u} \tag{5.14}$$

Finally, substituting for u, the solution for the density is

$$\rho = \rho_o \left(\frac{x_o}{x}\right)^2 \tag{5.15}$$

5.5 Bernoulli's Equation

Statement of the Problem: *The watering tube shown has a 90° bend and an outlet velocity given by*

$$\mathbf{u}_{out} = 5\mathbf{i} - 2\mathbf{j} \text{ ft/sec}$$

The pressure is atmospheric at the outlet, the fluid enters the tube at a speed of 7 ft/sec and the tube is $h = 10$ in high. What is the pressure at the inlet, p_{in}? Assume the flow is steady and irrotational throughout.

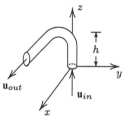

Figure 5.1: *Schematic of a watering tube.*

Solution: Water is an incompressible fluid, and we are given that the flow is steady and irrotational. The only evident body force acting is gravity, which is conservative. Thus, all of the necessary conditions are satisfied to justify use of Bernoulli's equation, so that

$$p_{in} + \frac{1}{2}\rho \mathbf{u}_{in} \cdot \mathbf{u}_{in} + \rho g z_{in} = p_{out} + \frac{1}{2}\rho \mathbf{u}_{out} \cdot \mathbf{u}_{out} + \rho g z_{out} \tag{5.16}$$

Hence, the inlet pressure is

$$p_{in} = p_{out} + \frac{1}{2}\rho \left(\mathbf{u}_{out} \cdot \mathbf{u}_{out} - \mathbf{u}_{in} \cdot \mathbf{u}_{in}\right) + \rho g \left(z_{out} - z_{in}\right) \tag{5.17}$$

54 CHAPTER 5. CONSERVATION OF MASS AND MOMENTUM

Now, the inlet lies at the origin of the coordinate system by design and the speed of the flow at the inlet is given to be 7 ft/sec so that

$$z_{in} = 0 \quad \text{and} \quad \mathbf{u}_{in} = 7\mathbf{k} \text{ ft/sec} \implies \mathbf{u}_{in} \cdot \mathbf{u}_{in} = 49 \text{ ft}^2/\text{sec}^2 \tag{5.18}$$

At the outlet, we have

$$z_{out} = 10 \text{ in} = \frac{5}{6} \text{ ft} \quad \text{and} \quad \mathbf{u}_{out} \cdot \mathbf{u}_{out} = (5\mathbf{i} - 2\mathbf{j}) \cdot (5\mathbf{i} - 2\mathbf{j}) = 29 \text{ ft}^2/\text{sec}^2 \tag{5.19}$$

Also, the outlet pressure is $p_{out} = 14.7$ psi $= 2116.8$ lb/ft^2, while the density of water is $\rho = 1.94$ slug/ft^3. Hence, combining Equations (5.17) through (5.19), the inlet pressure is given by

$$\begin{aligned}
p_{in} &= 2116.8 \, \frac{\text{lb}}{\text{ft}^2} + \frac{1}{2}\left(1.94 \, \frac{\text{slug}}{\text{ft}^3}\right)\left(29 \, \frac{\text{ft}^2}{\text{sec}^2} - 49 \, \frac{\text{ft}^2}{\text{sec}^2}\right) \\
&\quad + \left(1.94 \, \frac{\text{slug}}{\text{ft}^3}\right)\left(32.174 \, \frac{\text{ft}}{\text{sec}^2}\right)\left(\frac{5}{6} \text{ ft} - 0\right) \\
&= 2116.8 \, \frac{\text{lb}}{\text{ft}^2} - 19.4 \, \frac{\text{lb}}{\text{ft}^2} + 52.01 \, \frac{\text{lb}}{\text{ft}^2} = 2149.41 \, \frac{\text{lb}}{\text{ft}^2} \\
&= 14.93 \text{ psi}
\end{aligned} \tag{5.20}$$

5.6 Bernoulli's Equation—Quasi-Steady Flow

Statement of the Problem: *Consider a large tank filled with a liquid of density ρ, and open to the atmosphere as shown in Figure 5.2. Due to the rounding at the entrance to the pipe near the bottom of the tank, the flow can be considered irrotational. Also, the tank is so large relative to the dimensions of the pipe that the flow can be approximated as being steady. If the flow speed at Point 2 in the pipe is $U = 2\sqrt{gh}$, what is the pressure at that point? Express your answer in terms of ρ, g, h and atmospheric pressure, p_o.*

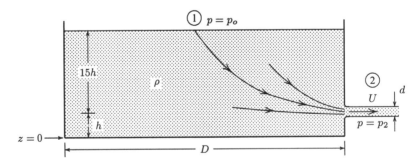

Figure 5.2: *A large cylindrical tank of diameter D with an attached pipe of diameter d; the tank and pipe are such that $d \ll D$.*

Solution: A liquid is an incompressible fluid, and we are given that the flow is steady and irrotational. The only evident body force acting is gravity, which is conservative. Thus, all of the necessary conditions are satisfied to justify use of Bernoulli's equation, so that:

$$p_1 + \frac{1}{2}\rho \mathbf{u}_1 \cdot \mathbf{u}_1 + \rho g z_1 = p_2 + \frac{1}{2}\rho \mathbf{u}_2 \cdot \mathbf{u}_2 + \rho g z_2 \tag{5.21}$$

5.7. EULER'S EQUATION

Hence, the pressure at Point 2 is

$$p_2 = p_1 + \frac{1}{2}\rho\left(\mathbf{u}_1 \cdot \mathbf{u}_1 - \mathbf{u}_2 \cdot \mathbf{u}_2\right) + \rho g\left(z_1 - z_2\right) \tag{5.22}$$

At the free surface, the pressure is atmospheric, the flow velocity is negligibly small and the vertical distance above the bottom of the tank is $16h$. Therefore, we have

$$p_1 = p_o, \quad \mathbf{u}_1 \approx \mathbf{0}, \quad z = 16h \tag{5.23}$$

In the pipe, we are given that the flow speed is $2\sqrt{gh}$ and $z = h$. Thus,

$$\mathbf{u}_2 = 2\sqrt{gh}\,\mathbf{i}, \quad z_2 = h \tag{5.24}$$

Substituting into Equation (5.22) gives

$$p_2 = p_o + \frac{1}{2}\rho\left(0 - 4gh\right) + \rho g(16h - h) = p_o + 13\rho gh \tag{5.25}$$

5.7 Euler's Equation

Statement of the Problem: *The velocity vector for a steady, incompressible flow is*

$$\mathbf{u} = \frac{U}{L^2}\left(yz\,\mathbf{i} + xz\,\mathbf{j} + xy\,\mathbf{k}\right)$$

where U and L are constant velocity and length scales, respectively. The fluid density is ρ and the pressure at $x = y = z = 0$ is p_o. Verify that this flow is irrotational and incompressible. Also, using Euler's equation, determine the pressure as a function of ρ, U, L, p_o, x, y and z. Assume there are no body forces acting. Compare your result with the pressure according to Bernoulli's equation.

Solution: First, to show that the flow is irrotational, we must verify that the curl of the velocity is zero. So,

$$\boldsymbol{\omega} = \frac{U}{L^2}\begin{vmatrix} \mathbf{i} & \mathbf{j} & \mathbf{k} \\ \frac{\partial}{\partial x} & \frac{\partial}{\partial y} & \frac{\partial}{\partial z} \\ yz & xz & xy \end{vmatrix} = \frac{U}{L^2}\left[\frac{\partial}{\partial y}(xy) - \frac{\partial}{\partial z}(xz)\right]\mathbf{i}$$

$$+ \frac{U}{L^2}\left[\frac{\partial}{\partial z}(yz) - \frac{\partial}{\partial x}(xy)\right]\mathbf{j} + \frac{U}{L^2}\left[\frac{\partial}{\partial x}(xz) - \frac{\partial}{\partial y}(yz)\right]\mathbf{k}$$

$$= \frac{U}{L^2}\left[(x-x)\mathbf{i} + (y-y)\mathbf{j} + (z-z)\mathbf{k}\right] = \mathbf{0} \tag{5.26}$$

Since the vorticity is zero, this velocity field is **irrotational**.

The flow is incompressible provided the divergence of the velocity vanishes. Hence,

$$\nabla \cdot \mathbf{u} = \frac{U}{L^2}\left(\mathbf{i}\frac{\partial}{\partial x} + \mathbf{j}\frac{\partial}{\partial y} + \mathbf{k}\frac{\partial}{\partial z}\right) \cdot (yz\,\mathbf{i} + xz\,\mathbf{j} + xy\,\mathbf{k})$$

$$= \frac{\partial}{\partial x}(yz) + \frac{\partial}{\partial y}(xz) + \frac{\partial}{\partial z}(xy) = 0 + 0 + 0 = 0 \tag{5.27}$$

which confirms that the flow is indeed **incompressible**.

In the absence of body forces, the three components of the Euler equation can be written as

$$-\frac{1}{\rho}\frac{\partial p}{\partial x} = u\frac{\partial u}{\partial x} + v\frac{\partial u}{\partial y} + w\frac{\partial u}{\partial z} \tag{5.28}$$

$$-\frac{1}{\rho}\frac{\partial p}{\partial y} = u\frac{\partial v}{\partial x} + v\frac{\partial v}{\partial y} + w\frac{\partial v}{\partial z} \tag{5.29}$$

$$-\frac{1}{\rho}\frac{\partial p}{\partial z} = u\frac{\partial w}{\partial x} + v\frac{\partial w}{\partial y} + w\frac{\partial w}{\partial z} \tag{5.30}$$

We begin by substituting the velocity components into the x component of the Euler equation, viz., Equation (5.28). There follows

$$\begin{aligned}-\frac{1}{\rho}\frac{\partial p}{\partial x} &= \frac{U^2}{L^4}\left[yz\frac{\partial}{\partial x}(yz) + xz\frac{\partial}{\partial y}(yz) + xy\frac{\partial}{\partial z}(yz)\right] \\ &= \frac{U^2}{L^4}\left[yz(0) + xz(z) + xy(y)\right] = \frac{U^2}{L^4}x\left(y^2 + z^2\right)\end{aligned} \tag{5.31}$$

Therefore, the pressure satisfies the following equation.

$$\frac{\partial p}{\partial x} = -\frac{\rho U^2}{L^4}x\left(y^2 + z^2\right) \tag{5.32}$$

So, integrating over x, we find

$$p(x, y, z) = -\frac{\rho U^2}{2L^4}x^2\left(y^2 + z^2\right) + f(y, z) \tag{5.33}$$

where $f(y, z)$ is a "function" of integration. Note that if this were an ordinary differential equation rather than a partial differential equation, there would be a "constant" of integration. Clearly, differentiating the function $f(y, z)$ appearing in Equation (5.33) with respect to x yields zero, as does the ordinary derivative of a constant.

Next, we substitute the velocity components into the y component of the Euler equation, Equation (5.29),

$$\begin{aligned}-\frac{1}{\rho}\frac{\partial p}{\partial y} &= \frac{U^2}{L^4}\left[yz\frac{\partial}{\partial x}(xz) + xz\frac{\partial}{\partial y}(xz) + xy\frac{\partial}{\partial z}(xz)\right] \\ &= \frac{U^2}{L^4}\left[yz(z) + xz(0) + xy(x)\right] = \frac{U^2}{L^4}y\left(x^2 + z^2\right)\end{aligned} \tag{5.34}$$

Hence, the pressure is given by

$$\frac{\partial p}{\partial y} = -\frac{\rho U^2}{L^4}y\left(x^2 + z^2\right) \tag{5.35}$$

Now, we can differentiate the equation for $p(x, y, z)$ in Equation (5.33) with respect to y to arrive at

$$\frac{\partial p}{\partial y} = -\frac{\rho U^2}{L^4}x^2 y + \frac{\partial f}{\partial y} \tag{5.36}$$

Combining Equations (5.35) and (5.36), we have

$$-\frac{\rho U^2}{L^4}x^2 y + \frac{\partial f}{\partial y} = -\frac{\rho U^2}{L^4}y\left(x^2 + z^2\right) \quad \Longrightarrow \quad \frac{\partial f}{\partial y} = -\frac{\rho U^2}{L^4}yz^2 \tag{5.37}$$

5.7. EULER'S EQUATION

Integrating over y and noting that f is at most a function of y and z, there follows:

$$f(y,z) = -\frac{\rho U^2}{2L^4} y^2 z^2 + g(z) \tag{5.38}$$

where $g(z)$ is another "function" of integration. Substituting Equation (5.38) into Equation (5.33) yields

$$p(x,y,z) = -\frac{\rho U^2}{2L^4} \left(x^2 y^2 + x^2 z^2 + y^2 z^2\right) + g(z) \tag{5.39}$$

Finally, we turn to the z component of the Euler equation. Substituting for the velocity components gives

$$-\frac{1}{\rho}\frac{\partial p}{\partial z} = \frac{U^2}{L^4}\left[yz\frac{\partial}{\partial x}(xy) + xz\frac{\partial}{\partial y}(xy) + xy\frac{\partial}{\partial z}(xy)\right]$$

$$= \frac{U^2}{L^4}\left[yz(y) + xz(x) + xy(0)\right] = \frac{U^2}{L^4}z\left(x^2 + y^2\right) \tag{5.40}$$

Hence, the pressure satisfies:

$$\frac{\partial p}{\partial z} = -\frac{\rho U^2}{L^4} z \left(x^2 + y^2\right) \tag{5.41}$$

Differentiating the equation for $p(x,y,z)$ in Equation (5.39) with respect to z tells us that

$$\frac{\partial p}{\partial z} = -\frac{\rho U^2}{L^4} z(x^2 + y^2) + \frac{dg}{dz} \tag{5.42}$$

Equating these two results for $\partial p/\partial z$ gives

$$-\frac{\rho U^2}{L^4} z(x^2 + y^2) + \frac{dg}{dz} = -\frac{\rho U^2}{L^4} z \left(x^2 + y^2\right) \quad \Longrightarrow \quad \frac{dg}{dz} = 0 \tag{5.43}$$

Thus, $g(z)$ is a true constant, and the pressure becomes

$$p(x,y,z) = -\frac{\rho U^2}{2L^4} \left(x^2 y^2 + x^2 z^2 + y^2 z^2\right) + \text{constant} \tag{5.44}$$

Finally, since $p(0,0,0) = p_o$, necessarily the constant is equal to p_o. Therefore, the solution for the pressure is

$$p(x,y,z) = p_o - \frac{\rho U^2}{2L^4} \left(x^2 y^2 + x^2 z^2 + y^2 z^2\right) \tag{5.45}$$

Since the flow is steady, incompressible, irrotational and subject to no body forces, all of the conditions required for Bernoulli's equation are satisfied. So, we should have

$$p(x,y,z) + \frac{1}{2}\rho \mathbf{u}(x,y,z) \cdot \mathbf{u}(x,y,z) = p_o + \underbrace{\frac{1}{2}\rho \mathbf{u}(0,0,0) \cdot \mathbf{u}(0,0,0)}_{=0} \tag{5.46}$$

Substituting the given velocity vector and rearranging terms yields

$$p(x,y,z) = p_o - \frac{\rho U^2}{2L^4} \left(x^2 y^2 + x^2 z^2 + y^2 z^2\right) \tag{5.47}$$

which is identical to Equation (5.45).

5.8 Euler's Equation—Rotating Tank

Statement of the Problem: *A rectangular tank partially filled with a liquid of density ρ is rotating about one of its sides as shown in Figure 5.3. The tank is open to the atmosphere and has been rotating for a time sufficient to establish rigid-body rotation. Determine the shape of the free surface on the plane passing through the axis of rotation if the minimum and maximum depths are as shown. Also, determine the angular-rotation rate, Ω, as a function of g and R.*

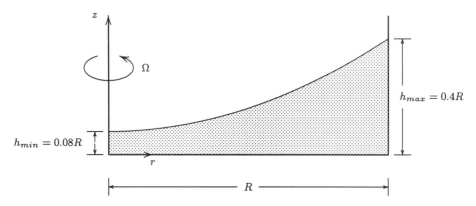

Figure 5.3: *Rotating tank in rigid-body rotation.*

Solution: For a fluid in rigid-body rotation, in the presence of a gravitational field aligned with the rotation axis, the Euler equation yields the following solution.

$$p + \rho g z - \frac{1}{2}\rho\Omega^2 r^2 = \text{constant} \tag{5.48}$$

Denoting atmospheric pressure by p_o and the shape of the free surface by $z = \zeta(r)$, applying this equation at radial distances r and 0, we have

$$p_o + \rho g \zeta(r) - \frac{1}{2}\rho\Omega^2 r^2 = p_o + \rho g \zeta(0) \tag{5.49}$$

But, from Figure 5.3, clearly $\zeta(0) = h_{min}$. So, canceling p_o from each side of Equation (5.49) and substituting for $\zeta(0)$, we arrive at the following.

$$\rho g \zeta(r) = \rho g h_{min} + \frac{1}{2}\rho\Omega^2 r^2 \tag{5.50}$$

Hence, solving for $\zeta(r)$, we arrive at

$$\zeta(r) = h_{min} + \frac{\Omega^2 r^2}{2g} \tag{5.51}$$

Finally, since $h_{min} = 0.08R$, substituting into Equation (5.51) tells us that the shape of the free surface is

$$\zeta(r) = 0.08R + \frac{\Omega^2 r^2}{2g} \tag{5.52}$$

5.9. PITOT-STATIC TUBE

We can solve for Ω by applying Equation (5.52) at $r = R$ where we are given the fact that $\zeta(R) = h_{max} = 0.4R$. Hence,

$$0.4R = 0.08R + \frac{\Omega^2 R^2}{2g} \implies \Omega^2 R^2 = 0.64gR \tag{5.53}$$

Solving for Ω, we find

$$\Omega = 0.8\sqrt{\frac{g}{R}} \tag{5.54}$$

5.9 Pitot-Static Tube

Statement of the Problem: *A pitot-static tube is placed in an incompressible air flow. The tube measures a stagnation pressure of p_{stag} = 104 kPa. If the flow speed is U = 75 m/sec, what static pressure, p_{stat}, will the tube measure? If the static pressure remains unchanged while the velocity decreases to 60 km/sec, what is the new value of the stagnation pressure? The density of the air is ρ = 1.2 kg/m^3.*

Solution: For a pitot-static tube, the velocity, U, is related to p_{stag}, p_{stat} and ρ by the following equation.

$$U = \sqrt{\frac{2(p_{stag} - p_{stat})}{\rho}} \tag{5.55}$$

Solving for p_{stat}, we have

$$p_{stat} = p_{stag} - \frac{1}{2}\rho U^2 \tag{5.56}$$

Now, substituting the given values of p_{stag}, p_{stat} and ρ into Equation (5.56), and noting that 1 Pa = 1 N/m^2 = 1 kg/(m·sec^2), we find

$$\begin{aligned} p_{stat} &= 104 \text{ kPa} - \frac{1}{2}\left(1.2 \text{ kg/m}^3\right)(75 \text{ m/sec})^2 \\ &= 104 \text{ kPa} - 3.375 \text{ kPa} = 100.625 \text{ kPa} \end{aligned} \tag{5.57}$$

Now, if we hold p_{stat} constant, the stagnation pressure is

$$p_{stag} = p_{stat} + \frac{1}{2}\rho U^2 \tag{5.58}$$

Again substituting given values:

$$\begin{aligned} p_{stag} &= 100.625 \text{ kPa} + \frac{1}{2}\left(1.2 \text{ kg/m}^3\right)(60 \text{ m/sec})^2 \\ &= 100.625 \text{ kPa} + 2.160 \text{ kPa} = 102.785 \text{ kPa} \end{aligned} \tag{5.59}$$

5.10 Galilean Invariance

Statement of the Problem: *An object is traveling through water at a speed U_1 = 20 m/sec. The flow speeds at Points 2 and 3 are U_2 = 12 m/sec and U_3 = 7 m/sec. If the pressure difference between Points 3 and 4 is half the difference between Points 2 and 3, what is the flow speed at Point 4, U_4?*

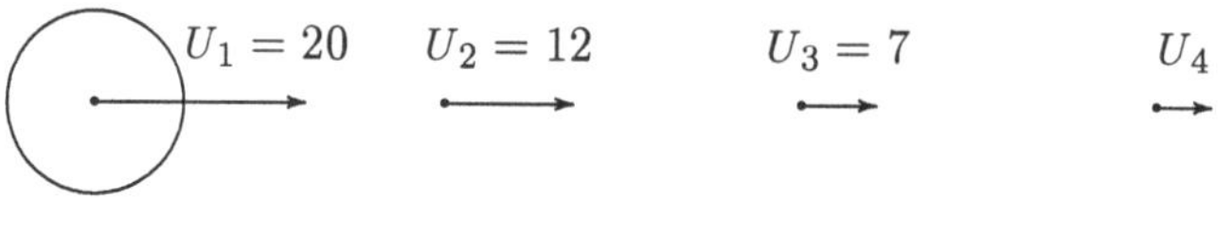

Figure 5.4: *Object translating in water.*

Solution: We must make a Galilean transformation in order to make the flow steady, wherefore we can use Bernoulli's equation. The velocities transform as indicated in Figure 5.5. Then, using Bernoulli's equation between Points 2 and 3,

$$p_2 + \frac{1}{2}\rho \left(U_2'\right)^2 = p_3 + \frac{1}{2}\rho \left(U_3'\right)^2 \implies p_3 - p_2 = \frac{1}{2}\rho \left[\left(U_2'\right)^2 - \left(U_3'\right)^2\right] \quad (5.60)$$

Similarly, using Bernoulli's equation between Points 3 and 4,

$$p_3 + \frac{1}{2}\rho \left(U_3'\right)^2 = p_4 + \frac{1}{2}\rho \left(U_4'\right)^2 \implies p_4 - p_3 = \frac{1}{2}\rho \left[\left(U_3'\right)^2 - \left(U_4'\right)^2\right] \quad (5.61)$$

Now, we are given

$$p_4 - p_3 = \frac{1}{2}(p_3 - p_2) \quad (5.62)$$

Combining these three equations yields

$$\frac{1}{2}\rho\left[\left(U_3'\right)^2 - \left(U_4'\right)^2\right] = \frac{1}{4}\rho\left[\left(U_2'\right)^2 - \left(U_3'\right)^2\right] \quad (5.63)$$

Multiplying through by $2/\rho$, there follows

$$\left(U_3'\right)^2 - \left(U_4'\right)^2 = \frac{1}{2}\left[\left(U_2'\right)^2 - \left(U_3'\right)^2\right] \quad (5.64)$$

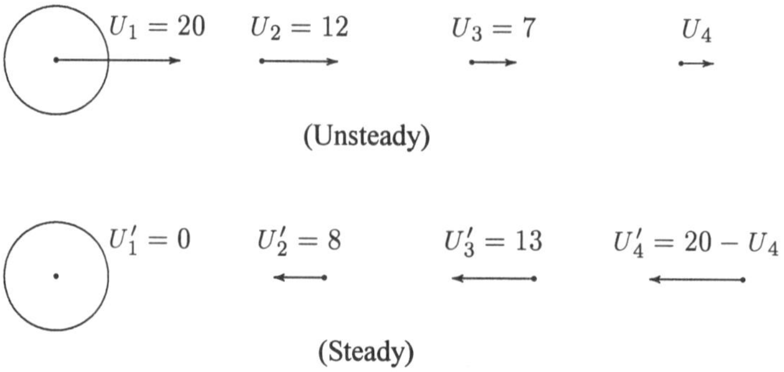

(Unsteady)

$U_1' = 0$ $U_2' = 8$ $U_3' = 13$ $U_4' = 20 - U_4$

(Steady)

Figure 5.5: *Galilean transformation details.*

5.10. GALILEAN INVARIANCE

Solving for U_4', we arrive at the following.

$$U_4' = \sqrt{\frac{3}{2}(U_3')^2 - \frac{1}{2}(U_2')^2} \tag{5.65}$$

From the given values,

$$U_3' = U_1 - U_3 = 13 \text{ m/sec} \quad \text{and} \quad U_2' = U_1 - U_2 = 8 \text{ m/sec} \tag{5.66}$$

Therefore, the transformed velocity at Point 4 is

$$U_4' = \sqrt{\frac{3}{2}(13 \text{ m/sec})^2 - \frac{1}{2}(8 \text{ m/sec})^2} = 14.88 \text{ m/sec} \tag{5.67}$$

Finally, transforming back to the frame in which the object is moving, U_4 is

$$U_4 = U_1 - U_4' = 20 \text{ m/sec} - 14.88 \text{ m/sec} = 5.12 \text{ m/sec} \tag{5.68}$$

Chapter 6

Control Volume Method

This chapter expands on the concepts introduced in Chapter 5, with the primary focus on using the integral form of the conservation laws for mass and momentum to analyze a finite control volume. Typically, we use the mass conservation equation to solve relatively simple problems involving inlet or exit velocities and/or area and density relations. In addition to supplying additional equations for the unknown quantities in a given problem, the momentum equation can be used for indirectly computing forces or pressures in terms of known mass and momentum fluxes.

For flows in which its use is justified, Bernoulli's equation can also be used to provide additional equations as needed to achieve a number of equations equal to the number of unknowns. The first seven problems in this chapter use the integral form of mass conservation for a control volume. The remaining problems use both mass and momentum conservation. The chapter closes with a three-dimensional problem.

There are four different types of control volumes, i.e., stationary, moving with constant velocity, deforming and accelerating. The easiest to analyze are stationary control volumes. They often include solid boundaries/objects. These types of problems are usually straightforward involving selection of a more-or-less obvious control volume, and applying the conservation laws to the selected volume.

When the control volume translates at constant velocity, a Galilean transformation replaces the moving control volume with a stationary one. Sometimes a control volume that changes in size, i.e., deforms, or that accelerates must be used. Although a bit more detailed, such control volume problems are handled in a manner similar to that used for stationary control volumes.

Section 6.1 Basic Control Volume Concepts: The problem in this section involves computation of unit normals, normal velocities, volume fluxes and the net pressure force on a specified control volume.

Section 6.2 Mass—Stationary Control Volume I: A simple stationary control volume problem that can be solved with the mass-conservation principle. The flow in this application is steady and requires computing an outlet velocity in terms of the known velocities at an inlet and at two other outlets.

Section 6.3 Mass—Stationary Control Volume II: This is also a stationary control volume problem that can be solved with the mass-conservation principle. However, this time the flow is unsteady and involves computing the rate of change of mass in the control volume.

Section 6.4 Mass—Translating Control Volume: The solution to this problem uses a moving control volume to analyze flow past a sphere advancing into a wind tunnel. A Galilean transformation makes use of a stationary control volume possible.

Section 6.5 Mass—Deforming Control Volume I: This is the first of two problems whose solution requires use of a deforming control volume. The problem in this section involves a sphere falling in a square tank of fluid.

Section 6.6 Mass—Deforming Control Volume II: This is another deforming control volume problem. Here, the application is to the flow beneath a circular disk that moves at constant velocity toward an infinite planar surface.

Section 6.7 Mass—Nonuniform Profiles: A simple straight-channel problem in which the inlet and outlet velocities are nonuniform. Part of the solution is a constant in the outlet velocity profile that guarantees mass conservation.

Section 6.8 Mass and Momentum—Stationary Control Volume I: This is the first of the momentum-conservation problems. It is a simple stationary control volume problem with all unit normals either horizontal or vertical.

Section 6.9 Mass and Momentum—Stationary Control Volume II: This is another stationary control volume, with the additional complication that one of the inlets lies at an angle to the horizontal.

Section 6.10 Mass and Momentum—Translating Control Volume: The problem in this section requires solving for the force on a hemisphere advancing into a stream of water. A Galilean transformation permits use of a stationary control volume to solve.

Section 6.11 Mass and Momentum—Rotational Flow: This is a control volume problem in which the flow is rotational. The mass- and momentum-conservation principles provide a sufficient number of equations to solve for inlet pressures and for the net force on the control volume.

Section 6.12 Mass and Momentum—Irrotational Flow: Application of the mass- and momentum-conservation principles in this problem does not yield enough equations to solve for the unknown quantities. Because its use is justified, we appeal to Bernoulli's equation to generate the extra equation needed to complete the solution.

Section 6.13 Mass and Momentum—Nonuniform Profiles: This is another pipe-flow problem that includes nonuniform velocity profiles. In addition to determining a constant in the outlet velocity profile, we determine the net friction force on the pipe walls as part of the solution.

Section 6.14 Mass and Momentum—Accelerating Control Volume: The problem in this section demonstrates the use of an accelerating control volume, with a cart being driven by a water jet and filling with water as it moves.

Section 6.15 Mass and Momentum—Three-Dimensional Flow: This problem involves fluid moving in all three directions, x, y, and z, to illustrate how the control volume method works in three dimensions. Aside from the complication of having to use the mass-conservation principle and all three components of the momentum-conservation equation, the problem is straightforward.

6.1 Basic Control Volume Concepts

Statement of the Problem: *Consider the branching pipe of circular cross section shown in Figure 6.1. Using the control volume indicated by the dashed contour, compute the following things.*

(a) *The unit normals,* $\mathbf{n}_1$, $\mathbf{n}_2$ *and* $\mathbf{n}_3$

(b) *The normal velocities,* $\mathbf{u}_1 \cdot \mathbf{n}_1$, $\mathbf{u}_2 \cdot \mathbf{n}_2$ *and* $\mathbf{u}_3 \cdot \mathbf{n}_3$

(c) *The volume fluxes,* $\mathbf{u}_1 \cdot \mathbf{n}_1 A_1$, $\mathbf{u}_2 \cdot \mathbf{n}_2 A_2$ *and* $\mathbf{u}_3 \cdot \mathbf{n}_3 A_3$

(d) *The net pressure force exerted by the surroundings on the control volume*

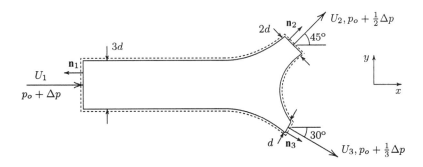

Figure 6.1: *Side view of a branching pipe.*

Solution: In all computations to follow, we cast our answers in terms of unit vectors $\mathbf{i}$ and $\mathbf{j}$ in the x and y directions, respectively. Also, keep in mind that this is a pipe so that cross-sectional areas must be computed accordingly.

6.1(a): By inspection, the three unit normals are

$$\mathbf{n}_1 = -\mathbf{i} \tag{6.1}$$

$$\mathbf{n}_2 = \mathbf{i}\cos 45° + \mathbf{j}\sin 45° = \frac{\sqrt{2}}{2}\mathbf{i} + \frac{\sqrt{2}}{2}\mathbf{j} \tag{6.2}$$

$$\mathbf{n}_3 = \mathbf{i}\cos 30° - \mathbf{j}\sin 30° = \frac{\sqrt{3}}{2}\mathbf{i} - \frac{1}{2}\mathbf{j} \tag{6.3}$$

6.1(b): At the inlet, the velocity vector is parallel to the unit normal and directed inward so that

$$\mathbf{u}_1 \cdot \mathbf{n}_1 = -U_1 \mathbf{n}_1 \cdot \mathbf{n}_1 = -U_1 \tag{6.4}$$

At both outlets, the velocity is parallel to the unit normal and directed outward, wherefore

$$\mathbf{u}_2 = U_2 \mathbf{n}_2 \cdot \mathbf{n}_2 = U_2 \quad \text{and} \quad \mathbf{u}_3 = U_3 \mathbf{n}_3 \cdot \mathbf{n}_3 = U_3 \tag{6.5}$$

6.1(c): Since the pipe has circular cross section, the areas at the inlet and the outlets are

$$A_1 = \frac{\pi}{4}(3d)^2 = \frac{9\pi}{4}d^2, \quad A_2 = \frac{\pi}{4}(2d)^2 = \pi d^2, \quad A_3 = \frac{\pi}{4}(d)^2 = \frac{\pi}{4}d^2 \tag{6.6}$$

Therefore, the volume fluxes are as follows.

$$\mathbf{u}_1 \cdot \mathbf{n}_1 A_1 = -\frac{9\pi}{4} U_1 d^2, \quad \mathbf{u}_2 \cdot \mathbf{n}_2 A_2 = \pi U_2 d^2, \quad \mathbf{u}_3 \cdot \mathbf{n}_3 A_3 = \frac{\pi}{4} U_3 d^2 \qquad (6.7)$$

6.1(d): For this problem, it is helpful to first take advantage of the fact that the net force exerted by the surroundings on the control volume is

$$\mathbf{F} = -\oiint_S p \mathbf{n} \, dS = -\oiint_S (p - p_o) \mathbf{n} \, dS \qquad (6.8)$$

where we make use of the fact that the closed surface integral of p_o is zero. Then, since the inlet and the outlets are the only parts of the control-volume surface on which the pressure differs from p_o, necessarily

$$\begin{aligned}
\mathbf{F} &= -(p_1 - p_o)\mathbf{n}_1 A_1 - (p_2 - p_o)\mathbf{n}_2 A_2 - (p_3 - p_o)\mathbf{n}_3 A_3 \\
&= -(\Delta p)(-\mathbf{i})\left(\frac{9\pi}{4}d^2\right) - \left(\frac{1}{2}\Delta p\right)\left(\frac{\sqrt{2}}{2}\mathbf{i} + \frac{\sqrt{2}}{2}\mathbf{j}\right)(\pi d^2) \\
&\quad - \left(\frac{1}{3}\Delta p\right)\left(\frac{\sqrt{3}}{2}\mathbf{i} - \frac{1}{2}\mathbf{j}\right)\left(\frac{\pi}{4}d^2\right) \\
&= \left(\frac{9}{4} - \frac{\sqrt{2}}{4} - \frac{\sqrt{3}}{24}\right)\pi d^2 \Delta p \, \mathbf{i} + \left(-\frac{\sqrt{2}}{4} + \frac{1}{24}\right)\pi d^2 \Delta p \, \mathbf{j} \qquad (6.9)
\end{aligned}$$

Therefore, the net force exerted by the surroundings on the control volume is

$$\mathbf{F} = (1.824\,\mathbf{i} - 0.312\,\mathbf{j})\,\pi d^2 \Delta p \qquad (6.10)$$

6.2 Mass—Stationary Control Volume I

Statement of the Problem: *A liquid of density ρ flows steadily through the piping system shown in Figure 6.2. The pipe cross sections are circular, and you can assume flow properties are uniform on all cross sections. Determine the velocity V as a function of U.*

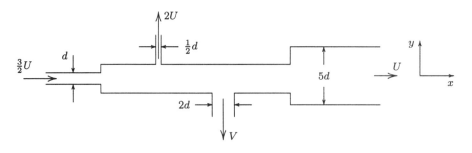

Figure 6.2: *Side view of a piping system.*

Solution: The first step in solving this problem is to select a suitable control volume. Since there is only one unknown quantity, viz., V, we need just one equation. Hence, the mass-conservation principle will suffice. Thus, the only issue influencing our choice of control

6.2. MASS—STATIONARY CONTROL VOLUME I

volume is that of making computation of mass fluxes across the control-volume boundaries as simple as possible. The control volume indicated by the dashed contour in Figure 6.3 satisfies this requirement.

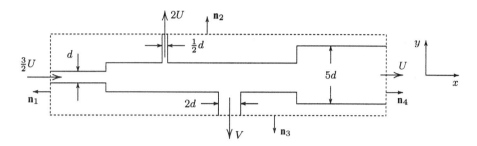

Figure 6.3: *Side view of the piping system and the control volume.*

We begin with the mass-conservation principle which tells us that

$$\iiint_V \frac{\partial \rho}{\partial t} dV + \oiint_S \rho \mathbf{u} \cdot \mathbf{n} \, dS = 0 \qquad (6.11)$$

Since the flow is steady, the first integral is zero, so that the net flux of mass out of the control volume is zero, i.e.,

$$\oiint_S \rho \mathbf{u} \cdot \mathbf{n} \, dS = 0 \qquad (6.12)$$

So, denoting the areas of the inlet and outlets by A_i, the integral yields

$$\underbrace{\rho \mathbf{u}_1 \cdot \mathbf{n}_1 A_1}_{Inlet} + \underbrace{\rho \mathbf{u}_2 \cdot \mathbf{n}_2 A_2 + \rho \mathbf{u}_3 \cdot \mathbf{n}_3 A_3 + \rho \mathbf{u}_4 \cdot \mathbf{n}_4 A_4}_{Outlets} = 0 \qquad (6.13)$$

At the inlet, the velocity vector is parallel to and points in the opposite direction of the outer unit normal, $\mathbf{n}_1$. All three outlet velocity vectors are parallel to and point in the same direction as the outer unit normal. Thus,

$$\mathbf{u}_1 \cdot \mathbf{n}_1 = -\frac{3}{2}U, \quad \mathbf{u}_2 \cdot \mathbf{n}_2 = 2U, \quad \mathbf{u}_3 \cdot \mathbf{n}_3 = V, \quad \mathbf{u}_4 \cdot \mathbf{n}_4 = U \qquad (6.14)$$

Also, the areas are

$$A_1 = \frac{\pi}{4}d^2, \quad A_2 = \frac{\pi}{16}d^2, \quad A_3 = \pi d^2, \quad A_4 = \frac{25\pi}{4}d^2 \qquad (6.15)$$

Combining Equations (6.13), (6.14) and (6.15) yields

$$-\frac{3}{8}\pi \rho U d^2 + \frac{1}{8}\pi \rho U d^2 + \pi \rho V d^2 + \frac{25}{4}\pi \rho U d^2 = 0 \qquad (6.16)$$

Note that the term at the inlet has a minus sign, while the outlet terms all have plus signs. This reflects the fact that the closed surface integral represents the net flux of fluid *out of* the control volume. At the inlet, fluid enters the control volume, corresponding to a negative flux out of the volume. For positive values of U and V, fluid leaves the control volume at the outlets, corresponding to a positive flux out. Solving for V, we find

$$V = -6U \qquad (6.17)$$

The fact that V is negative means that fluid enters the main pipe from "Outlet" 3.

6.3 Mass—Stationary Control Volume II

Statement of the Problem: *Compute the rate of change of the mass of the (incompressible) fluid contained in the tank shown in Figure 6.4. Express your answer as a function of fluid density, ρ, inlet flow velocity, U, and pipe diameter, d. Determine whether the tank is filling or emptying.*

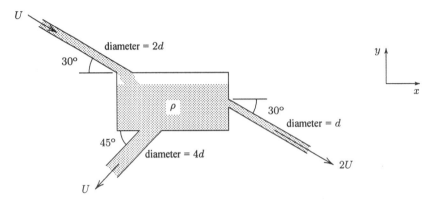

Figure 6.4: *Tank and piping system—all pipes have circular cross section.*

Solution: Since all velocities are given, the only unknown quantity is the rate of change of mass, $\dot{m}$, of the fluid within the control volume. Thus, we need a single equation so that the mass-conservation principle is sufficient. The first step in formulating the solution is to select a suitable control volume. The only issue influencing our choice of control volume lies in making computation of mass fluxes across the control-volume boundaries as simple as possible. For the problem at hand, this means the control-volume surface should be normal to the velocity vectors at all points where fluid passes through the surface. The control volume indicated by the dashed contour in Figure 6.5 satisfies this requirement.

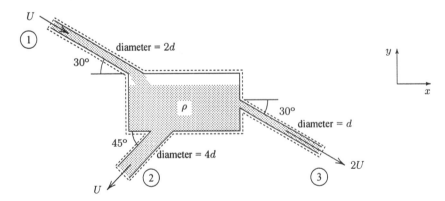

Figure 6.5: *Control-volume boundaries for the tank and piping system.*

Our starting point is the mass-conservation principle, viz.,

$$\frac{\partial}{\partial t} \iiint_V \rho \, dV + \oiint_S \rho \mathbf{u} \cdot \mathbf{n} \, dS = 0 \tag{6.18}$$

6.4. MASS—TRANSLATING CONTROL VOLUME

Note that in writing Equation (6.18), we take advantage of the fact that the control volume is stationary so that the partial derivative appears ahead of the volume integral [cf. Equation (6.11)]. Since the volume integral of the density is the mass of the fluid in the control volume, m, we can rewrite Equation (6.18) as follows.

$$\frac{dm}{dt} + \oiint_S \rho \mathbf{u} \cdot \mathbf{n}\, dS = 0 \tag{6.19}$$

So, denoting the areas of the inlet and outlets by A_i, the integral yields

$$\dot{m} + \underbrace{\rho \mathbf{u}_1 \cdot \mathbf{n}_1 A_1}_{Inlet} + \underbrace{\rho \mathbf{u}_2 \cdot \mathbf{n}_2 A_2 + \rho \mathbf{u}_3 \cdot \mathbf{n}_3 A_3}_{Outlets} = 0 \tag{6.20}$$

At the inlet, the velocity vector is parallel to and points in the opposite direction of the outer unit normal, $\mathbf{n}_1$. Both outlet velocity vectors are parallel to and point in the same direction as the outer unit normal. Thus,

$$\mathbf{u}_1 \cdot \mathbf{n}_1 = -U, \quad \mathbf{u}_2 \cdot \mathbf{n}_2 = U, \quad \mathbf{u}_3 \cdot \mathbf{n}_3 = 2U \tag{6.21}$$

Also, the areas are

$$A_1 = \frac{\pi}{4}(2d)^2 = \pi d^2, \quad A_2 = \frac{\pi}{4}(4d)^2 = 4\pi d^2, \quad A_3 = \frac{\pi}{4}d^2 \tag{6.22}$$

Combining Equations (6.20), (6.21) and (6.22) yields

$$\dot{m} - \pi \rho U d^2 + 4\pi \rho U d^2 + \frac{1}{2}\pi \rho U d^2 = 0 \tag{6.23}$$

As in the preceding section, the term at the inlet has a minus sign, while the outlet terms all have plus signs. This reflects the fact that the closed surface integral represents the net flux of fluid *out of* the control volume. At the inlet, fluid enters the control volume, corresponding to a negative flux out of the volume. Fluid leaves the control volume at the outlets, corresponding to a positive flux out. Solving for $\dot{m}$, we find

$$\dot{m} = -\frac{7}{2}\pi \rho U d^2 \tag{6.24}$$

The fact that $\dot{m}$ is negative means that the tank is emptying.

6.4 Mass—Translating Control Volume

Statement of the Problem: *In order to destroy Blofeld's headquarters, James Bond wants to launch a spherical object filled with explosives into an air-conditioning duct of square cross section. As shown in Figure 6.6, Q has designed the sphere to move at the same speed as the air, U, but in the opposite direction. The maximum flow speed, U_{max}, in the duct will occur on the cross section at the center plane of the sphere. In order for the sphere to move undetected due to the noise associated with its motion, the maximum flow speed must not exceed the undisturbed duct speed by more than 10%. Determine the maximum diameter, D, of the sphere as a function of the duct width, H.*

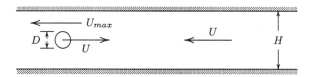

Figure 6.6: *Sphere advancing into a square duct—side view.*

Solution: The first step in solving this problem is to select a suitable control volume. Since there is only one unknown quantity, i.e., the diameter of the sphere that yields $U_{max} = 1.1U$, we need just one equation. Hence, the mass-conservation principle will provide the equation we need to solve this problem. Now, the indicated motion is unsteady because the sphere is moving. We can make this a steady-flow problem that can be analyzed with a stationary control volume by making a Galilean transformation. In the transformed problem, the sphere is at rest and the air approaches the sphere with a speed of $2U$ to the left. Also, the flow velocity between the sphere and duct walls transforms to $U_{max} + U$, again directed to the left. The control volume indicated by the dashed contour in Figure 6.7 is ideal for developing the solution as the outer unit normals are everywhere parallel to the flow velocity where fluid crosses the control-volume surface.

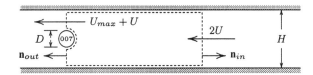

Figure 6.7: *Sphere at rest with a stationary control volume.*

We begin with the mass-conservation principle which tells us that

$$\iiint_V \frac{\partial \rho}{\partial t} dV + \oiint_S \rho \mathbf{u} \cdot \mathbf{n} \, dS = 0 \qquad (6.25)$$

Since the flow is steady in the transformed frame of Figure 6.7, the first integral is zero so that the net flux of mass out of the control volume is zero, i.e.,

$$\oiint_S \rho \mathbf{u} \cdot \mathbf{n} \, dS = 0 \qquad (6.26)$$

Assuming velocities are uniform on cross sections, the integral yields

$$\underbrace{\rho \mathbf{u}_{in} \cdot \mathbf{n}_{in} A_{in}}_{\text{Inlet}} + \underbrace{\rho \mathbf{u}_{out} \cdot \mathbf{n}_{out} A_{out}}_{\text{Outlet}} = 0 \qquad (6.27)$$

At the inlet, the velocity vector is parallel to and points in the opposite direction of the outer unit normal, $\mathbf{n}_{in}$. The outlet velocity vector is parallel to and points in the same direction as the outer unit normal. Thus,

$$\mathbf{u}_{in} \cdot \mathbf{n}_{in} = -2U \quad \text{and} \quad \mathbf{u}_{out} \cdot \mathbf{n}_{out} = U_{max} + U \qquad (6.28)$$

Also, the areas are

$$A_{in} = H^2 \quad \text{and} \quad A_{out} = H^2 - \frac{\pi}{4} D^2 \qquad (6.29)$$

6.5. MASS—DEFORMING CONTROL VOLUME I

Substituting Equations (6.28) and (6.29) into Equation (6.27) gives

$$-2\rho U H^2 + \rho (U_{max} + U)\left(H^2 - \frac{\pi}{4}D^2\right) = 0 \tag{6.30}$$

Of course, the term at the inlet has a minus sign, while the outlet term has a plus sign. This reflects the fact that the closed surface integral represents the net flux of fluid *out of* the control volume. At the inlet, fluid enters the control volume, corresponding to a negative flux out of the volume. At the outlet, fluid leaves the control volume, which corresponds to a positive flux out. Solving for D/H, there follows

$$\frac{D}{H} = \sqrt{\frac{4}{\pi}\frac{U_{max}/U - 1}{U_{max}/U + 1}} \tag{6.31}$$

Finally, substituting $U_{max}/U = 1.1$, the maximum diameter of the explosive sphere is

$$D = 0.246 H \tag{6.32}$$

6.5 Mass—Deforming Control Volume I

Statement of the Problem: *A spherical ball of diameter d is dropped into a tank with square cross section of width h as shown below. The tank is filled with an incompressible fluid of density ρ. Using a deforming control volume, determine the ball's diameter if the mean upward velocity of the fluid between the ball and the tank walls is one tenth the downward velocity of the ball.*

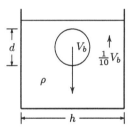

Figure 6.8: *Spherical ball falling into a tank of square cross section.*

Solution: Since the geometry of this problem is continuously changing with time, the motion is inherently unsteady. Such problems are conveniently analyzed using a deforming control volume. Consider the control volume whose upper boundary moves with the ball as shown in Figure 6.9. For this control volume, we must use the mass-conservation principle in the following form.

$$\frac{\partial}{\partial t}\iiint_V \rho\, dV + \oiint_S \rho\, (\mathbf{u}_{abs} - \mathbf{u}_{cv})\cdot \mathbf{n}\, dS = 0 \tag{6.33}$$

Since the upper boundary of the control volume moves with the same velocity as the ball, in a time Δt, the height of the control volume changes by $V_b \Delta t$. Until the ball reaches the bottom of the tank, the change in height of the control volume is the same across the entire width of the tank. Hence, because the tank cross section is square with side h, the volume change is given by

$$\Delta V = -h^2 V_b \Delta t \quad \Longrightarrow \quad \frac{dV}{dt} = -V_b h^2 \tag{6.34}$$

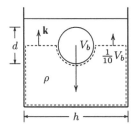

Figure 6.9: *Spherical ball, tank and control volume.*

Since the fluid is incompressible, the unsteady term, i.e., the first term in Equation (6.33), thus becomes

$$\frac{\partial}{\partial t}\iiint_V \rho\,dV = \rho\frac{dV}{dt} = -\rho V_b h^2 \tag{6.35}$$

Turning to the mass-flux integral, i.e., the second term in Equation (6.33), we observe that the only part of the control volume across which fluid flows is the horizontal plane between the ball and the tank walls. On this part of the control-volume surface, the absolute velocity of the fluid and the control-volume velocity are

$$\mathbf{u}_{abs} = \frac{1}{10}V_b\mathbf{k} \quad \text{and} \quad \mathbf{u}_{cv} = -V_b\mathbf{k} \tag{6.36}$$

Also, the outer unit normal on this part of the control volume is $\mathbf{n} = \mathbf{k}$. Therefore, because the area between the tank and the sphere is the difference between the tank cross-sectional area, h^2, and the area of the sphere's projection on a horizontal plane, $\pi d^2/4$, we have

$$\oiint_S \rho\,(\mathbf{u}_{abs}-\mathbf{u}_{cv})\cdot\mathbf{n}\,dS = \iint_A \rho\left[\frac{1}{10}V_b\mathbf{k}-(-V_b\mathbf{k})\right]\cdot\mathbf{k}\,dA = \frac{11}{10}\rho V_b\left(h^2-\frac{\pi}{4}d^2\right) \tag{6.37}$$

Finally, substituting Equations (6.35) and (6.37) into Equation (6.33) yields

$$-\rho V_b h^2 + \frac{11}{10}\rho V_b\left(h^2-\frac{\pi}{4}d^2\right) = 0 \quad\Longrightarrow\quad \frac{1}{10}\rho V_b h^2 = \frac{11\pi}{40}\rho V_b d^2 \tag{6.38}$$

Solving for d, we find

$$d = \frac{2h}{\sqrt{11\pi}} \quad\Longrightarrow\quad d = 0.34h \tag{6.39}$$

6.6 Mass—Deforming Control Volume II

Statement of the Problem: *A disk of diameter d is moving at speed U and is approaching a stationary disk, also of diameter d as shown in Figure 6.10. The fluid between the disks is incompressible and has density ρ. Using a deforming cylindrical control volume between the disks (indicated by the dashed lines), determine the radial velocity, U_r, of the fluid. Express your answer for U_r as a function of U, radial distance, r, and instantaneous distance between the disks, ℓ.*

Solution: For the indicated control volume, we must use the mass-conservation principle in the following form.

$$\frac{\partial}{\partial t}\iiint_V \rho\,dV + \oiint_S \rho\,(\mathbf{u}_{abs}-\mathbf{u}_{cv})\cdot\mathbf{n}\,dS = 0 \tag{6.40}$$

6.6. MASS—DEFORMING CONTROL VOLUME II

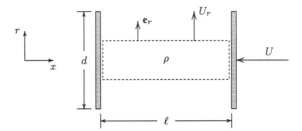

Figure 6.10: *Fluid motion between two disks with a deforming control volume.*

Since the right boundary of the control volume moves with the same velocity as the disk, in a time Δt, the height of the cylindrical control volume changes by $\Delta \ell = U \Delta t$. Hence, because the control-volume cross section is a circle with diameter r, the volume change is given by

$$\Delta V = -\pi r^2 \Delta \ell = -\pi r^2 U \Delta t \quad \Longrightarrow \quad \frac{dV}{dt} = -\pi U r^2 \qquad (6.41)$$

Since the fluid is incompressible, the unsteady term, i.e., the first term in Equation (6.40), thus becomes

$$\frac{\partial}{\partial t} \iiint_V \rho \, dV = \rho \frac{dV}{dt} = -\pi \rho U r^2 \qquad (6.42)$$

Turning to the mass-flux integral, i.e., the second term in Equation (6.40), we observe that the only part of the control volume across which fluid flows is the cylindrical surface between the disks. The fluid is being forced out in the radial direction. Unlike the problem in the preceding section, there is no fluid moving across the moving part of the bounding surface of the control volume since $\mathbf{u}_{abs} = \mathbf{u}_{cv}$ there. Here, the part of the bounding surface across which fluid passes is stationary. On this part of the control-volume surface, the absolute velocity of the fluid and the control-volume velocity are as follows.

$$\mathbf{u}_{abs} = U_r \mathbf{e}_r \quad \text{and} \quad \mathbf{u}_{cv} = \mathbf{0} \qquad (6.43)$$

Note that $\mathbf{e}_r$ is a unit vector in the radial direction. Also, the outer unit normal on this part of the control volume is $\mathbf{n} = \mathbf{e}_r$. Therefore, assuming U_r does not vary with z and that the left plate lies at $z = z_o$, the mass-flux integral is

$$\oiint_S \rho \left(\mathbf{u}_{abs} - \mathbf{u}_{cv} \right) \cdot \mathbf{n} \, dS = \int_{z_o}^{z_o + \ell} \int_0^{2\pi} \rho \left[U_r \mathbf{e}_r - \mathbf{0} \right] \cdot \mathbf{e}_r \, r d\theta dz = 2\pi \rho U_r r \ell \qquad (6.44)$$

Finally, substituting Equations (6.42) and (6.44) into Equation (6.40) gives

$$-\pi \rho U r^2 + 2\pi U_r r \ell = 0 \qquad (6.45)$$

Solving for U_r, we find

$$U_r = \frac{U}{2} \frac{r}{\ell} \qquad (6.46)$$

As a concluding comment, in reality, the radial velocity will depend upon both r and z. Our solution is still valid, although the computed $U_r(r)$ is actually the average value of the true radial velocity, $u_r(r, z)$, where

$$U_r(r) = \frac{1}{\ell} \int_{z_o}^{z_o + \ell} u_r(r, z) \, dz \qquad (6.47)$$

6.7 Mass—Nonuniform Profiles

Statement of the Problem: *The velocity profile in a channel changes from U_1 to U_2, where*

$$U_1 = 2U\left[\frac{y}{h} - \frac{1}{2}\left(\frac{y}{h}\right)^2\right] \quad \text{and} \quad U_2 = K\left[\frac{y}{h} + \left(\frac{y}{h}\right)^2 - \left(\frac{y}{h}\right)^3\right]$$

for $0 \leq y \leq h$, and is symmetric about the centerline at $y = h$. Assuming the flow is steady and incompressible, solve for the constant K.

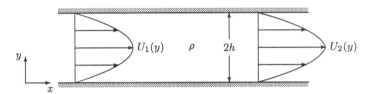

Figure 6.11: *Flow in a channel.*

Solution: Because the flow is symmetric about the centerline, we select the control volume indicated by the dashed contour in Figure 6.12. The control volume includes the lower half of the channel, and the unit normals are parallel to the inlet and outlet velocities. We begin with the mass-conservation principle which tells us that

$$\iiint_V \frac{\partial \rho}{\partial t} dV + \oiint_S \rho \mathbf{u} \cdot \mathbf{n}\, dS = 0 \tag{6.48}$$

Since the flow is steady, the first integral is zero, so that the net flux of mass out of the control volume is zero, i.e.,

$$\oiint_S \rho \mathbf{u} \cdot \mathbf{n}\, dS = 0 \tag{6.49}$$

At the inlet, the velocity vector is parallel to and points in the opposite direction of the outer unit normal so that $\mathbf{u}_1 \cdot \mathbf{n}_1 = -U_1$. The outlet velocity vector is parallel to and points in the same direction as the outer unit normal so that $\mathbf{u}_2 \cdot \mathbf{n}_2 = U_2$. There can be no flow across the centerline and the channel wall, of course, since $v = 0$ and $\mathbf{n} = \mathbf{j}$ wherefore $\mathbf{u} \cdot \mathbf{n} = 0$. Thus,

$$-\int_0^h \rho U_1(y)\, dy + \int_0^h \rho U_2(y)\, dy = 0 \tag{6.50}$$

Substituting the given velocity profiles and dividing through by ρ yields:

$$-\int_0^h 2U\left[\frac{y}{h} - \frac{1}{2}\left(\frac{y}{h}\right)^2\right] dy + \int_0^h K\left[\frac{y}{h} + \left(\frac{y}{h}\right)^2 - \left(\frac{y}{h}\right)^3\right] dy = 0 \tag{6.51}$$

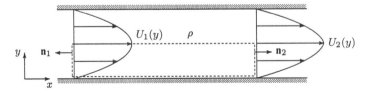

Figure 6.12: *Control volume for flow in a channel.*

6.8. MASS AND MOMENTUM—STATIONARY CONTROL VOLUME I

This equation can be rearranged to read as follows.

$$2Uh \int_0^h \left[\frac{y}{h} - \frac{1}{2}\left(\frac{y}{h}\right)^2\right] d\left(\frac{y}{h}\right) = Kh \int_0^h \left[\frac{y}{h} + \left(\frac{y}{h}\right)^2 - \left(\frac{y}{h}\right)^3\right] d\left(\frac{y}{h}\right) \qquad (6.52)$$

Next, it is convenient to change integration variables according to $\eta \equiv y/h$, wherefore

$$2Uh \int_0^1 \left[\eta - \frac{1}{2}\eta^2\right] d\eta = Kh \int_0^1 \left[\eta + \eta^2 - \eta^3\right] d\eta \qquad (6.53)$$

Dividing through by h and evaluating the integrals gives

$$2U \left[\frac{1}{2}\eta^2 - \frac{1}{6}\eta^3\right]_{\eta=0}^{\eta=1} = K \left[\frac{1}{2}\eta^2 + \frac{1}{3}\eta^3 - \frac{1}{4}\eta^4\right]_{\eta=0}^{\eta=1} \qquad (6.54)$$

or,

$$2U \left[\frac{1}{2} - \frac{1}{6}\right] = K \left[\frac{1}{2} + \frac{1}{3} - \frac{1}{4}\right] \implies \frac{2}{3}U = \frac{7}{12}K \qquad (6.55)$$

Finally, solving for K, the solution is

$$K = \frac{8}{7}U \qquad (6.56)$$

6.8 Mass and Momentum—Stationary Control Volume I

Statement of the Problem: *An incompressible jet of fluid strikes a circular plate and spreads radially. The jet diameter is d, the fluid density is ρ and the plate diameter is D. Using the control volume indicated by the dashed contour in Figure 6.13, compute the thickness, t, of the radially-spreading jet, and the force, F, required to hold the plate in place. You may assume the flow is steady, inviscid and that no body forces are present. Also, assume properties are uniform on all cross sections.*

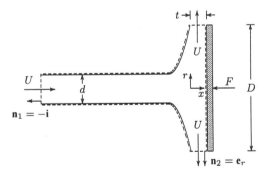

Figure 6.13: *Axisymmetric jet impinging on a disk—side view.*

Solution: This problem involves two unknown quantities, viz., the thickness of the radially-spreading jet, t, and the force required to hold the plate in place, F. Thus, in addition to the mass-conservation principle, we must appeal to the momentum-conservation equation to establish a sufficient number of equations to solve (i.e., 2). By symmetry, there are no forces in the circumferential or radial directions. Hence, we will use only the x component of the momentum equation.

Mass Conservation: Since the flow is steady, the mass-conservation principle is

$$\oiint_S \rho \mathbf{u} \cdot \mathbf{n}\, dS = 0 \tag{6.57}$$

Expanding the closed-surface integral, we have

$$\underbrace{\int_0^{2\pi}\int_0^{d/2} \rho \mathbf{u}_1 \cdot \mathbf{n}_1 r\, dr\, d\theta}_{\text{Incident jet}} + \underbrace{\int_0^{2\pi}\int_0^{t} \rho \mathbf{u}_2 \cdot \mathbf{n}_2 \left(\frac{D}{2}\right) dx\, d\theta}_{\text{Radially-spreading jet}} = 0 \tag{6.58}$$

Note that the incident jet has a circular cross section of diameter d, so that the differential surface area is $r\, dr\, d\theta$. We thus integrate from 0 to 2π for θ and 0 to $d/2$ for r. The radially-spreading jet passes through a cylindrical surface of height t and diameter D so that $dS = dx\, R\, d\theta$, where $R = D/2$. Hence, we integrate from 0 to 2π for θ and 0 to t for x.

Now, as shown in Figure 6.13, the velocity in the incident jet is $\mathbf{u}_1 = U\mathbf{i}$, while the outer unit normal is $\mathbf{n}_1 = -\mathbf{i}$. For the radially-spreading jet, we have $\mathbf{u}_2 = U\mathbf{e}_r$ and $\mathbf{n}_2 = \mathbf{e}_r$. Therefore, we have

$$\mathbf{u}_1 \cdot \mathbf{n}_1 = -U \quad \text{and} \quad \mathbf{u}_2 \cdot \mathbf{n}_2 = U \tag{6.59}$$

Hence, the mass-conservation equation becomes

$$-2\pi\rho U \int_0^{d/2} r\, dr + \pi\rho U D \int_0^t dx = 0 \quad \Longrightarrow \quad -\frac{\pi}{4}\rho U d^2 + \pi\rho U D t = 0 \tag{6.60}$$

Solving for the thickness t, we arrive at

$$t = \frac{d^2}{4D} \tag{6.61}$$

x-Momentum Conservation: The x component of the momentum-conservation equation for steady flow is

$$\oiint_S \rho u\, (\mathbf{u} \cdot \mathbf{n})\, dS = -\mathbf{i} \cdot \oiint_S (p - p_o)\, \mathbf{n}\, dS \tag{6.62}$$

where we take the dot product of the pressure integral with $\mathbf{i}$ to extract its x component. We can evaluate the momentum-flux integral [the integral on the left-hand side of Equation (6.62)] as follows.

$$\oiint_S \rho u\, (\mathbf{u} \cdot \mathbf{n})\, dS = \underbrace{\int_0^{2\pi}\int_0^{d/2} \rho U(\mathbf{u}_1 \cdot \mathbf{n}_1) r\, dr\, d\theta}_{\text{Incident jet}} + \underbrace{\int_0^{2\pi}\int_0^{t} \rho(0)(\mathbf{u}_2 \cdot \mathbf{n}_2)\left(\frac{D}{2}\right) dx\, d\theta}_{\text{Radially-spreading jet}}$$

$$= \rho U(-U)\frac{\pi}{4}d^2 + 0 = -\frac{\pi}{4}\rho U^2 d^2 \tag{6.63}$$

Turning to the pressure integral, we have

$$\oiint_S (p - p_o)\, \mathbf{n}\, dS = \underbrace{\iint_A (p - p_o)\, \mathbf{n}\, dS}_{\text{Non-plate}} + \underbrace{\int_0^{2\pi}\int_0^{D/2} (p - p_o)\, \mathbf{i}\, r\, dr\, d\theta}_{\text{Plate}} \tag{6.64}$$

6.9. MASS AND MOMENTUM—STATIONARY CONTROL VOLUME II

Now, the pressure at the interface between the jet and the surroundings is atmospheric, i.e., $p = p_o$. Also, since a jet is a classical case of a *thin shear layer*, the pressure on cross sections is also atmospheric. Thus, the only part of the control-volume surface on which the pressure differs from the atmospheric value is the plate. Since we are neglecting viscous effects, the force on the plate is simply the integral of p over its surface. Also, since the pressure is atmospheric on the other side of the plate, the net force is the integral of $(p - p_o)$. Since the plate is stationary, this force must exactly balance the force required to hold it in place, F, so that

$$\oint_S (p - p_o)\,\mathbf{n}\,dS = \mathbf{0} + \int_0^{2\pi}\int_0^{D/2} (p - p_o)\,\mathbf{i}\,r\,dr\,d\theta = F\,\mathbf{i} \tag{6.65}$$

Combining Equations (6.62), (6.63) and (6.65) yields

$$-\frac{\pi}{4}\rho U^2 d^2 = -\mathbf{i}\cdot(F\,\mathbf{i}) = -F \tag{6.66}$$

Therefore, the force required to hold the plate in place is

$$F = \frac{\pi}{4}\rho U^2 d^2 \tag{6.67}$$

6.9 Mass and Momentum—Stationary Control Volume II

Statement of the Problem: *An incompressible fluid flows in a pipe with branches as shown below. The flow is steady, inviscid and no body forces are acting. Solve for U_1 and the pressure difference $p_4 - p_1$ as a function of p_2, ρ, U and atmospheric pressure, p_o. Assume velocity and pressure are constant on all cross sections, all of which are circular. Also assume the fluid exerts no net force on the pipe in the x direction.*

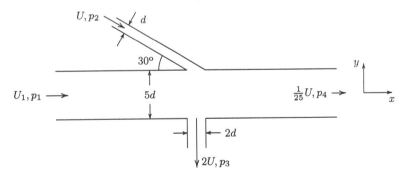

Figure 6.14: Pipe with multiple branches—side view.

Solution: There are two unknown quantities in this problem, which are the velocity U_1 and the pressure p_4. So, in addition to the mass-conservation principle, we must appeal to the momentum-conservation equation to establish the pair of equations needed to solve. It will suffice to use the x component of the momentum equation.

The optimum control volume lies entirely outside of the pipe system and has its bounding surface normal to all inlets and outlets. This ensures that the pressure differs from atmospheric pressure only at the inlets and outlets, and that the inlet/outlet velocity vectors are all parallel to the outer unit normals. Figure 6.15 shows the control volume as a dashed contour.

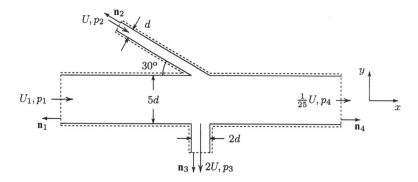

Figure 6.15: *Pipe with multiple branches and control volume.*

Mass Conservation: Because the flow is steady, the mass-conservation principle is

$$\oiint_S \rho \mathbf{u} \cdot \mathbf{n} \, dS = 0 \tag{6.68}$$

Expanding the closed-surface integral and noting that velocities are constant on all cross sections, we obtain

$$\rho \mathbf{u}_1 \cdot \mathbf{n}_1 \frac{\pi}{4}(5d)^2 + \rho \mathbf{u}_2 \cdot \mathbf{n}_2 \frac{\pi}{4}d^2 + \rho \mathbf{u}_3 \cdot \mathbf{n}_3 \frac{\pi}{4}(2d)^2 + \rho \mathbf{u}_4 \cdot \mathbf{n}_4 \frac{\pi}{4}(5d)^2 = 0 \tag{6.69}$$

As shown in Figure 6.15, the velocities are all parallel to the outer unit normals. Therefore, we have

$$\mathbf{u}_1 \cdot \mathbf{n}_1 = -U_1, \quad \mathbf{u}_2 \cdot \mathbf{n}_2 = -U, \quad \mathbf{u}_3 \cdot \mathbf{n}_3 = 2U, \quad \mathbf{u}_4 \cdot \mathbf{n}_4 = \frac{1}{25}U \tag{6.70}$$

Hence, the mass-conservation equation becomes

$$-\frac{25\pi}{4}\rho U_1 d^2 - \frac{\pi}{4}\rho U d^2 + 2\pi \rho U d^2 + \frac{\pi}{4}\rho U d^2 = 0 \tag{6.71}$$

Solving for the inlet velocity, U_1, we arrive at

$$U_1 = \frac{8}{25}U \tag{6.72}$$

x-Momentum Conservation: The x component of the momentum-conservation equation for steady flow is

$$\oiint_S \rho u \left(\mathbf{u} \cdot \mathbf{n} \right) dS = -\mathbf{i} \cdot \oiint_S (p - p_o) \mathbf{n} \, dS \tag{6.73}$$

where we take the dot product of the pressure integral with $\mathbf{i}$ to extract its x component. We also subtract atmospheric pressure from p to simplify the analysis. If we fail to do this and work with the pressure, p, rather than $p - p_o$, we must include the integral of p_o over all parts of the bounding surface of the control volume except the inlets and outlets, where the pressure is equal to a value other than p_o. We are taking advantage of the fact that the closed surface integral of p_o is exactly zero, so that the pressure integral with $(p - p_o)$ involves only contributions from the inlets and outlets where p differs from p_o.

6.9. MASS AND MOMENTUM—STATIONARY CONTROL VOLUME II

We can evaluate the momentum-flux integral [the integral on the left-hand side of Equation (6.73)] as follows (remember that we focus on flow in the x direction).

$$\begin{aligned}
\oiint_S \rho u \, (\mathbf{u} \cdot \mathbf{n}) \, dS &= \rho U_1 (\mathbf{u}_1 \cdot \mathbf{n}_1) \frac{\pi}{4}(5d)^2 + \rho U \cos 30° (\mathbf{u}_2 \cdot \mathbf{n}_2) \frac{\pi}{4} d^2 \\
&\quad + \rho (0)(\mathbf{u}_3 \cdot \mathbf{n}_3) \frac{\pi}{4}(2d)^2 + \frac{1}{25} \rho U (\mathbf{u}_4 \cdot \mathbf{n}_4) \frac{\pi}{4}(5d)^2 \\
&= -\frac{25\pi}{4} \rho U_1^2 d^2 - 0.866 \frac{\pi}{4} \rho U^2 d^2 + \frac{\pi}{100} \rho U^2 d^2 \\
&= -\frac{25\pi}{4} \rho U_1^2 d^2 - 0.826 \frac{\pi}{4} \rho U^2 d^2 \qquad (6.74)
\end{aligned}$$

Combining Equations (6.72) and (6.74) gives

$$\oiint_S \rho u \, (\mathbf{u} \cdot \mathbf{n}) \, dS = -\frac{25\pi}{4} \rho \left(\frac{8}{25} U\right)^2 d^2 - 0.826 \frac{\pi}{4} \rho U^2 d^2 = -3.386 \frac{\pi}{4} \rho U^2 d^2 \qquad (6.75)$$

Turning to the pressure integral, we have

$$\begin{aligned}
\oiint_S (p - p_o) \mathbf{n} \, dS &= (p_1 - p_o)(-\mathbf{i}) \frac{\pi}{4}(5d)^2 + (p_2 - p_o)(-\mathbf{i} \cos 30° + \mathbf{j} \sin 30°) \frac{\pi}{4} d^2 \\
&\quad + (p_3 - p_o)(-\mathbf{j}) \frac{\pi}{4}(2d)^2 + (p_4 - p_o)(\mathbf{i}) \frac{\pi}{4}(5d)^2 \\
&= -\frac{25\pi}{4}(p_1 - p_o) d^2 \mathbf{i} + \frac{\pi}{4}(p_2 - p_o) d^2 (-0.866 \mathbf{i} + 0.5 \mathbf{j}) \\
&\quad - \pi (p_3 - p_o) d^2 \mathbf{j} + \frac{25\pi}{4}(p_4 - p_o) d^2 \mathbf{i} \qquad (6.76)
\end{aligned}$$

So, taking the dot product of this equation with $-\mathbf{i}$, all contributions proportional to $\mathbf{j}$ drop out, and what remains is

$$-\mathbf{i} \cdot \oiint_S (p - p_o) \mathbf{n} \, dS = -\frac{25\pi}{4}(p_4 - p_1) d^2 + 0.866 \frac{\pi}{4}(p_2 - p_o) d^2 \qquad (6.77)$$

Combining Equations (6.73), (6.75) and (6.77) yields

$$-3.386 \frac{\pi}{4} \rho U^2 d^2 = -\frac{25\pi}{4}(p_4 - p_1) d^2 + 0.866 \frac{\pi}{4}(p_2 - p_o) d^2 \qquad (6.78)$$

Finally, solving for $p_4 - p_1$, we conclude that

$$p_4 - p_1 = 0.035 \, (p_2 - p_o) + 0.135 \rho U^2 \qquad (6.79)$$

Observe that the pressure at the lower outlet, p_3, does not appear in the final result. This is true because the pressure acts normal to the control volume surface, so that the contribution from this outlet acts entirely in the y direction. Since we have used the x-momentum equation to develop our answer, we should not expect forces in the y direction to play a role in determining the pressure difference, $p_4 - p_1$.

6.10 Mass and Momentum—Translating Control Volume

Statement of the Problem: *An object in the shape of a hemisphere of diameter $2d$ is advancing into a tube of diameter $3d$. The hemisphere is moving to the left at a speed U while an incompressible fluid of density ρ flows to the right at a speed $2U$. The flow speed and pressure across a plane coincident with the base of the hemisphere are U_2 and p_2, where $p_2 = p_\infty - 16\rho U^2$. Assuming the pressure on the base of the hemisphere is constant and equal to p_2, compute the flow speed U_2 and the net force, F, on the hemisphere. Assume the flow velocity is uniform on all cross sections.*

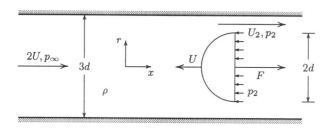

Figure 6.16: *A hemisphere advancing into a tube—center-plane view.*

Solution: The first step in solving this problem is to decide what equations are needed to solve. There are two unknown quantities, U_2 and F, so that we must use the mass-conservation principle and the x component of the momentum equation. The motion is unsteady because the hemisphere is moving. We can make this a steady-flow problem that can be analyzed with a stationary control volume by making a Galilean transformation. In the transformed problem, the hemisphere is at rest and the fluid approaches the hemisphere with a speed of $3U$ to the right. Also, the flow velocity between the hemisphere and tube wall transforms to $U + U_2$, again directed to the right. The control volume indicated by the dashed contour in Figure 6.17 is ideal for developing the solution as the outer unit normals are everywhere parallel to the flow velocity where fluid crosses the control-volume surface.

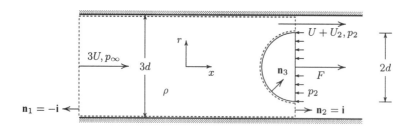

Figure 6.17: *Hemisphere, tube and control volume after Galilean transformation.*

Mass Conservation: Because the flow is steady in the transformed coordinate frame, the equation for conservation of mass is

$$\oiint_S \rho \mathbf{u} \cdot \mathbf{n}\, dS = 0 \qquad (6.80)$$

6.10. MASS AND MOMENTUM—TRANSLATING CONTROL VOLUME

Expanding the closed-surface integral gives

$$\underbrace{\rho \mathbf{u}_1 \cdot \mathbf{n}_1 \frac{\pi}{4}(3d)^2}_{Inlet} + \underbrace{\rho \mathbf{u}_2 \cdot \mathbf{n}_2 \frac{\pi}{4}\left[(3d)^2 - (2d)^2\right]}_{Outlet} = 0 \qquad (6.81)$$

Now, as shown in Figure 6.17, because the velocities are all parallel to the outer unit normals, we have

$$\mathbf{u}_1 \cdot \mathbf{n}_1 = -3U, \quad \mathbf{u}_2 \cdot \mathbf{n}_2 = U + U_2 \qquad (6.82)$$

Hence, the mass-conservation equation becomes

$$-\frac{27\pi}{4}\rho U d^2 + \frac{5\pi}{4}\rho (U + U_2) d^2 = 0 \qquad (6.83)$$

Solving for the outlet velocity, U_2, we arrive at

$$U_2 = \frac{22}{5} U \qquad (6.84)$$

x-Momentum Conservation: The x component of the momentum-conservation equation for steady flow is

$$\oiint_S \rho u (\mathbf{u} \cdot \mathbf{n}) \, dS = -\mathbf{i} \cdot \oiint_S p \mathbf{n} \, dS \qquad (6.85)$$

where we take the dot product of the pressure integral with $\mathbf{i}$ to extract its x component. We can evaluate the momentum-flux integral, i.e., the integral on the left-hand side of Equation (6.85), in the following manner.

$$\begin{aligned}\oiint_S \rho u (\mathbf{u} \cdot \mathbf{n}) \, dS &= (3\rho U)(\mathbf{u}_1 \cdot \mathbf{n}_1)\frac{\pi}{4}(3d)^2 + \rho(U+U_2)(\mathbf{u}_2 \cdot \mathbf{n}_2)\frac{\pi}{4}\left[(3d)^2 - (2d)^2\right] \\ &= -\frac{81}{4}\pi\rho U^2 d^2 + \frac{5}{4}\pi\rho \underbrace{(U+U_2)^2}_{=\left(\frac{27}{5}U\right)^2} d^2 = \frac{81\pi}{5}\rho U^2 d^2 \end{aligned} \qquad (6.86)$$

Focusing on the pressure integral, the pressure on the tube walls acts radially and thus makes no contribution to the balance of forces in the x direction. Thus, we have

$$\begin{aligned}-\mathbf{i}\cdot\oiint_S p\mathbf{n}\,dS &= \underbrace{-\mathbf{i}\cdot p_\infty(-\mathbf{i})\frac{\pi}{4}(3d)^2}_{Inlet} - \underbrace{\mathbf{i}\cdot p_2(\mathbf{i})\frac{\pi}{4}\left[(3d)^2 - (2d)^2\right]}_{Outlet} - \underbrace{\mathbf{i}\cdot\iint_A p\mathbf{n}\,dS}_{Hemisphere\ front} \\ &= \frac{9\pi}{4}p_\infty d^2 - \frac{5\pi}{4}p_2 d^2 - F_{front}\end{aligned} \qquad (6.87)$$

where F_{front} denotes the force on the front part of the hemisphere. Then, substituting the given p_2, we find

$$\begin{aligned}-\mathbf{i}\cdot\oiint_S p\mathbf{n}\,dS &= \frac{9\pi}{4}p_\infty d^2 - \frac{5\pi}{4}\left(p_\infty - 16\rho U^2\right) d^2 - F_{front} \\ &= \pi p_\infty d^2 + 20\pi\rho U^2 d^2 - F_{front}\end{aligned} \qquad (6.88)$$

Substituting Equations (6.86) and (6.88) into Equation (6.85) yields

$$\frac{81\pi}{5}\rho U^2 d^2 = \pi p_\infty d^2 + 20\pi\rho U^2 d^2 - F_{front} \qquad (6.89)$$

82   CHAPTER 6. CONTROL VOLUME METHOD

Therefore, the force on the front of the hemisphere is

$$F_{front} = \frac{19\pi}{5}\rho U^2 d^2 + \pi p_\infty d^2 \tag{6.90}$$

The net force on the hemisphere is

$$F = F_{front} - F_{back} \tag{6.91}$$

where F_{back} is the integral of the pressure over the base of the hemisphere. Since we are given that the pressure is constant and equal to p_2 on the back face of the hemisphere, we have

$$F_{back} = p_2 \frac{\pi}{4}(2d)^2 = \pi p_2 d^2 = -16\pi\rho U^2 d^2 + \pi p_\infty d^2 \tag{6.92}$$

Combining Equations (6.90), (6.91) and (6.92), the contributions proportional to p_∞ cancel and the net force on the hemisphere is

$$F = \frac{99\pi}{5}\rho U^2 d^2 \tag{6.93}$$

6.11 Mass and Momentum—Rotational Flow

Statement of the Problem: *Consider steady, incompressible flow through the device shown. Body forces can be ignored and the flow is strongly rotational. The flow enters with a known flow speed U, and the areas of the inlets and outlets are as shown. The flow exhausts to the atmosphere at the outlets where the flow speed is V. The force required to hold the device in place is*

$$\mathbf{R} = R_x \mathbf{i} + R_y \mathbf{j}$$

Assuming $R_y = 0$, solve for V, the pressure differential, Δp, and R_x. Also, assume flow properties are constant on all cross sections.

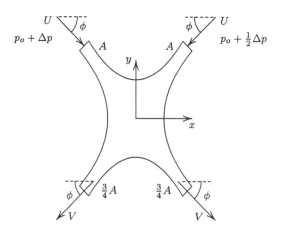

Figure 6.18: *Flow through a fluid-mechanical device.*

Solution: There are three unknown quantities in this problem, viz., V, Δp and R_x. We can appeal to the mass-conservation principle for one equation. The x and y components of

6.11. MASS AND MOMENTUM—ROTATIONAL FLOW

the momentum-conservation principle provide the other two equations. Since the problem is stated in terms of the force required to hold the device in place, i.e., the reaction force, we select a control volume external to the device. Figure 6.19 shows the control volume as a dashed contour.

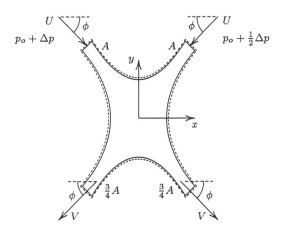

Figure 6.19: *Fluid-mechanical device and control volume.*

Mass Conservation: We are given the fact that the flow is steady. Hence, the equation for conservation of mass is

$$\oiint_S \rho \mathbf{u} \cdot \mathbf{n} \, dS = 0 \tag{6.94}$$

The control volume has been chosen so that the outer unit normals are parallel to the velocity vectors on all inlets and outlets. Thus, at the inlets and outlets we have

$$\mathbf{u} \cdot \mathbf{n} = \begin{cases} -U, & \text{Upper left} \\ -U, & \text{Upper right} \\ V, & \text{Lower right} \\ V, & \text{Lower left} \end{cases} \tag{6.95}$$

Hence, expanding the closed-surface integral gives

$$\underbrace{\rho(-UA)}_{Upper\ left} + \underbrace{\rho(-UA)}_{Upper\ right} + \underbrace{\rho\left(\frac{3}{4}VA\right)}_{Lower\ right} + \underbrace{\rho\left(\frac{3}{4}VA\right)}_{Lower\ left} = 0 \tag{6.96}$$

Solving for the outlet velocity, V, we find

$$V = \frac{4}{3}U \tag{6.97}$$

x-Momentum Conservation: The x component of the momentum-conservation equation for steady flow is

$$\oiint_S \rho u \left(\mathbf{u} \cdot \mathbf{n}\right) dS = -\mathbf{i} \cdot \oiint_S (p - p_o) \mathbf{n} \, dS + R_x \tag{6.98}$$

where we take the dot product of the pressure integral with **i** to extract its x component. We can evaluate the momentum-flux integral, i.e., the integral on the left-hand side of Equation (6.98), as follows.

$$\oiint_S \rho u \, (\mathbf{u} \cdot \mathbf{n}) \, dS = \underbrace{(\rho U \cos \phi)(-UA)}_{Upper\ left} + \underbrace{(-\rho U \cos \phi)(-UA)}_{Upper\ right}$$

$$+ \underbrace{(\rho V \cos \phi)\left(\frac{3}{4}VA\right)}_{Lower\ right} + \underbrace{(-\rho V \cos \phi)\left(\frac{3}{4}VA\right)}_{Lower\ left} = 0 \quad (6.99)$$

Turning to the pressure integral, the pressure differs from p_o only at the inlets. Thus,

$$-\mathbf{i} \cdot \oiint_S (p - p_o) \, \mathbf{n} \, dS = \underbrace{\Delta p A \cos \phi}_{Upper\ left} - \underbrace{\frac{1}{2}\Delta p A \cos \phi}_{Upper\ right} = \frac{1}{2}\Delta p A \cos \phi \quad (6.100)$$

Substituting Equations (6.99) and (6.100) into Equation (6.98) yields

$$0 = \frac{1}{2}\Delta p A \cos \phi + R_x \quad \Longrightarrow \quad R_x = -\frac{1}{2}\Delta p A \cos \phi \quad (6.101)$$

y-Momentum Conservation: The y component of the momentum-conservation equation is

$$\oiint_S \rho v \, (\mathbf{u} \cdot \mathbf{n}) \, dS = -\mathbf{j} \cdot \oiint_S (p - p_o) \, \mathbf{n} \, dS + R_y \quad (6.102)$$

where we take the dot product of the pressure integral with **j** to extract its y component. As with the x-momentum equation, we begin by focusing first on the momentum-flux integral.

$$\oiint_S \rho v \, (\mathbf{u} \cdot \mathbf{n}) \, dS = \underbrace{(-\rho U \sin \phi)(-UA)}_{Upper\ left} + \underbrace{(-\rho U \sin \phi)(-UA)}_{Upper\ right}$$

$$+ \underbrace{(-\rho V \sin \phi)\left(\frac{3}{4}VA\right)}_{Lower\ right} + \underbrace{(-\rho V \cos \phi)\left(\frac{3}{4}VA\right)}_{Lower\ left}$$

$$= 2\rho U^2 A \sin \phi - \frac{3}{2}\rho V^2 A \sin \phi \quad (6.103)$$

But, from Equation (6.97), we know that $V = \frac{4}{3}U$. Therefore,

$$\oiint_S \rho v \, (\mathbf{u} \cdot \mathbf{n}) \, dS = \left[2 - \frac{3}{2}\left(\frac{4}{3}\right)^2\right] \rho U^2 A \sin \phi = -\frac{2}{3}\rho U^2 A \sin \phi \quad (6.104)$$

Again, noting that the pressure differs from p_o only at the inlets, expanding the closed-surface integral of the pressure, we find

$$-\mathbf{j} \cdot \oiint_S (p - p_o) \, \mathbf{n} \, dS = \underbrace{-\Delta p A \sin \phi}_{Upper\ left} - \underbrace{\frac{1}{2}\Delta p A \sin \phi}_{Upper\ right} = -\frac{3}{2}\Delta p A \sin \phi \quad (6.105)$$

6.12. MASS AND MOMENTUM—IRROTATIONAL FLOW

Combining Equations (6.102), (6.104) and (6.105), and noting that we are given $R_y = 0$, there follows

$$-\frac{2}{3}\rho U^2 A \sin\phi = -\frac{3}{2}\Delta p A \sin\phi \qquad (6.106)$$

Therfore, solving for the pressure differential, Δp, we have

$$\Delta p = \frac{4}{9}\rho U^2 \qquad (6.107)$$

Finally, substituting this value of Δp into Equation (6.101), the x component of the force required to hold the device in place is

$$R_x = -\frac{2}{9}\rho U^2 A \cos\phi \qquad (6.108)$$

6.12 Mass and Momentum—Irrotational Flow

Statement of the Problem: *Consider steady, incompressible flow into a 180° bend that springs a leak as shown. Body forces can be ignored, and you may assume the flow is irrotational. The cross-sectional areas at the inlet and the outlet are the same and equal to A, while the area of the jet of fluid emanating from the leak is $\frac{1}{25}A$. Assume flow properties are constant on all cross sections, and that p_o is atmospheric pressure. Solve for the jet velocity, U_j, the pressure differentials, Δp_1 and Δp_2, and the force exerted by the fluid on the bend.*

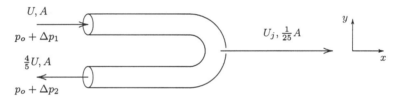

Figure 6.20: *Flow into a 180° bend with a leak.*

Solution: As with the problem in the preceding section, we must determine four unknown quantities. They are U_j, Δp_1, Δp_2 and the force on the bend. From the symmetry of the geometry, it is obvious that the force has only an x component, which we will call F. The mass-conservation principle yields one equation. Another equation follows from the x component of the momentum-conservation principle. (The y component simply confirms that the force in the y direction is zero.) This leaves us short two equations.

From the statement of the problem, the flow is steady, incompressible, irrotational and free of body forces. Hence, all of the conditions required for Bernoulli's equation to be valid are satisfied for this flow. So, we can obtain two additional equations by using Bernoulli's equation twice.

It is convenient to use a control volume external to the device. Figure 6.21 shows the control volume as a dashed contour.

Mass Conservation: Because the flow is steady, the mass-conservation equation simplifies to

$$\oiint_S \rho \mathbf{u}\cdot\mathbf{n}\,dS = 0 \qquad (6.109)$$

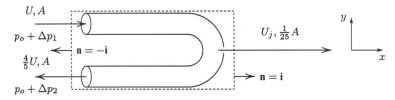

Figure 6.21: *Control volume for flow into a 180° bend with a leak.*

The control volume has been chosen so that the outer unit normals are parallel to the velocity vectors at the inlet, outlet and the jet. Thus, we have

$$\mathbf{u} \cdot \mathbf{n} = \begin{cases} -U, & \text{Inlet} \\ \frac{4}{5}U, & \text{Outlet} \\ U_j, & \text{Jet} \end{cases} \qquad (6.110)$$

So, the closed-surface integral of the mass flux becomes

$$\underbrace{\rho(-UA)}_{Inlet} + \underbrace{\rho\left(\frac{1}{25}U_jA\right)}_{Jet} + \underbrace{\rho\left(\frac{4}{5}UA\right)}_{Outlet} = 0 \qquad (6.111)$$

Solving for the jet velocity, U_j, we find

$$U_j = 5U \qquad (6.112)$$

Momentum Conservation: For steady flow, the x component of the momentum-conservation equation assumes the following form.

$$\oiint_S \rho u\,(\mathbf{u} \cdot \mathbf{n})\, dS = -\mathbf{i} \cdot \oiint_S (p - p_o)\,\mathbf{n}\, dS + R_x \qquad (6.113)$$

The quantity R_x is a reaction force, which is the force required to hold the bend in place. Also, we take the dot product of the pressure integral with $\mathbf{i}$ to extract its x component, and we subtract atmospheric pressure from p to simplify the analysis. If we fail to do this and work with the pressure, p, rather than $p - p_o$, we must include the integral of p_o over all parts of the bounding surface of the control volume except the inlets and outlets, where the pressure is equal to a value other than p_o (the jet exhausts to the atmosphere so that the pressure in the jet is p_o). We are taking advantage of the fact that the closed surface integral of p_o is exactly zero, so that the pressure integral with $(p - p_o)$ involves only contributions from the inlets and outlets where p differs from p_o.

Our first step is to evaluate the momentum-flux integral, i.e., the integral on the left-hand side of Equation (6.113). We have

$$\begin{aligned}
\oiint_S \rho u\,(\mathbf{u}\cdot\mathbf{n})\,dS &= \underbrace{(\rho U)(-UA)}_{Inlet} + \underbrace{(\rho U_j)\left(\frac{1}{25}U_jA\right)}_{Jet} + \underbrace{\left(-\frac{4}{5}\rho U\right)\left(\frac{4}{5}UA\right)}_{Outlet} \\
&= -\frac{41}{25}\rho U^2 A + \frac{1}{25}\rho U_j^2 A \\
&= -\frac{41}{25}\rho U^2 A + \frac{1}{25}\rho(5U)^2 A = -\frac{16}{25}\rho U^2 A \qquad (6.114)
\end{aligned}$$

6.12. MASS AND MOMENTUM—IRROTATIONAL FLOW

Note that we use Equation (6.112) to eliminate U_j in the last line of Equation (6.114).

Turning to the pressure integral, the pressure differs from p_o only at the inlet and the outlet. Thus,

$$-\mathbf{i} \cdot \oiint_S (p - p_o)\mathbf{n}\, dS = \underbrace{\Delta p_1 A}_{Inlet} + \underbrace{\Delta p_2 A}_{Outlet} = (\Delta p_1 + \Delta p_2)A \quad (6.115)$$

Combining Equations (6.113), (6.114) and (6.115) yields

$$-\frac{16}{25}\rho U^2 A = (\Delta p_1 + \Delta p_2)A + R_x \quad (6.116)$$

Solving for R_x, we arrive at the following.

$$R_x = -\frac{16}{25}\rho U^2 A - (\Delta p_1 + \Delta p_2)A \quad (6.117)$$

Bernoulli's Equation: We can use Bernoulli's equation twice to determine the pressure differentials, Δp_1 and Δp_2. First, applying the equation between the inlet and the jet, we find

$$p_o + \Delta p_1 + \frac{1}{2}\rho U^2 = p_o + \frac{1}{2}\rho(5U)^2 = p_o + \frac{25}{2}\rho U^2 \quad (6.118)$$

Solving for Δp_1 yields

$$\Delta p_1 = 12\rho U^2 \quad (6.119)$$

Similarly, we apply the equation between the outlet and the jet, wherefore

$$p_o + \Delta p_2 + \frac{1}{2}\rho\left(\frac{4}{5}U\right)^2 = p_o + \frac{1}{2}\rho(5U)^2 \implies \Delta p_2 = \left(\frac{25}{2} - \frac{8}{25}\right)\rho U^2 \quad (6.120)$$

Hence, the pressure differential at the outlet is

$$\Delta p_2 = \frac{609}{50}\rho U^2 \quad (6.121)$$

Substituting Equations (6.119) and (6.121) into Equation (6.117), the reaction force is

$$R_x = -\frac{16}{25}\rho U^2 A - \left(12 + \frac{609}{50}\right)\rho U^2 A = -\frac{1241}{50}\rho U^2 A \quad (6.122)$$

Finally, the sum of the force exerted by the fluid on the bend, F, and the force required to hold the bend in place is zero, i.e., $F + R_x = 0$, so that the force is

$$F = -R_x = \frac{1241}{50}\rho U^2 A = 24.82\rho U^2 A \quad (6.123)$$

Because F is positive, it acts to the right, so that the force vector is given by

$$\mathbf{F} = 24.82\rho U^2 A\, \mathbf{i} \quad (6.124)$$

6.13 Mass and Momentum—Nonuniform Profiles

Statement of the Problem: *The velocity profile in a channel changes from U_1 to U_2, where*

$$U_1 = U\left[1 - \left(\frac{y}{h}\right)^2\right] \quad \text{and} \quad U_2 = AU\left[1 - \left(\frac{y}{h}\right)^4\right]$$

for $-h \leq y \leq h$. Also, the pressure difference between the outlet and inlet of the channel is $p_2 - p_1 = \frac{1}{15}\rho U^2$, and the pressure is constant on cross sections. Assuming the flow is steady and incompressible, solve for the constant A, and the net viscous force per unit width on the channel, F_τ.

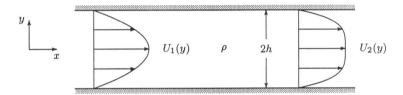

Figure 6.22: *Flow in a channel.*

Solution: To solve this problem, we select a rectangular control volume that is coincident with the inner walls of the channel, as indicated by the dashed contour in Figure 6.23. This control volume is convenient as the unit normals are parallel to the inlet and outlet velocities.

Mass Conservation: We begin with the mass-conservation principle for steady flow, which tells us that

$$\oiint_S \rho \mathbf{u} \cdot \mathbf{n} \, dS = 0 \tag{6.125}$$

At the inlet, the velocity vector is parallel to and points in the opposite direction of the outer unit normal so that $\mathbf{u}_1 \cdot \mathbf{n}_1 = -U_1$. The outlet velocity vector is parallel to and points in the same direction as the outer unit normal so that $\mathbf{u}_2 \cdot \mathbf{n}_2 = U_2$. Thus,

$$-\underbrace{\int_{-h}^{h} \rho\, U_1(y)\, dy}_{Inlet} + \underbrace{\int_{-h}^{h} \rho\, U_2(y)\, dy}_{Outlet} = 0 \tag{6.126}$$

Note that because this is a two-dimensional problem, we regard all of our results as being "per unit width." Thus, integrating over the inlet and outlet cross sections means simply

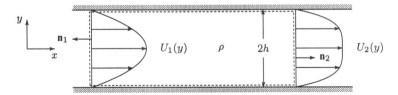

Figure 6.23: *Control volume for flow in a channel.*

6.13. MASS AND MOMENTUM—NONUNIFORM PROFILES

integrating over y. Equivalently, we can regard the channel as having unit width so that the cross-sectional area at the inlet, for example, is $2h \cdot 1$, where the factor '1' makes the area dimensionally correct. Either way, all of our results will have factors of h present where we would normally expect to find a quantity of dimensions (length)^2.

Substituting the given velocity profiles and dividing through by ρ yields

$$-\int_{-h}^{h} U\left[1 - \left(\frac{y}{h}\right)^2\right] dy + \int_{-h}^{h} AU\left[1 - \left(\frac{y}{h}\right)^4\right] dy = 0 \quad (6.127)$$

This equation can be rearranged to read as follows.

$$Uh \int_{-h}^{h} \left[1 - \left(\frac{y}{h}\right)^2\right] d\left(\frac{y}{h}\right) = AUh \int_{-h}^{h} \left[1 - \left(\frac{y}{h}\right)^4\right] d\left(\frac{y}{h}\right) \quad (6.128)$$

Next, it is convenient to change integration variables according to $\eta \equiv y/h$. Also, dividing through by Uh, we find

$$\int_{-1}^{1} \left[1 - \eta^2\right] d\eta = A \int_{-1}^{1} \left[1 - \eta^4\right] d\eta \quad (6.129)$$

Evaluating the integrals gives

$$\left[\eta - \frac{1}{3}\eta^3\right]_{\eta=-1}^{\eta=1} = A \left[\eta - \frac{1}{5}\eta^5\right]_{\eta=-1}^{\eta=1} \quad (6.130)$$

or,

$$\left[\left(1 - \frac{1}{3}\right) - \left(-1 + \frac{1}{3}\right)\right] = A \left[\left(1 - \frac{1}{5}\right) - \left(-1 + \frac{1}{5}\right)\right] \implies \frac{4}{3} = \frac{8}{5}A \quad (6.131)$$

Finally, solving for A, the solution is

$$A = \frac{5}{6} \quad (6.132)$$

Momentum Conservation: For steady flow, the x component of the momentum-conservation equation assumes the following form.

$$\oiint_S \rho u (\mathbf{u} \cdot \mathbf{n}) \, dS = -\mathbf{i} \cdot \oiint_S p \mathbf{n} \, dS - F_\tau \quad (6.133)$$

The quantity F_τ is the net viscous force that acts on the channel walls. We include this force with a minus sign to reflect the fact that it opposes the motion. Also, we take the dot product of the pressure integral with $\mathbf{i}$ to extract its x component. Our first step is to evaluate the momentum-flux integral, i.e.,

$$\begin{aligned}\oiint_S \rho u (\mathbf{u} \cdot \mathbf{n}) \, dS &= \underbrace{\int_{-h}^{h} (\rho U_1)(-U_1) \, dy}_{Inlet} + \underbrace{\int_{-h}^{h} (\rho U_2)(U_2) \, dy}_{Outlet} \\ &= -\int_{-h}^{h} \rho U^2 \left[1 - \left(\frac{y}{h}\right)^2\right]^2 dy + \int_{-h}^{h} \frac{25}{36} \rho U^2 \left[1 - \left(\frac{y}{h}\right)^4\right]^2 dy \\ &= -\rho U^2 h \int_{-h}^{h} \left[1 - \left(\frac{y}{h}\right)^2\right]^2 d\left(\frac{y}{h}\right) \\ &\quad + \frac{25}{36} \rho U^2 h \int_{-h}^{h} \left[1 - \left(\frac{y}{h}\right)^4\right]^2 d\left(\frac{y}{h}\right)\end{aligned} \quad (6.134)$$

Note that we use Equation (6.132) to replace A^2 by 25/36 in the second line of Equation (6.134). As above, we again change integration variables according to $\eta \equiv y/h$. Then, the momentum-flux integral can be evaluated as follows.

$$
\begin{aligned}
\oiint_S \rho u \left(\mathbf{u} \cdot \mathbf{n}\right) dS &= -\rho U^2 h \int_{-1}^{1} \left(1 - \eta^2\right)^2 d\eta + \frac{25}{36} \rho U^2 h \int_{-1}^{1} \left(1 - \eta^4\right)^2 d\eta \\
&= -\rho U^2 h \int_{-1}^{1} \left(1 - 2\eta^2 + \eta^4\right) d\eta + \frac{25}{36} \rho U^2 h \int_{-1}^{1} \left(1 - 2\eta^4 + \eta^8\right) d\eta \\
&= -\rho U^2 h \left[\eta - \frac{2}{3}\eta^3 + \frac{1}{5}\eta^5\right]_{\eta=-1}^{\eta=1} + \frac{25}{36} \rho U^2 h \left[\eta - \frac{2}{5}\eta^5 + \frac{1}{9}\eta^9\right]_{\eta=-1}^{\eta=1} \\
&= -\rho U^2 h \left[\left(1 - \frac{2}{3} + \frac{1}{5}\right) - \left(-1 + \frac{2}{3} - \frac{1}{5}\right)\right] \\
&\quad + \frac{25}{36} \rho U^2 h \left[\left(1 - \frac{2}{5} + \frac{1}{9}\right) - \left(-1 + \frac{2}{5} - \frac{1}{9}\right)\right] \\
&= -\rho U^2 h \left[\frac{16}{15}\right] + \frac{25}{36} \rho U^2 h \left[\frac{64}{45}\right] \\
&= -\frac{32}{405} \rho U^2 h
\end{aligned} \qquad (6.135)
$$

The pressure integral is straightforward and simplifies to

$$-\mathbf{i} \cdot \oiint_S p \mathbf{n} \, dS = (p_1 - p_2) h = -\frac{1}{15} \rho U^2 h \qquad (6.136)$$

Combining Equations (6.133), (6.135) and (6.136) gives

$$-\frac{32}{405} \rho U^2 h = -\frac{1}{15} \rho U^2 h - F_\tau \qquad (6.137)$$

Solving for F_τ, we arrive at the following.

$$F_\tau = \frac{1}{81} \rho U^2 h \qquad (6.138)$$

As noted above in the discussion of this problem's two-dimensional geometry, this force has dimensions of force per unit length. This is the desired result, i.e., F_τ is the viscous force per unit width of the channel.

6.14 Mass and Momentum—Accelerating Control Volume

Statement of the Problem: *A small cart receives water from a jet as shown in Figure 6.24. The speed and cross-sectional area of the jet are U_j and A, respectively. At time $t = 0$, the cart is at rest and its mass is m_o. Assume the motion is such that aerodynamic forces on the cart are negligible, i.e., ignore viscosity and assume $p \approx p_o$ at the cart surface, where p_o is atmospheric pressure. The only force acting on the cart is rolling friction, $\mu m g$, where μ is the coefficient of rolling friction, m is the instantaneous mass of the cart and g is gravitational acceleration. Assume the motion of the fluid inside the cart is steady as seen by an observer riding on the cart. Finally, let $U(t)$ denote the instantaneous speed of the cart.*

6.14. MASS AND MOMENTUM—ACCELERATING CONTROL VOLUME

(a) Using the translating (and accelerating) control volume indicated by the dashed contour, develop differential equations for m and U that describe the motion of the cart.

(b) Assuming the surface is frictionless so that $\mu = 0$, combine the equations developed in Part (a) into a differential equation involving dU/dm.

(c) Solve the equation derived in Part (b) and determine the velocity when $m(t) = 5m_o$.

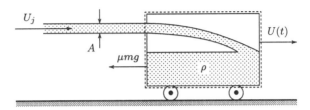

Figure 6.24: *Cart driven by a water jet.*

Solution: We solve this problem using a control volume that is fixed with the cart. Since the cart is accelerating, so is the control volume.

6.14(a): Conservation of mass tells us

$$\frac{d}{dt} \iiint_V \rho \, dV + \oiint_S \rho \mathbf{u} \cdot \mathbf{n} \, dS = 0 \qquad (6.139)$$

By definition, the mass of the cart, including the water it receives, is m, and the only place fluid crosses the control-volume surface is through the opening where the water jet enters the cart. Let the positive x direction be to the right. The absolute velocity of the fluid in the jet is $\mathbf{u}_{abs} = U_j \mathbf{i}$, while the control-volume velocity is $\mathbf{u}_{cv} = U(t)\mathbf{i}$. Hence, the relative velocity of the water jet as it crosses the control-volume surface, and the outer unit normal are

$$\mathbf{u} = (U_j - U)\mathbf{i} \quad \text{and} \quad \mathbf{n} = -\mathbf{i} \qquad (6.140)$$

Thus,

$$m \equiv \iiint_V \rho \, dV \quad \text{and} \quad \oiint_S \rho \mathbf{u} \cdot \mathbf{n} \, dS = -\rho(U_j - U)A \qquad (6.141)$$

Combining Equations (6.139) and (6.141), we have

$$\frac{dm}{dt} - \rho(U_j - U)A = 0 \qquad (6.142)$$

Turning to momentum, the conservation principle tells us that

$$\iiint_V \frac{\partial}{\partial t}(\rho u_{abs}) \, dV + \oiint_S \rho u_{abs} (\mathbf{u} \cdot \mathbf{n}) \, dS = -\mathbf{i} \cdot \oiint_S (p - p_o) \mathbf{n} \, dS - \mu m g \qquad (6.143)$$

where the absolute velocity, as noted above, is

$$\mathbf{u}_{abs} = (U + u)\mathbf{i} \qquad (6.144)$$

and u is the velocity as seen by an observer on the cart. Now, the unsteady term is

$$\iiint_V \frac{\partial}{\partial t}(\rho u_{abs}) \, dV = \iiint_V \frac{\partial}{\partial t}(\rho U) \, dV + \iiint_V \frac{\partial}{\partial t}(\rho u) \, dV \qquad (6.145)$$

Since we are given that the flow inside the cart is steady as seen by an observer riding on the cart, the last integral on the right-hand side of this equation vanishes. Also, since the volume does not change in size or shape, $\partial/\partial t$ commutes with the integral. Thus,

$$\iiint_V \frac{\partial}{\partial t}(\rho u_{abs}) \, dV = \frac{d}{dt} \iiint_V \rho U \, dV = \frac{d}{dt}(mU) \tag{6.146}$$

The only part of the control-volume boundary across which momentum passes is the jet. Hence,

$$\oiint_S \rho u_{abs}(\mathbf{u} \cdot \mathbf{n}) \, dS = \rho U_j \left[-(U_j - U) A \right] = -\rho U_j (U_j - U) A \tag{6.147}$$

Finally, for the chosen control volume, ignoring aerodynamic forces means the net pressure integral is zero. Hence, combining Equations (6.143), (6.146) and (6.147), momentum conservation simplifies to

$$\frac{d}{dt}(mU) - \rho U_j (U_j - U) A = -\mu m g \tag{6.148}$$

We can expand the first term on the left-hand side of the momentum equation according to

$$\frac{d}{dt}(mU) = m \frac{dU}{dt} + U \frac{dm}{dt} = m \frac{dU}{dt} + \rho U (U_j - U) A \tag{6.149}$$

where we make use of the mass-conservation Equation (6.142) to eliminate dm/dt. Substituting into the momentum equation, the differential equations for m and U are as follows.

$$m \frac{dU}{dt} = \rho (U_j - U)^2 A - \mu m g \tag{6.150}$$

$$\frac{dm}{dt} = \rho (U_j - U) A \tag{6.151}$$

6.14(b): If the surface is frictionless so that $\mu = 0$, then the differential equations for U and m are

$$m \frac{dU}{dt} = \rho (U_j - U)^2 A \quad \text{and} \quad \frac{dm}{dt} = \rho (U_j - U) A \tag{6.152}$$

Dividing the momentum equation by the mass equation, we find

$$m \frac{dU}{dm} = U_j - U \tag{6.153}$$

6.14(c): First, we rewrite Equation (6.153) as

$$\frac{dU}{U_j - U} = \frac{dm}{m} \tag{6.154}$$

Integrating this equation, there follows

$$-\ell n (U_j - U) = \ell n m + \text{constant} \tag{6.155}$$

Using the fact that $U(0) = 0$ and $m(0) = m_o$, the constant is

$$\text{constant} = -\ell n U_j - \ell n m_o \tag{6.156}$$

6.15. MASS AND MOMENTUM—THREE-DIMENSIONAL FLOW

Hence, the solution for $U(t)$ as a function of $m(t)$ is

$$-\ell n\left[\frac{U_j - U(t)}{U_j}\right] = \ell n\left[\frac{m(t)}{m_o}\right] \implies U_j - U(t) = U_j \frac{m_o}{m(t)} \quad (6.157)$$

Therefore, the cart velocity is

$$U(t) = U_j\left[1 - \frac{m_o}{m(t)}\right] \quad (6.158)$$

Finally, when $m(t) = 5m_o$, we have

$$U(t) = U_j\left[1 - \frac{1}{5}\right] = \frac{4}{5}U_j \quad (6.159)$$

6.15 Mass and Momentum—Three-Dimensional Flow

Statement of the Problem: *A pipe has a 90° bend and exhausts to the atmosphere. Ahead of the bend, the pipe diameter is d, and it decreases to $\frac{1}{2}d$ at the outlet. In terms of Cartesian coordinates, the inlet and outlet velocities are*

$$\mathbf{u}_1 = -U\mathbf{i} \quad \text{and} \quad \mathbf{u}_2 = \frac{V}{\sqrt{2}}(\mathbf{j} + \mathbf{k})$$

The flow is steady, incompressible, strongly rotational, and the pressure at the inlet is

$$p_1 = p_o + 3\rho U^2$$

where ρ is the fluid density and p_o is atmospheric pressure. Ignoring body forces, solve for V and determine the force required to hold the pipe in place.

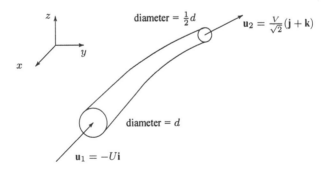

Figure 6.25: *Pipe with a 90° bend.*

Solution: This problem involves four unknown quantities, viz., the outlet velocity, V, and the three components of the force required to hold the pipe in place, R_x, R_y and R_z. Thus, in addition to the mass-conservation principle, we must appeal to the full three-dimensional momentum-conservation equation to obtain the three additional equations that we must have in order to solve the problem. It will prove to be most convenient to work with the momentum equation in vector form.

The first step is to select a control volume. Since the goal is to compute the force required to hold the pipe in place, i.e., the reaction force, the obvious choice is a control volume coincident with the outer surface of the pipe. When we do this, we are obligated to include a reaction force, which is the force required to hold the control volume in place. Also, the control-volume surface should be such that the outer unit normals are parallel to the flow velocity vectors at the inlet and the outlet. Figure 6.26 shows the control volume as a dashed contour.

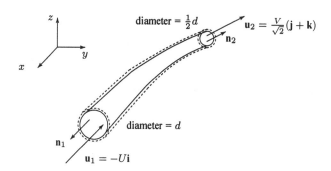

Figure 6.26: *Pipe and control volume.*

First, note that the outer unit normals at the inlet and outlet are

$$\mathbf{n}_1 = \mathbf{i} \quad \text{and} \quad \mathbf{n}_2 = \frac{1}{\sqrt{2}}(\mathbf{j} + \mathbf{k}) \tag{6.160}$$

By design, the inlet and outlet velocities are parallel to the unit normals, wherefore

$$\mathbf{u}_1 \cdot \mathbf{n}_1 = -U \quad \text{and} \quad \mathbf{u}_2 \cdot \mathbf{n}_2 = V \tag{6.161}$$

Mass Conservation: Since the flow is steady, the mass-conservation principle is

$$\oiint_S \rho \mathbf{u} \cdot \mathbf{n}\, dS = 0 \tag{6.162}$$

Expanding the closed-surface integral, we have

$$\underbrace{\rho \mathbf{u}_1 \cdot \mathbf{n}_1 \frac{\pi}{4} d^2}_{\text{Inlet}} + \underbrace{\rho \mathbf{u}_2 \cdot \mathbf{n}_2 \frac{\pi}{4}\left(\frac{d}{2}\right)^2}_{\text{Outlet}} = 0 \tag{6.163}$$

Now, taking advantage of Equation (6.161), we have

$$-\frac{\pi}{4}\rho U d^2 + \frac{\pi}{16}\rho V d^2 = 0 \quad \Longrightarrow \quad V = 4U \tag{6.164}$$

Momentum Conservation: The momentum-conservation equation for steady flow in vector form is

$$\oiint_S \rho \mathbf{u}\,(\mathbf{u} \cdot \mathbf{n})\, dS = -\oiint_S (p - p_o)\,\mathbf{n}\, dS + \mathbf{R} \tag{6.165}$$

6.15. MASS AND MOMENTUM—THREE-DIMENSIONAL FLOW

We can evaluate the momentum-flux integral as follows.

$$\begin{aligned}
\oint_S \rho \mathbf{u}(\mathbf{u}\cdot\mathbf{n})\,dS &= \underbrace{\iint_A (-\rho U\,\mathbf{i})(-U)\,dA}_{\text{Inlet}} + \underbrace{\iint_A \frac{\rho V}{\sqrt{2}}(\mathbf{j}+\mathbf{k})(V)\,dA}_{\text{Outlet}} \\
&= \frac{\pi}{4}\rho U^2 d^2\,\mathbf{i} + \frac{\pi}{16\sqrt{2}}\rho V^2 d^2(\mathbf{j}+\mathbf{k}) \\
&= \frac{\pi}{4}\rho U^2 d^2\,\mathbf{i} + \frac{\pi}{\sqrt{2}}\rho U^2 d^2(\mathbf{j}+\mathbf{k}) \\
&= \frac{\pi}{4}\rho U^2 d^2\left(\mathbf{i} + 2\sqrt{2}\,\mathbf{j} + 2\sqrt{2}\,\mathbf{k}\right)
\end{aligned} \qquad (6.166)$$

In the third line of Equation (6.166), we have used the fact that $V = 4U$ [Equation (6.164)].

Turning to the pressure integral, we note first that the pressure at the outlet is atmospheric, i.e., $p = p_o$. Also, it is atmospheric on the outer surface of the pipe. The only part of the control-volume surface on which the pressure differs from the atmospheric value is at the inlet. Thus,

$$\oint_S (p-p_o)\mathbf{n}\,dS = \underbrace{\iint_A (p_1 - p_o)\mathbf{i}\,dS}_{\text{Inlet}} = \frac{\pi}{4}(p_1 - p_o)d^2\,\mathbf{i} \qquad (6.167)$$

But, we are given $p_1 = p_o + 3\rho U^2$, so that the pressure integral is

$$\oint_S (p-p_o)\mathbf{n}\,dS = \frac{3\pi}{4}\rho U^2 d^2\,\mathbf{i} \qquad (6.168)$$

Combining Equations (6.165), (6.166) and (6.168) yields

$$\frac{\pi}{4}\rho U^2 d^2\left(\mathbf{i} + 2\sqrt{2}\,\mathbf{j} + 2\sqrt{2}\,\mathbf{k}\right) = -\frac{3\pi}{4}\rho U^2 d^2\,\mathbf{i} + \mathbf{R} \qquad (6.169)$$

Solving for $\mathbf{R}$, the force required to hold the pipe in place is

$$\mathbf{R} = \pi\rho U^2 d^2\left(\mathbf{i} + 2\sqrt{2}\,\mathbf{j} + 2\sqrt{2}\,\mathbf{k}\right) \qquad (6.170)$$

Chapter 7

Conservation of Energy

In this chapter, we turn to the energy-conservation principle, focusing on both integral and differential forms. Typical problems using the integral form deal with control volumes, and are similar to the types of problems considered in Chapter 6. The differential form of the energy-conservation equation finds application in flow through pipes, piping systems and open channels. When we analyze pipe and open-channel flows, we use concepts from the field of hydraulics. Although empirical in its essence, hydraulics provides useful design techniques that can be expected to be accurate to within 10% to 20%, depending on the complexity of the application. The problems in this chapter cover the following topics.

Section 7.1 Thermodynamics: A simple problem in thermodynamics that uses basic thermodynamic concepts to compute heat and work.

Section 7.2 Energy—Integral Form: This is a problem that uses the integral form of the energy equation to determine the power delivered by a pump.

Section 7.3 Energy—Differential Form: The solution to this pipe-flow problem uses the differential form of the energy-conservation equation. It includes computation of kinetic-energy correction factors and the total head loss in the pipe.

Section 7.4 Moody Diagram: Using the Moody diagram, this problem involves calculation of friction factors and head loss for various pipes.

Section 7.5 Flow Through a Turbine: The problem in this section uses the differential form of the energy equation to analyze a hydroelectric plant.

Section 7.6 Minor Losses I: This is the first of two pipe-flow problems involving both major and minor losses. The solution makes use of the Moody diagram to determine the friction factor.

Section 7.7 Minor Losses II: This is the second of two pipe-flow problems involving both major and minor losses. In this problem, we use the Colebrook equation to determine the friction factor.

Section 7.8 Pipe System: The problem in this section uses the differential form of the energy equation to analyze a pipe system. The solution uses the Moody diagram to determine friction factors.

Section 7.9 Free-Surface Waves: Approximating a stream of water as a (very wide) open-channel flow, this problem analyzes motion of surface waves in the stream.

Section 7.10 Open-Channel Flow I: This is the first of two open-channel flow problems. It focuses on a rectangular channel, and can be solved directly, i.e., without iteration.

Section 7.11 Open-Channel Flow II: This is the second of two open-channel flow problems. It involves flow in a channel of triangular cross section, and requires an iterative solution.

Section 7.12 Hydraulic Jump: Using the hydraulic-jump relations, the solution to this problem shows how to determine jump properties in terms of the Froude number.

7.1 Thermodynamics

Statement of the Problem: *In a polytropic process, the pressure and density of a perfect gas are related by* $p = p_1(\rho/\rho_1)^{(n+1)/2}$, *where n is a constant and subscript '1' denotes initial condition.*

(a) *In the initial state, we have* $V_1 = 1000 \, ft^3$ *of air with a total mass of* $m = 1$ *slug and a pressure of* $p_1 = 14.7$ *psi. The air undergoes a reversible, polytropic process in which the air pressure and temperature change to* $p_2 = 20$ *psi and* $T_2 = 65°F$, *respectively. What is the value of the constant n?*

(b) *Determine the work done by the surroundings on the gas, W, where* $W \equiv \int_{V_1}^{V_2} p \, dV$.

(c) *Determine the heat transferred from the surroundings to the gas, Q. Express your answer in Btu.*

Solution: Since the equation relating state variables for this process involves pressure and density, and pressures are given but not densities, we must determine ρ_1 and ρ_2. First, by definition,

$$\rho_1 = \frac{m}{V_1} = \frac{1 \text{ slug}}{1000 \text{ ft}^3} = 0.001 \, \frac{\text{slug}}{\text{ft}^3} \qquad (7.1)$$

Then, using the perfect-gas law for air with a perfect-gas constant of $R = 1716$ ft·lb/(slug·°R), we find

$$\rho_2 = \frac{p_2}{RT_2} = \frac{\left(20 \text{ lb/in}^2\right)\left(144 \text{ in}^2/\text{ft}^2\right)}{(1716 \text{ ft} \cdot \text{lb/slug/°R})\left[(65 + 459.67)° \text{R}\right]} = 0.0032 \, \frac{\text{slug}}{\text{ft}^3} \qquad (7.2)$$

7.1(a): To compute the constant n, note that we are given

$$\frac{p_2}{p_1} = \left(\frac{\rho_2}{\rho_1}\right)^{(n+1)/2} \quad \Longrightarrow \quad n = 2\frac{\ell n \, (p_2/p_1)}{\ell n \, (\rho_2/\rho_1)} - 1 \qquad (7.3)$$

Hence, n is

$$n = 2\frac{\ell n\left[(20 \text{ psi})/(14.7 \text{ psi})\right]}{\ell n\left[\left(0.0032 \text{ slug/ft}^3\right)/\left(0.001 \text{ slug/ft}^3\right)\right]} - 1 = -0.47 \qquad (7.4)$$

7.1(b): To determine the work done, it is convenient to express p as a function of volume, V. Since the density is $\rho = m/V$, necessarily

$$p = p_1 \left(\frac{\rho}{\rho_1}\right)^{(n+1)/2} = p_1 \left(\frac{V_1}{V}\right)^{(n+1)/2} \qquad (7.5)$$

7.1. THERMODYNAMICS

Thus, the work done by the surroundings on the gas, W, is

$$W = \int_{V_1}^{V_2} p \, dV = p_1 V_1^{(n+1)/2} \int_{V_1}^{V_2} \frac{dV}{V^{(n+1)/2}} = \frac{2p_1 V_1^{(n+1)/2}}{1-n} V^{(1-n)/2} \bigg|_{V=V_1}^{V=V_2} \tag{7.6}$$

Therefore, the work is

$$W = \frac{2p_1 V_1^{(n+1)/2}}{1-n} \left[V_2^{(1-n)/2} - V_1^{(1-n)/2} \right] \tag{7.7}$$

We can rearrange this by noting first that $p_1 V_1^{(n+1)/2} = p_2 V_2^{(n+1)/2}$, wherefore

$$W = \frac{2}{1-n}(p_2 V_2 - p_1 V_1) \tag{7.8}$$

Then, since $V = m/\rho$, we have

$$W = \frac{2m}{1-n}\left(\frac{p_2}{\rho_2} - \frac{p_1}{\rho_1}\right) \tag{7.9}$$

Hence, using the given values, the work is

$$W = \frac{2(1 \text{ slug})}{1+0.47}\left[\frac{(20 \text{ lb/in}^2)(144 \text{ in}^2/\text{ft}^2)}{0.0032 \text{ slug/ft}^3} - \frac{(14.7 \text{ lb/in}^2)(144 \text{ in}^2/\text{ft}^2)}{0.001 \text{ slug/ft}^3}\right]$$
$$= -1.66 \cdot 10^6 \text{ ft} \cdot \text{lb} \tag{7.10}$$

7.1(c): From the first law of thermodynamics,

$$\Delta E = Q - W \quad \Longrightarrow \quad Q = \Delta E + W \tag{7.11}$$

Now, we know that

$$\Delta E = m(e_2 - e_1) = mc_v(T_2 - T_1) \tag{7.12}$$

So, using the perfect-gas law, the change in energy is

$$\Delta E = m\frac{c_v}{R}\left(\frac{p_2}{\rho_2} - \frac{p_1}{\rho_1}\right) \tag{7.13}$$

Therefore, using the fact that, for a calorically-perfect gas $c_v = R/(\gamma-1)$, the heat transferred from the surroundings to the gas is

$$Q = \underbrace{m\frac{c_v}{R}\left(\frac{p_2}{\rho_2} - \frac{p_1}{\rho_1}\right)}_{\Delta E} + \underbrace{\frac{2m}{1-n}\left(\frac{p_2}{\rho_2} - \frac{p_1}{\rho_1}\right)}_{W} = \left[\frac{1}{\gamma-1} + \frac{2}{1-n}\right] m \left(\frac{p_2}{\rho_2} - \frac{p_1}{\rho_1}\right) \tag{7.14}$$

Combining Equations (7.9) and (7.14), we arrive at the following relation between Q and W.

$$Q = \left[\frac{1}{\gamma-1} + \frac{2}{1-n}\right]\frac{1-n}{2}W = \left[1 + \frac{1-n}{2(\gamma-1)}\right]W \tag{7.15}$$

For the given values, Q is

$$Q = \left[1 + \frac{1 + 0.47}{2(1.4 - 1)}\right](-1.66 \cdot 10^6 \text{ ft} \cdot \text{lb}) = -4.71 \cdot 10^6 \text{ ft} \cdot \text{lb} \quad (7.16)$$

Finally, noting that 1 Btu = 778 ft·lb, we have

$$Q = -6054 \text{ Btu} \quad (7.17)$$

The fact that Q is negative indicates that heat transfer is from the gas to the surroundings.

7.2 Energy—Integral Form

Statement of the Problem: *Your yard drainage system is backing up. To relieve the problem, you install a pump to help drain the water from the primary pipe whose diameter is d. As shown in Figure 7.1, the outlet pipe has a diameter of $2d$. The pump supplies energy to the flow such that the upstream absolute pressure is p and the downstream pressure is $4p$, the steady mass-flow rate is $\dot{m}$, and the temperature increases by ΔT. Also, you may neglect effects of gravity. What is the power, P, delivered by the pump to the flow? Express your answer for the power in terms of $\dot{m}$, p, d, water density, ρ, and specific-heat coefficient, c_v.*

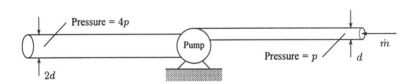

Figure 7.1: *Drainage system with assistance of a pump.*

Solution: There are two unknowns in this problem, namely, the outlet velocity, u_2, and the power, P, delivered by the pump to the flow. One equation follows from the mass-conservation principle. Using the momentum equation, while adding more equations, also introduces additional unknowns in the form of the reaction-force components. Hence, there is no point in appealing to the momentum equation. Rather, our second equation comes from the energy-conservation principle, which explicitly includes the power, P. To solve, we use the control volume indicated in Figure 7.2.

Since the flow is steady, the mass-conservation principle is

$$\oiint_S \rho \mathbf{u} \cdot \mathbf{n} \, dS = 0 \quad (7.18)$$

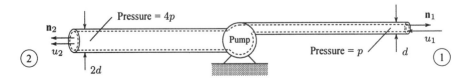

Figure 7.2: *Drainage system, pump and control volume.*

7.2. ENERGY—INTEGRAL FORM

Let u_1 and u_2 denote inlet and outlet velocity, respectively. Expanding the closed-surface integral, we find

$$\underbrace{\rho\left[-u_1\frac{\pi}{4}d^2\right]}_{\text{Inlet}} + \underbrace{\rho\left[u_2\frac{\pi}{4}(2d)^2\right]}_{\text{Outlet}} = 0 \implies \frac{\pi}{4}\rho u_1 d^2 = \pi\rho u_2 d^2 \qquad (7.19)$$

Now, we are given that the mass-flow at the inlet (and hence at the outlet) is $\dot{m}$, and by definition, we have $\dot{m} = \frac{\pi}{4}\rho u_1 d^2$. Thus, the velocities at the inlet and outlet are

$$u_1 = \frac{4\dot{m}}{\pi\rho d^2} \quad \text{and} \quad u_2 = \frac{\dot{m}}{\pi\rho d^2} \qquad (7.20)$$

Energy Conservation: For steady flow, the exact energy conservation principle is

$$\dot{Q} - \dot{W}_s = \oiint_S \rho\left[h + \frac{1}{2}\mathbf{u}\cdot\mathbf{u} + gz\right](\mathbf{u}\cdot\mathbf{n})\,dS \qquad (7.21)$$

We are given that gravitational effects are negligible. Also, the difference between $\dot{Q}$ and $\dot{W}_s$ is the power, P, delivered by the pump to the flow. Therefore, the energy-conservation equation simplifies to

$$P = \oiint_S \rho\left[h + \frac{1}{2}\mathbf{u}\cdot\mathbf{u}\right](\mathbf{u}\cdot\mathbf{n})\,dS \qquad (7.22)$$

Since fluid crosses the control volume surface only at the inlet and the outlet, expanding the closed-surface integral yields

$$\begin{aligned} P &= \underbrace{\rho\left[h_1 + \frac{1}{2}u_1^2\right]\left[-u_1\frac{\pi}{4}d^2\right]}_{\text{Inlet}} + \underbrace{\rho\left[h_2 + \frac{1}{2}u_2^2\right]\left[u_2\frac{\pi}{4}(2d)^2\right]}_{\text{Outlet}} \\ &= \dot{m}\left[h_2 - h_1 + \frac{1}{2}\left(u_2^2 - u_1^2\right)\right] \end{aligned} \qquad (7.23)$$

Now, by definition of enthalpy, $h = c_v T + p/\rho$, so that

$$h_2 - h_1 = c_v \Delta T + \frac{4p}{\rho} - \frac{p}{\rho} = c_v \Delta T + 3\frac{p}{\rho} \qquad (7.24)$$

Hence, appealing to Equation (7.20) to eliminate u_1 and u_2, the power delivered by the pump to the flow is

$$\begin{aligned} P &= \dot{m}\left[c_v\Delta T + 3\frac{p}{\rho} + \frac{1}{2}\left(u_2^2 - u_1^2\right)\right] \\ &= \dot{m}\left[c_v\Delta T + 3\frac{p}{\rho} - \frac{15}{2}\left(\frac{\dot{m}}{\pi\rho d^2}\right)^2\right] \end{aligned} \qquad (7.25)$$

7.3 Energy—Differential Form

Statement of the Problem: *Consider steady, laminar flow of an incompressible fluid with density ρ through the circular pipe of radius R shown below. At the inlet the velocity is uniform and equal to U. At the outlet, the velocity is*

$$U_{out} = U_{max}\left(1 - \frac{r^2}{R^2}\right)$$

where r is radial distance from the centerline. The change in elevation, Δz, and the change in pressure between inlet and outlet, Δp, are

$$\Delta z = z_{out} - z_{in} = \frac{U^2}{12g} \quad \text{and} \quad \Delta p = p_{out} - p_{in} = -\frac{2}{3}\rho U^2$$

Determine U_{max} and the head loss, h_L.

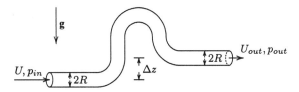

Figure 7.3: *Laminar flow through a pipe with a bend.*

Solution: There are two unknowns in this problem, namely, the outlet maximum velocity, U_{max}, and the head loss, h_L. One equation follows from the mass-conservation principle. As discussed in the preceding section, using the momentum equation adds an unknown reaction force, and thus does not close our system of equations. Since the head loss is required as part of the solution, we use the differential form of the energy equation. To solve, we select the control volume indicated in Figure 7.4.

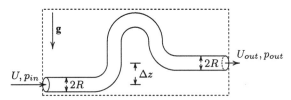

Figure 7.4: *Control volume for laminar flow through a pipe with a bend.*

We are given the fact that the flow is steady. Hence, the mass-conservation principle is

$$\oiint_S \rho \mathbf{u} \cdot \mathbf{n}\, dS = 0 \tag{7.26}$$

Expanding the closed-surface integral gives

$$\underbrace{-\rho U \pi R^2}_{\text{Inlet}} + \underbrace{\int_0^{2\pi}\int_0^R \rho U_{max}\left(1 - \frac{r^2}{R^2}\right) r\, dr\, d\theta}_{\text{Outlet}} = 0 \tag{7.27}$$

7.3. ENERGY—DIFFERENTIAL FORM

Integrating over θ, dividing through by $\rho \pi R^2$, and rearranging terms yields

$$U_{max} \int_0^R \left(1 - \frac{r^2}{R^2}\right) \frac{r}{R} d\left(\frac{r}{R}\right) = \frac{1}{2} U \quad (7.28)$$

Then, changing integration variables according to $\eta \equiv r/R$, we have

$$U_{max} \int_0^1 \left(1 - \eta^2\right) \eta d\eta = \frac{1}{2} U \quad (7.29)$$

Finally, evaluating the integral, there follows

$$U_{max} \underbrace{\left[\frac{1}{2}\eta^2 - \frac{1}{4}\eta^4\right]_{\eta=0}^{\eta=1}}_{=1/4} = \frac{1}{2} U \implies U_{max} = 2U \quad (7.30)$$

In order to use the differential form of the energy-conservation equation, we will need the kinetic-energy correction factors at the inlet, α_{in}, and the outlet, α_{out}. By definition,

$$\alpha \equiv \frac{1}{A} \iint_A \left(\frac{u}{\bar{u}}\right)^3 dA \quad (7.31)$$

Since the velocity is uniform at the inlet, i.e., $u_{in}(r) = U$, clearly the average value, $\bar{u}$, is also equal to U. Thus, by inspection, the value of α_{in} is

$$\alpha_{in} = 1 \quad (7.32)$$

Now, since the radius of the pipe is constant, clearly the average velocity at the outlet is also U. That is, the mass flux is constant on all cross sections, and $\dot{m} = \pi \rho \bar{u} R^2 = \pi \rho U R^2$. The kinetic-energy correction factor at the outlet of the pipe can be computed in the following manner.

$$\begin{aligned}
\alpha_{out} &= \frac{1}{\pi R^2} \int_0^{2\pi} \int_0^R \left[\frac{U_{max}\left(1 - r^2/R^2\right)}{U}\right]^3 r \, dr \, d\theta \\
&= \frac{2\pi}{\pi R^2} \left(\frac{U_{max}}{U}\right)^3 \int_0^R \left(1 - \frac{r^2}{R^2}\right)^3 r \, dr \\
&= 16 \int_0^1 \left(1 - \frac{r^2}{R^2}\right)^3 \frac{r}{R} d\left(\frac{r}{R}\right) \\
&= 16 \int_0^1 \left(1 - \eta^2\right)^3 \eta \, d\eta \\
&= 16 \int_0^1 \left(1 - 3\eta^2 + 3\eta^4 - \eta^6\right) \eta \, d\eta \\
&= 16 \left(\frac{1}{2}\eta^2 - \frac{3}{4}\eta^4 + \frac{1}{2}\eta^6 - \frac{1}{8}\eta^8\right)\bigg|_{\eta=0}^{\eta=1} \\
&= 2 \quad (7.33)
\end{aligned}$$

Because there are no turbines or pumps attached to our pipe, the differential form of the energy equation is

$$\frac{p_{in}}{\rho g} + \alpha_{in} \frac{\bar{u}_{in}^2}{2g} + z_{in} = \frac{p_{out}}{\rho g} + \alpha_{out} \frac{\bar{u}_{out}^2}{2g} + z_{out} + h_L \quad (7.34)$$

Solving for the head loss gives

$$h_L = \frac{p_{in} - p_{out}}{\rho g} + \frac{1}{2g}\left(\alpha_{in}\bar{u}_{in}^2 - \alpha_{out}\bar{u}_{out}^2\right) + z_{in} - z_{out} \qquad (7.35)$$

As discussed above, $\bar{u}_{in} = \bar{u}_{out} = U$, while α_{in} and α_{out} are given by Equations (7.32) and (7.33), respectively. Using the given pressure and elevation changes, the head loss between the pipe inlet and outlet is

$$h_L = \frac{2}{3}\frac{U^2}{g} + \frac{1}{2g}\left(U^2 - 2U^2\right) - \frac{U^2}{12g} = \frac{U^2}{12g} \qquad (7.36)$$

7.4 Moody Diagram

Statement of the Problem: *In designing a piping system, you have a choice of pipes made of copper, steel and cast iron. You also have of choice of using a pipe diameter of either $D = 6$ cm or 10 cm. The system calls for a straight section of length $L = 50$ m, through which a liquid of kinematic viscosity $\nu = 2.25 \cdot 10^{-7}$ m^2/sec will flow. Also, the average flow velocity will be $\bar{u} = 7.5$ m/sec, regardless of the pipe diameter. Using the Moody diagram, determine the friction factors for these six different pipes. Also, compute the head loss for the pipes with the minimum and maximum friction factors.*

Solution: In order to use the Moody diagram, i.e., Figure B.10, we must know the Reynolds number, $Re_D = \bar{u}D/\nu$, and the ratio of roughness height, k_s, to pipe diameter, D. Now, the Reynolds number per unit length is

$$\frac{\bar{u}}{\nu} = \frac{7.5 \text{ m/sec}}{2.25 \cdot 10^{-7} \text{ m}^2/\text{sec}} = 3.33 \cdot 10^7 \text{ m}^{-1} \qquad (7.37)$$

Therefore, the Reynolds number based on D is

$$Re_D = \begin{cases} 2.00 \cdot 10^6, & D = 6 \text{ cm} \\ 3.33 \cdot 10^6, & D = 10 \text{ cm} \end{cases} \qquad (7.38)$$

From the inset in the Moody diagram, we find

$$k_s = \begin{cases} 0.0015 \text{ mm}, & \text{copper} \\ 0.0460 \text{ mm}, & \text{steel} \\ 0.2600 \text{ mm}, & \text{cast iron} \end{cases} \qquad (7.39)$$

Appealing to the Moody diagram, the friction factors for the 6 different pipes are as follows.

Pipe material	D(cm)	k_s/D	f
Copper	6	$2.5 \cdot 10^{-5}$	0.0112
Copper	10	$1.5 \cdot 10^{-5}$	0.0102
Steel	6	$7.7 \cdot 10^{-4}$	0.0186
Steel	10	$4.6 \cdot 10^{-4}$	0.0166
Cast Iron	6	$4.3 \cdot 10^{-3}$	0.0291
Cast Iron	10	$2.6 \cdot 10^{-3}$	0.0252

The minimum friction factor occurs for a 10 cm copper pipe. In this case the head loss is

$$h_{L_{min}} = f\frac{\bar{u}^2}{2g}\frac{L}{D} = (0.0102)\frac{(7.5 \text{ m/sec})^2}{2\left(9.81 \text{ m/sec}^2\right)}\frac{50 \text{ m}}{0.10 \text{ m}} = 14.6 \text{ m} \qquad (7.40)$$

The maximum friction factor occurs for a 6 cm cast-iron pipe, and the head loss is

$$h_{L_{max}} = f\frac{\overline{u}^2}{2g}\frac{L}{D} = (0.0291)\frac{(7.5 \text{ m/sec})^2}{2\left(9.81 \text{ m/sec}^2\right)}\frac{50 \text{ m}}{0.06 \text{ m}} = 69.5 \text{ m} \qquad (7.41)$$

7.5 Flow Through a Turbine

Statement of the Problem: *Consider the hydroelectric plant depicted in Figure 7.5. The head loss between Points 1 and 2 is $h_L = \frac{1}{25}H$. If the head supplied to the turbine is $h_t = \frac{9}{10}H$, what is the Froude number (based on H) of the water emitted by the turbine? In developing your answer, assume the kinetic-energy correction factor, α, is 1.08 throughout the flow, and that the water emitted by the turbine exhausts to the atmosphere.*

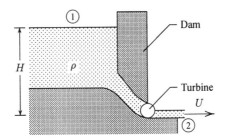

Figure 7.5: *Schematic of a hydroelectric plant.*

Solution: We can start by using the differential form of the energy equation to relate flow properties between Points 1 and 2. The key to this problem is to recognize that the reservoir is sufficiently large for a hydroelectric plant that we can ignore any motion that might occur at Point 1. Because the flow exhausts to the atmosphere after it passes through the turbine, and since the water at Point 1 lies at a free surface, the pressure is atmospheric at both Points 1 and 2.

Applying the differential form of the energy equation between Points 1 and 2, we have

$$\frac{p_1}{\rho g} + \alpha_1\frac{\overline{u}_1^2}{2g} + z_1 = \frac{p_2}{\rho g} + \alpha_2\frac{\overline{u}_2^2}{2g} + z_2 + h_t + h_L \qquad (7.42)$$

Now, as discussed above, the pressure is atmospheric at both Points 1 and 2 so that $p_1 = p_2$. Also, $\overline{u}_1 \approx 0$ and $\overline{u}_2 = U$, while $z_1 = H$ and $z_2 = 0$. Finally, we are given $\alpha \approx 1.08$, $h_L = \frac{1}{25}H$ and $h_t = \frac{9}{10}H$. Hence,

$$H = 1.08\frac{U^2}{2g} + \frac{9}{10}H + \frac{1}{25}H = 0.54\frac{U^2}{g} + 0.94H \qquad (7.43)$$

Regrouping terms, there follows

$$\frac{U^2}{gH} = \frac{1}{9} \qquad (7.44)$$

Hence, the Froude number based on H for this flow is

$$Fr_H = \frac{U}{\sqrt{gH}} = \frac{1}{3} \qquad (7.45)$$

7.6 Minor Losses I

Statement of the Problem: *Determine the head loss in the following section of pipe. The pipe has a 50° beveled inlet with a length-to-diameter ratio of $L/D = 0.10$. Also, it has a threaded union and a sudden expansion where the pipe diameter becomes $\frac{3}{2}D$. The length $\ell = 25D$, the surface roughness is $k_s/D = 0.001$ and the Reynolds number is*

$$Re_D = \frac{\overline{u}D}{\nu} = 10^6$$

What percentages of the total head loss are due to major and minor losses? Use the Moody diagram to determine the friction factor.

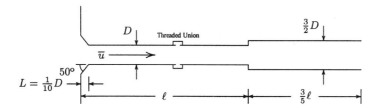

Figure 7.6: *Segment of a piping system—side view.*

Solution: Letting $\overline{u}_2$ denote the velocity in the pipe after the sudden expansion, the total head loss is

$$h_L = \frac{\overline{u}^2}{2g}\left[f\frac{\ell}{D} + K_{inlet} + K_{union} + K_{expansion} + \frac{\overline{u}_2^2}{\overline{u}^2}f_2\frac{\frac{3}{5}\ell}{\frac{3}{2}D}\right] \quad (7.46)$$

Since the flow is steady, conservation of mass tells us the mass-flow rate, $\dot{m}$, is constant on all cross sections. Thus,

$$\dot{m} = \frac{\pi}{4}\rho\overline{u}D^2 = \frac{\pi}{4}\rho\overline{u}_2\left(\frac{3}{2}D\right)^2 \quad \Longrightarrow \quad \overline{u}_2 = \frac{4}{9}\overline{u} \quad (7.47)$$

Therefore, we can rewrite the total head loss as follows.

$$h_L = \frac{\overline{u}^2}{2g}\left[\underbrace{\left(f + \frac{32}{405}f_2\right)\frac{\ell}{D}}_{major\ losses} + \underbrace{K_{inlet} + K_{union} + K_{expansion}}_{minor\ losses}\right] \quad (7.48)$$

Major losses: In order to determine the major losses from the straight pipe sections, we must find the friction factors, f and f_2. We can do this using the Moody diagram, i.e., Figure B.10. First, we must know the Reynolds number based on local velocity and local pipe diameter and the ratio of k_s to local pipe diameter. We are given $Re_D = 10^6$ and $k_s/D = 0.001$. Hence, after the expansion,

$$Re_{D_2} = \frac{\overline{u}_2 D_2}{\nu} = \frac{\frac{4}{9}\overline{u}\frac{3}{2}D}{\nu} = \frac{2}{3}Re_D \quad (7.49)$$

7.6. MINOR LOSSES I

Also, we have

$$\frac{k_s}{D_2} = \frac{k_s}{\frac{3}{2}D} = \frac{2}{3}\frac{k_s}{D} \tag{7.50}$$

So, after the expansion, we have $Re_{D_2} = 6.67 \cdot 10^5$ and $k_s/D_2 = 0.000667$. Using the Moody diagram, the friction factors are

$$f \approx 0.0200 \quad \text{and} \quad f_2 \approx 0.0185 \tag{7.51}$$

Therefore, the major losses are

$$h_L^{(major)} = \frac{\bar{u}^2}{2g}\left(f + \frac{32}{405}f_2\right)\frac{\ell}{D} = \frac{\bar{u}^2}{2g}\left(0.200 + \frac{32}{405}\cdot 0.0185\right)(25) = 0.54\frac{\bar{u}^2}{2g} \tag{7.52}$$

Minor losses: There are three different types of minor loss in this section of pipe. *The first minor loss* is at the inlet. Inspection of Figure B.9 shows that for a 50° beveled inlet with $L/D = 0.10$, the loss coefficient is

$$K_{inlet} = 0.28 \tag{7.53}$$

The second minor loss is at the threaded union. Again, inspection of Figure B.9 indicates that

$$K_{union} = 0.08 \tag{7.54}$$

The third minor loss occurs at the sudden expansion. Using Equation (B.6), we have the following.

$$K_{expansion} \approx \left[1 - \left(\frac{D}{\frac{3}{2}D}\right)^2\right]^2 = \left[1 - \frac{4}{9}\right]^2 = \frac{25}{81} = 0.31 \tag{7.55}$$

So, combining Equations (7.48), (7.52), (7.53), (7.54) and (7.55), the total head loss for this piping system is

$$h_L = \frac{\bar{u}^2}{2g}[0.54 + 0.28 + 0.08 + 0.31)] = 1.21\frac{\bar{u}^2}{2g} \tag{7.56}$$

Finally, the minor-loss contributions are

$$h_L^{(minor)} = h_L - h_L^{(major)} = 1.21\frac{\bar{u}^2}{2g} - 0.54\frac{\bar{u}^2}{2g} = 0.67\frac{\bar{u}^2}{2g} \tag{7.57}$$

Computing the percentages of major and minor losses yields

$$\frac{h_L^{(major)}}{h_L} = \frac{0.54\bar{u}^2/2g}{1.21\bar{u}^2/2g} = 0.45 \tag{7.58}$$

and

$$\frac{h_L^{(minor)}}{h_L} = \frac{0.67\bar{u}^2/2g}{1.21\bar{u}^2/2g} = 0.55 \tag{7.59}$$

Thus, major losses account for 45% of the total head loss while minor losses are responsible for 55%. Minor losses can actually dominate in a piping system when, as in this example, the straight sections are relatively short.

7.7 Minor Losses II

Statement of the Problem: *Water flows into a 90° bend with a radius of curvature $\mathcal{R}$. The pipe is circular with diameter D, has surface roughness of $k_s/D = 0.001$, and the Reynolds number is $Re_D = \bar{u}D/\nu = 10^6$. If the radius of curvature is $\mathcal{R} = 7D$ and the lengths of pipe before and after the bend are $\ell = 10D$, what is the head loss? Use the Colebrook formula to determine the friction factor.*

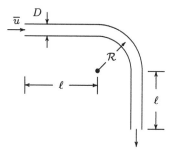

Figure 7.7: *Flow into a 90° bend – side view.*

Solution: The head loss for this pipe consists of major losses in the straight sections and a minor loss in the bend. Focusing first on the minor loss, Equation (B.7) tells us the loss coefficient for flow through a 90° bend with surface roughness is

$$K = \frac{\pi}{2} f \frac{\mathcal{R}}{D} + \Delta K \qquad (7.60)$$

where ΔK follows from Figure B.8. For $\mathcal{R}/D = 7$ and $k_s/D = 0.001$, the figure indicates

$$\Delta K \approx 0.13 \qquad (7.61)$$

Turning to the major loss, since the flow is steady, conservation of mass tells us the mass-flow rate, $\dot{m} = \frac{\pi}{4}\rho\bar{u}D^2$, is constant on all cross sections. Thus, since the diameter of the pipe is constant, the average flow velocity, $\bar{u}$, is constant. Therefore, using the fact that $\ell = 10D$, we can write the total head loss as follows.

$$h_L = \underbrace{2f\frac{\bar{u}^2}{2g}\frac{\ell}{D}}_{\substack{major \\ loss}} + \underbrace{\frac{\bar{u}^2}{2g}\left(\frac{\pi}{2}f\frac{\mathcal{R}}{D} + \Delta K\right)}_{\substack{minor \\ loss}} = \frac{\bar{u}^2}{2g}\left[\left(20 + \frac{\pi}{2}\frac{\mathcal{R}}{D}\right)f + 0.13\right] \qquad (7.62)$$

Finally, in order to determine the major loss from the straight pipe sections and the contribution to the minor loss that is proportional to $\mathcal{R}/D$, we must find the friction factor, f. We can do this using the Colebrook formula, i.e., Equation (B.10). First, we must know the Reynolds number and k_s/D. We are given $Re_D = 10^6$ and $k_s/D = 0.001$. Now, the Colebrook equation is

$$\frac{1}{\sqrt{f}} = -2\log_{10}\left(\frac{k_s/D}{3.7} + \frac{2.51}{Re_D\sqrt{f}}\right) \qquad (7.63)$$

Denoting the right-hand side of the Colebrook equation by *RHS*, a straightforward computation yields the following.

7.8. PIPE SYSTEM

f	$1/\sqrt{f}$	RHS	% Error
0.0196	7.1429	7.1307	0.17
0.0197	7.1247	7.1307	-0.08
0.0198	7.1067	7.1307	-0.34

Hence, the friction factor is

$$f \approx 0.0197 \qquad (7.64)$$

Substituting for f in Equation (7.62), we find

$$h_L = \frac{\bar{u}^2}{2g}\left[\left(20+\frac{7\pi}{2}\right)(0.0197)+0.13\right] = 0.74\frac{\bar{u}^2}{2g} \qquad (7.65)$$

7.8 Pipe System

Statement of the Problem: *Consider the pipe system shown in Figure 7.8. Water at 68° F flows from one large reservoir to another. Both reservoirs are open to the atmosphere and the pipes are made of steel. If d = 4 inches and $\bar{u}_2$ = 16 ft/sec, what must the value of $\bar{u}_1$ be? You may ignore minor losses.*

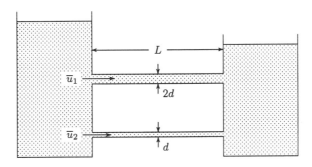

Figure 7.8: *Pipe system between two reservoirs – side view.*

Solution: Because the two pipes are in parallel, the head losses in each pipe are the same. Since minor losses can be ignored, h_L is entirely due to the major loss so that

$$h_L = f\frac{\bar{u}^2}{2g}\frac{L}{d} \qquad (7.66)$$

Denoting properties of Pipe 1 by subscript 1 and similarly for Pipe 2, we have

$$f_1\frac{\bar{u}_1^2}{2g}\frac{L_1}{d_1} = f_2\frac{\bar{u}_2^2}{2g}\frac{L_2}{d_2} \implies \bar{u}_1 = \bar{u}_2\sqrt{\frac{f_2}{f_1}\frac{L_2}{L_1}\frac{d_1}{d_2}} \qquad (7.67)$$

From Figure 7.8, $d_1/d_2 = 2$ and $L_2/L_1 = 1$. Therefore,

$$\bar{u}_1 = \bar{u}_2\sqrt{2\frac{f_2}{f_1}} \qquad (7.68)$$

Now, we must determine the friction factors for the two pipes. First, we must know the Reynolds number based on local velocity and local pipe diameter and the ratio of k_s to local pipe diameter. For water at 68° F, reference to Equation (A.5) shows that the kinematic viscosity is $\nu = 1.08 \cdot 10^{-5}$ ft^2/sec. Hence, since 4 in = 1/3 ft, the Reynolds numbers in the two pipes are

$$Re_{d_1} = \frac{\overline{u}_1 d_1}{\nu} = \frac{\overline{u}_1 (2/3 \text{ ft})}{1.08 \cdot 10^{-5} \text{ ft}^2/\text{sec}} = 61728 \overline{u}_1 \quad (7.69)$$

Also, we have

$$Re_{d_2} = \frac{\overline{u}_2 d_2}{\nu} = \frac{(16 \text{ ft/sec})(1/3 \text{ ft})}{1.08 \cdot 10^{-5} \text{ ft}^2/\text{sec}} = 4.94 \cdot 10^5 \quad (7.70)$$

Since $\overline{u}_1$ is part of the solution, we will have to solve iteratively.

We have sufficient information to determine the friction factor in Pipe 2. Because the pipe is made of steel, inspection of the insert in the Moody diagram [Figure B.10] shows that the roughness height is $k_s = 1.5 \cdot 10^{-4}$ ft. Therefore, we have

$$\frac{k_s}{d_2} = \frac{1.5 \cdot 10^{-4} \text{ ft}}{1/3 \text{ ft}} = 0.00045 \quad (7.71)$$

Hence using either the Moody diagram or the Colebrook formula, Equation (B.10), we find

$$f_2 = 0.0174 \quad (7.72)$$

Combining Equations (7.68), (7.69) and (7.72), the Reynolds number in Pipe 1 is

$$Re_{d_1} = 61728 \overline{u}_2 \sqrt{2 \frac{f_2}{f_1}} = \frac{1.84 \cdot 10^5}{\sqrt{f_1}} \quad (7.73)$$

Also, the roughness-height to diameter ratio is

$$\frac{k_s}{d_1} = \frac{1.5 \cdot 10^{-4} \text{ ft}}{2/3 \text{ ft}} = 0.000225 \quad (7.74)$$

Finally, to solve for f_1, we proceed by trial and error:

f_{guess}	Re_{d_1}	f_1	% Error
0.0150	$1.50 \cdot 10^6$	0.0147	2.04
0.0149	$1.51 \cdot 10^6$	0.0147	1.36
0.0148	$1.51 \cdot 10^6$	0.0147	0.68
0.0147	$1.52 \cdot 10^6$	0.0147	0.00
0.0146	$1.52 \cdot 10^6$	0.0147	-0.68

Hence, the friction factor is

$$f_1 \approx 0.0147 \quad (7.75)$$

Substituting for f_1 in Equation (7.68), the velocity in Pipe 1 is

$$\overline{u}_1 = 24.62 \text{ ft/sec} \quad (7.76)$$

7.9 Free-Surface Waves

Statement of the Problem: *From an overhead bridge, you drop a stone into a wide canal that is 10 m deep. You notice that the resulting ripples on the surface of the water propagate only in the downstream direction. Having studied open-channel flow, you know that this means the flow is either critical or supercritical. Based on what you have learned, what is the minimum speed of the flow in the canal? What is the corresponding speed of the ripples traveling in the downstream direction?*

Solution: The waves travel upstream and downstream at velocities u_{up} and u_{down}, respectively, where

$$u_{up} = \sqrt{gy} - \overline{u} \quad \text{and} \quad u_{down} = \sqrt{gy} + \overline{u} \tag{7.77}$$

Hence, the minimum speed of the flow in the canal, $\overline{u}_{min}$, which corresponds to $u_{up} = 0$, is given by

$$\overline{u}_{min} = \sqrt{gy} \tag{7.78}$$

Correspondingly, the velocity of the ripples traveling in the downstream direction is

$$u_{down} = \overline{u}_{min} + \sqrt{gy} = 2\sqrt{gy} \tag{7.79}$$

Since $y = 10$ m, the minimum velocity of the flow in the canal is

$$\overline{u}_{min} = \sqrt{\left(9.81 \text{ m/sec}^2\right)(10 \text{ m})} = 9.90 \text{ m/sec} \tag{7.80}$$

and the velocity of the ripples traveling downstream is

$$u_{down} = 2\sqrt{\left(9.81 \text{ m/sec}^2\right)(10 \text{ m})} = 19.81 \text{ m/sec} \tag{7.81}$$

7.10 Open-Channel Flow I

Statement of the Problem: *The rectangular open channel shown in Figure 7.9 has width $b = 3$ ft and depth $y = 5$ ft. The bottom slope is $S_o = 0.0005$ and the channel is made of brass. Using the Chézy-Manning formula, determine the specific energy and whether the flow is subcritical or supercritical.*

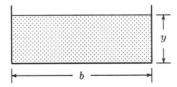

Figure 7.9: *Uniform flow in a rectangular open channel.*

Solution: We begin with the Chézy-Manning equation written in terms of volume-flow rate, Q, which is of the following form.

$$Q = \frac{\chi}{n} A S_o^{1/2} R_h^{2/3}, \quad \chi = 1.49 \text{ ft}^{1/3}/\text{sec} \tag{7.82}$$

Appealing to Figure B.2, the area and hydraulic radius of a rectangular cross section are

$$A = by \quad \text{and} \quad R_h = \frac{by}{b+2y} \tag{7.83}$$

Hence, for the given values,

$$A = (3\text{ ft})(5\text{ ft}) = 15\text{ ft}^2 \quad \text{and} \quad R_h = \frac{(3\text{ ft})(5\text{ ft})}{(3\text{ ft})+2(5\text{ ft})} = \frac{15}{13}\text{ ft} \tag{7.84}$$

Finally, for brass, reference to Table B.1 indicates that the Manning roughness coefficient is

$$n = 0.011 \pm 0.002 \tag{7.85}$$

So, using the nominal value of $n = 0.011$, the volume-flow rate is

$$Q = \frac{\left(1.49\text{ ft}^{1/3}/\text{sec}\right)}{0.011}(15\text{ ft}^2)\sqrt{0.0005}\left(\frac{15}{13}\right)^{2/3} = 50.0\text{ ft}^3/\text{sec} \tag{7.86}$$

We can determine if the flow is subcritical or supercritical by computing the Froude number. If it is less than 1, the flow is subcritical, while a value in excess of 1 means the flow is supercritical. The Froude number for this flow is most conveniently computed in terms of the critical depth, y_c, i.e.,

$$Fr = \left(\frac{y_c}{y}\right)^{3/2} \tag{7.87}$$

Also, the critical depth is defined as follows.

$$y_c \equiv \left(\frac{Q^2}{gb^2}\right)^{1/3} = \left[\frac{(50.0\text{ ft}^3/\text{sec})^2}{(32.174\text{ ft/sec}^2)(3\text{ ft})^2}\right]^{1/3} = 2.05\text{ ft} \tag{7.88}$$

The channel depth is $y = 5$ ft. Hence, the Froude number is

$$Fr = \left(\frac{2.05\text{ ft}}{5\text{ ft}}\right)^{3/2} = 0.26 \tag{7.89}$$

Therefore, the flow is **subcritical**.

Finally, the specific energy, E, is

$$\frac{E}{y_c} = \frac{y}{y_c} + \frac{1}{2}\left(\frac{y_c}{y}\right)^2 \implies E = y + \frac{y_c^3}{2y^2} \tag{7.90}$$

Hence, the numerical value of the specific energy is

$$E = 5\text{ ft} + \frac{(2.05\text{ ft})^3}{2(5\text{ ft})^2} = 5.172\text{ ft} \tag{7.91}$$

7.11 Open-Channel Flow II

Statement of the Problem: *The triangular open channel shown in Figure 7.10 has depth $y = 1.25$ m. The bottom slope is $S_o = 0.005$ and the channel is made of cast iron. Using the Chézy-Manning formula, determine the angle α that will give a volume-flow rate, Q, of 3 m³/sec. Solve for α by trial and error (or Newton's iterations if you are familiar with the method) to the nearest tenth of a degree.*

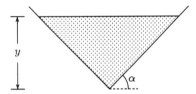

Figure 7.10: *Uniform flow in a triangular open channel.*

Solution: We begin with the Chézy-Manning equation written in terms of volume-flow rate, Q, which is of the following form.

$$Q = \frac{\chi}{n} A S_o^{1/2} R_h^{2/3}, \qquad \chi = 1.00 \text{ m}^{1/3}/\text{sec} \tag{7.92}$$

Appealing to Figure B.2, the area and hydraulic radius of a triangular cross section are

$$A = y^2 \cot \alpha \qquad \text{and} \qquad R_h = \frac{1}{2} y \cos \alpha \tag{7.93}$$

Hence, the volume-flow rate is

$$Q = \frac{\chi}{n} y^2 \cot \alpha \, S_o^{1/2} \left(\frac{1}{2} y \cos \alpha \right)^{2/3} = \frac{\chi S_o^{1/2} y^2}{n} \left(\frac{y}{2} \right)^{2/3} \cot \alpha \cos^{2/3} \alpha \tag{7.94}$$

Rearranging terms, we find

$$\cos^{2/3} \alpha = \frac{Qn}{\chi S_o^{1/2} y^2} \left(\frac{2}{y} \right)^{2/3} \tan \alpha \implies \cos^{5/3} \alpha = \frac{Qn}{\chi S_o^{1/2}} \left(\frac{2}{y^4} \right)^{2/3} \sin \alpha \tag{7.95}$$

Finally, for cast iron, reference to Table B.1 indicates that the Manning roughness coefficient is

$$n = 0.013 \pm 0.003 \tag{7.96}$$

So, using the nominal value of $n = 0.013$, the coefficient appearing in Equation (7.95) is

$$\frac{Qn}{\chi S_o^{1/2}} \left(\frac{2}{y^4} \right)^{2/3} = \frac{(3 \text{ m}^3/\text{sec})(0.013)}{(1.00 \text{ m}^{1/3}/\text{sec}) \sqrt{0.005}} \left[\frac{2}{(1.25 \text{ m})^4} \right]^{2/3} = 0.4829 \tag{7.97}$$

Therefore, the equation we must solve to determine the angle α is

$$\cos^{5/3} \alpha = 0.4829 \sin \alpha \tag{7.98}$$

This is a transcendental equation that must be solved iteratively. We can solve either by trial and error or with a more systematic method such as Newton's iterations. Either way, it is worthwhile to rewrite Equation (7.98) as follows.

$$f(\alpha) = \cos^{5/3} \alpha - 0.4829 \sin \alpha = 0 \tag{7.99}$$

Trial and error: The following table lists several values of $f(\alpha)$ as a function of α.

α (degrees)	$f(\alpha)$
54.5	$1.106 \cdot 10^{-2}$
54.9	$2.522 \cdot 10^{-3}$
55.0	$3.923 \cdot 10^{-4}$
55.1	$-1.735 \cdot 10^{-3}$
55.5	$-1.023 \cdot 10^{-2}$

Thus, the angle of the triangular channel is

$$\alpha \approx 55.0° \tag{7.100}$$

Newton's iterations: First, note that

$$f'(\alpha) = -\frac{5}{3} \sin\alpha \cos^{2/3}\alpha - 0.4829 \cos\alpha \tag{7.101}$$

Using Newton's iterations, we expand according to

$$f(\alpha + \Delta\alpha) = f(\alpha) + f'(\alpha)\Delta\alpha + \cdots = 0 \quad \Longrightarrow \quad \Delta\alpha \approx -\frac{f(\alpha)}{f'(\alpha)} \tag{7.102}$$

So, we make an initial guess of $\alpha = 45°$. Hence,

$$f(45°) = (0.707107)^{5/3} - 0.4829(0.707107) = 0.21977$$
$$f'(45°) = -\frac{5}{3}(0.707107)(0.707107)^{2/3} - 0.4829(0.707107) = -1.27685$$
$$\Delta\alpha = -\frac{0.21977}{-1.27685} = 0.17212 \text{ radian} = 9.8617°$$

Thus, as the next guess, we try $\alpha = 54.8617°$. Then,

$$f(54.8617°) = (0.575552)^{5/3} - 0.4829(0.817765) = 3.3377 \cdot 10^{-3}$$
$$f'(54.8617°) = -\frac{5}{3}(0.817765)(0.575552)^{2/3} - 0.4829(0.575552) = -1.22098$$
$$\Delta\alpha = -\frac{3.3377 \cdot 10^{-3}}{-1.22098} = 2.7336 \cdot 10^{-3} \text{ radian} = 0.1566°$$

For the third guess, we try $\alpha = 55.0183°$. Then,

$$f(55.0183°) = (0.573315)^{5/3} - 0.4829(0.819335) = 3.1538 \cdot 10^{-6}$$
$$f'(54.8617°) = -\frac{5}{3}(0.819335)(0.573315)^{2/3} - 0.4829(0.573315) = -1.21926$$
$$\Delta\alpha = -\frac{3.1538 \cdot 10^{-6}}{-1.21926} = 2.5867 \cdot 10^{-6} \text{ radian} = 0.0001°$$

wherefore $\alpha = 55.0184°$. Since the correction is less than 10^{-5}, we can consider the solution to be converged to a high degree of accuracy. Therefore, as determined above by trial and error,

$$\alpha \approx 55.0° \tag{7.103}$$

7.12 Hydraulic Jump

Statement of the Problem: *Water flows down a steep hill in a wide rectangular open channel as shown. A hydraulic jump occurs, abruptly increasing the channel depth from y_1 to y_2. The volume-flow rate is $Q = 40$ m³/sec, the channel width (out of the page) is $b = 2$ m and the depth before the jump is $y_1 = 1$ m. Compute the depth after the jump, and the Froude numbers upstream and downstream of the jump.*

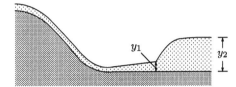

Figure 7.11: *Hydraulic jump in a rectangular open channel.*

Solution: Because the flow is steady and incompressible, the mass-flow rate, and hence the volume-flow rate, is constant. So, since the cross-sectional area of a rectangular channel is $A = by$, the volume-flow rate is

$$Q = \overline{u}_1 b y_1 \quad \Longrightarrow \quad \overline{u}_1 = \frac{Q}{by_1} \tag{7.104}$$

From the given values,

$$\overline{u}_1 = \frac{40 \text{ m}^3/\text{sec}}{(2 \text{ m})(1 \text{ m})} = 20 \text{ m/sec} \tag{7.105}$$

Now, by definition, the Froude number ahead of the hydraulic jump is

$$Fr_1 = \frac{\overline{u}_1}{\sqrt{gy_1}} = \frac{20 \text{ m/sec}}{\sqrt{\left(9.81 \text{ m/sec}^2\right)(1 \text{ m})}} = 6.39 \tag{7.106}$$

Hence, the Froude number after the hydraulic jump is

$$Fr_2 = Fr_1 \left[\frac{2}{\sqrt{1 + 8Fr_1^2} - 1}\right]^{3/2} = (6.39) \left[\frac{2}{\sqrt{1 + 8(6.39)^2} - 1}\right]^{3/2} = 0.26 \tag{7.107}$$

Also, using another of the hydraulic-jump relations,

$$\frac{y_2}{y_1} = \frac{\sqrt{1 + 8Fr_1^2} - 1}{2} = \left[\frac{\sqrt{1 + 8(6.39)^2} - 1}{2}\right] = 8.55 \tag{7.108}$$

Therefore, the depth downstream of the hydraulic jump is

$$y_2 = 8.55 y_1 = (8.55)(1 \text{ m}) = 8.55 \text{ m} \tag{7.109}$$

Chapter 8

One-Dimensional Compressible Flow

The first 7 chapters have focused almost exclusively on incompressible flows. In this chapter, we turn our attention to compressible flows. These are flows in which density variations cannot be ignored. This occurs most commonly in gases because, except under extreme conditions, liquid flows are incompressible. We restrict our attention to steady, inviscid, one-dimensional applications, and defer complicating effects of viscosity, heat transfer and two-dimensionality to Chapter 15. The common theme of the types of problems discussed is the use of the isentropic-flow and normal-shock relations. The tables of Appendix C can be used for gases with a specific-heat ratio, γ, of 1.4. Algebraic formulas must be used for other gases. The chapter begins with problems illustrating how the speed of sound depends on altitude and how flow properties are related in isentropic flows. The rest of the chapter focuses on problems with shock waves and on the Laval nozzle. The problems are as follows.

Section 8.1 Speed of Sound: This is a simple problem illustrating how the speed of sound depends upon altitude.

Section 8.2 Isentropic Flow I: A computation of the wind-tunnel test-section temperature when the reservoir temperature, Mach number and specific-heat ratio are given.

Section 8.3 Isentropic Flow II: This problem involves isentropic flow of a gas whose specific-heat ratio, γ, must be deduced from what is given. Since γ is unknown, we must use the isentropic-flow *relations* to calculate flow properties rather than the *tables* of Appendix C.

Section 8.4 Normal Shock I: A straightforward problem involving a normal shock wave. Since the gas is air, the normal-shock tables of Appendix C can be used.

Section 8.5 Normal Shock II: A second problem involving a normal shock wave. Since the gas is air, the normal-shock tables of Appendix C can be used.

Section 8.6 Laval Nozzle I: This is the first of two Laval-nozzle problems. It has a shock wave near the nozzle exit. The gas is air so that the isentropic-flow and normal-shock tables can be used. We are given specific conditions and required to determine if the nozzle meets a given constraint.

Section 8.7 Laval Nozzle II: This is the second of two Laval-nozzle problems, and includes most of the features present in elementary one-dimensional compressible flows. This time, the gas has $\gamma \neq 1.4$, so that the tables cannot be used.

8.1 Speed of Sound

Statement of the Problem: *Imagine that you launch a small rocket from a beach somewhere in the Los Angeles area. At the same time, a friend launches a similar rocket in Denver, which is a mile above sea level. If both rockets have a Mach number of 0.7, whose rocket is traveling faster? Assume conditions in the atmosphere are given by the U. S. Standard Atmosphere with a sea-level temperature of $T_o = 78°\, F$.*

Solution: Since we are given the Mach number, the speed of each rocket is proportional to the speed of sound. There are two key observations needed to effect solution of this problem. The first is that the speed of sound for a perfect gas is proportional to $\sqrt{T}$, where T is the absolute temperature of the gas. The second is that, according to the U. S. Standard Atmosphere, temperature is a linearly decreasing function of altitude.

From the definition of Mach number, the speed of each rocket is

$$U = M a = M \sqrt{\gamma R T} \qquad (8.1)$$

In the second equality we use the fact that the speed of sound of air is $a = \sqrt{\gamma R T}$, where $\gamma = 1.4$ is the specific-heat ratio, $R = 1716$ ft·lb/(slug·°R) is the perfect-gas constant, and T is absolute temperature in degrees Rankine.

Also, according to the U. S. Standard Atmosphere, the temperature of the atmosphere is

$$T = T_o - \alpha z \qquad (8.2)$$

where T_o is specified to be 78° F and $\alpha = 18.85°$ R/mi is the adiabatic lapse rate. Therefore, the speed of each rocket is

$$U = M \sqrt{\gamma R (T_o - \alpha z)} \qquad (8.3)$$

Now, we must convert the surface temperature from degrees Fahrenheit to degrees Rankine, i.e.,

$$T_o = 78°\, \text{F} + 459.67°\, \text{R} = 537.67°\, \text{R} \qquad (8.4)$$

Hence, for the given conditions,

$$\begin{aligned} U &= 0.7 \sqrt{(1.4) \left(1716 \, \frac{\text{ft} \cdot \text{lb}}{\text{slug}°\text{R}}\right) \left[537.67°\, \text{R} - \left(18.85 \, \frac{°\text{R}}{\text{mi}}\right) z\right]} \\ &= 795.6 \sqrt{1 - \frac{z}{28.5 \, \text{mi}}} \, \frac{\text{ft}}{\text{sec}} \end{aligned} \qquad (8.5)$$

Therefore, on the beach in Los Angeles, which is at sea level so that $z = 0$, the rocket's velocity is

$$U_{Los\ Angeles} = 796 \, \frac{\text{ft}}{\text{sec}} \qquad (8.6)$$

Similarly, in Denver where $z = 1$ mi, substitution into Equation (8.5) gives

$$U_{Denver} = 782 \, \frac{\text{ft}}{\text{sec}} \qquad (8.7)$$

Therefore, the rocket travels faster in Los Angeles.

8.2 Isentropic Flow I

Statement of the Problem: *Hypersonic experiments are being done in a wind tunnel with a reservoir temperature of 300° F. If the test-section Mach number is 6, determine the temperature in the test section when the working gas is air ($\gamma = 1.40$) and Helium ($\gamma = 1.66$).*

Solution: By definition, the total temperature for Mach 6 flow is

$$\frac{T_t}{T} = 1 + \frac{\gamma - 1}{2}M^2 = 1 + 18(\gamma - 1) = 18\gamma - 17 \implies T = \frac{T_t}{18\gamma - 17} \tag{8.8}$$

Then, note that the total temperature is $T_t = 459.67°\,\text{R} + 300°\,\text{F} = 759.67°\,\text{R}$. Thus,

$$T_{air} = \frac{759.67°\,\text{R}}{18(1.4) - 17} = 92.64°\,\text{R} = -367.0°\,\text{F} \tag{8.9}$$

$$T_{He} = \frac{759.67°\,\text{R}}{18(1.66) - 17} = 58.98°\,\text{R} = -400.7°\,\text{F} \tag{8.10}$$

8.3 Isentropic Flow II

Statement of the Problem: *The ratio of the total density to the static density of a gas is $\rho_t/\rho = 5.02$. The Mach number is $M = 2.4$ and the flow is isentropic.*

(a) Assuming the gas is one of those listed in Table A.1, which gas is it?

(b) If the total temperature is $T_t = 600°\,C$, what is the static temperature, T?

(c) If the total pressure is $p_t = 1050\,kPa$, find the static pressure, p.

Solution: Because we do not know the value of γ, we cannot use the isentropic-flow tables of Appendix C, which are valid only for $\gamma = 1.4$. Rather, we must use the isentropic relations in algebraic form. For this problem, the appropriate relations are

$$\rho_t = \rho\left[1 + \frac{\gamma - 1}{2}M^2\right]^{1/(\gamma-1)}, \quad T_t = T\left[1 + \frac{\gamma - 1}{2}M^2\right], \quad p_t = p\left(\frac{\rho_t}{\rho}\right)^\gamma \tag{8.11}$$

8.3(a): Appealing to the first of Equations (8.11), we have

$$\frac{\rho_t}{\rho} = \left[1 + \frac{\gamma - 1}{2}M^2\right]^{1/(\gamma-1)} \tag{8.12}$$

Taking the natural logarithm of both sides of this equation and rearranging terms yields

$$\gamma = 1 + \frac{\ln\left(1 + \frac{\gamma-1}{2}M^2\right)}{\ln(\rho_t/\rho)} \tag{8.13}$$

This is an implicit equation for γ that cannot be solved in closed form. Rather, we must solve iteratively. To do so, it is convenient to rewrite Equation (8.13) as follows.

$$\gamma - f(\gamma) = 0 \quad \text{where} \quad f(\gamma) \equiv 1 + \frac{\ln\left(1 + \frac{\gamma-1}{2}M^2\right)}{\ln(\rho_t/\rho)} \tag{8.14}$$

Then, substituting the given values of $M = 2.4$ and $\rho_t/\rho = 5.02$ into Equation (8.14), the equation we must solve to determine γ is

$$\gamma - f(\gamma) = 0 \quad \text{where} \quad f(\gamma) \equiv 1 + 0.6198 \ln(2.88\gamma - 1.88) \tag{8.15}$$

Finally, to solve numerically, we proceed by trial and error (although we omit the details here, you can use Newton's iterations if you are familiar with the method—see Section 7.11):

γ	$f(\gamma)$	$\gamma - f(\gamma)$
1.60	1.622	-0.022
1.65	1.654	-0.004
1.66	1.660	0.000
1.67	1.666	0.004
1.70	1.684	0.016

Hence, the specific-heat ratio is

$$\gamma = 1.66 \tag{8.16}$$

Inspection of Table A.1 indicates that the gas is **helium**.

8.3(b): Appealing to the second of Equations (8.11), the total temperature is

$$T_t = T\left[1 + \frac{\gamma-1}{2}M^2\right] = T\left(\frac{\rho_t}{\rho}\right)^{\gamma-1} \tag{8.17}$$

Thus, the static temperature is

$$T = \frac{T_t}{(\rho_t/\rho)^{\gamma-1}} \tag{8.18}$$

Now, the total temperature is $T_t = 600°\text{C} = 873.16\text{ K}$, so that the static temperature is

$$T = \frac{873.16 \text{ K}}{(5.02)^{0.66}} = 301.04 \text{ K} = 27.88°\text{ C} \tag{8.19}$$

Therefore, to the nearest degree, we have

$$T = 28°\text{ C} \tag{8.20}$$

8.3(c): As noted in the last of Equations (8.11), the total pressure is given by

$$p_t = p\left(\frac{\rho_t}{\rho}\right)^{\gamma} \quad \Longrightarrow \quad p = \frac{p_t}{(\rho_t/\rho)^{\gamma}} \tag{8.21}$$

Using the given values,

$$p = \frac{1050 \text{ kPa}}{(5.02)^{1.66}} = 72.11 \text{ kPa} \tag{8.22}$$

Therefore, to the nearest kPa, the static pressure is

$$p = 72 \text{ kPa} \tag{8.23}$$

8.4 Normal Shock I

Statement of the Problem: *In the process of calibrating a wind tunnel, an engineer inadvertently places her Pitot tube and other instrumentation so far into the test section that the devices lie downstream of a normal shock. She realizes her oversight when the measured flow properties correspond to subsonic flow. Rather than move the instruments and rerun the test, she opts to simply use the normal-shock tables. The flow downstream of the shock has Mach number, $M_2 = 0.45$, static temperature, $T_2 = 15.8°$ C and static pressure, $p_2 = 40$ kPa. Find the Mach number, M_1, total temperature, T_{t_1}, and total pressure, p_{t_1}, ahead of the shock.*

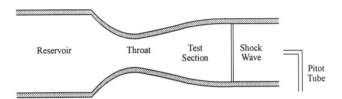

Figure 8.1: *Schematic of a supersonic wind tunnel.*

Solution: Since we are dealing with air, which has $\gamma = 1.4$, we can use the normal-shock tables of Appendix C. Since the given value of the Mach number behind the shock is $M_2 = 0.45$, using linear interpolation in the normal-shock tables, we have

M_1	M_2	p_2/p_1	T_2/T_1	Source
3.50	0.4512	14.13	3.315	Table
3.53	0.4500	14.39	3.359	Interpolated
3.60	0.4474	14.95	3.454	Table

Therefore, the Mach number ahead of the shock is

$$M_1 = 3.53 \tag{8.24}$$

Also, the static pressure and static temperature ahead of the shock, which we will need in order to compute total conditions, are

$$p_1 = \frac{p_2}{p_2/p_1} = \frac{40 \text{ kPa}}{14.39} = 2.78 \text{ kPa} \tag{8.25}$$

$$T_1 = \frac{T_2}{T_2/T_1} = \frac{(15.8 + 273.16) \text{ K}}{3.359} = 86.0 \text{ K} \approx -187° \text{ C} \tag{8.26}$$

Ahead of the shock, the flow is isentropic. Hence, we use the isentropic-flow tables of Appendix C to determine total conditions. Using linear interpolation gives

M_1	p_1/p_t	T_1/T_t	Source
3.50	$0.1311 \cdot 10^{-1}$	0.2899	Table
3.53	$0.1259 \cdot 10^{-1}$	0.2865	Interpolation
3.60	$0.1138 \cdot 10^{-1}$	0.2784	Table

Therefore, using the static pressure and static temperature computed in Equations (8.25) and (8.26), the total conditions are as follows.

$$p_t = \frac{p_1}{p_1/p_t} = \frac{2.78 \text{ kPa}}{0.01259} = 220.8 \text{ kPa} \tag{8.27}$$

$$T_t = \frac{T_1}{T_1/T_t} = \frac{86.0 \text{ K}}{0.2865} = 300.17 \text{ K} \approx 27.0° \text{ C} \tag{8.28}$$

In summary, this shows that the test-section flow conditions for this wind tunnel are

$$M_1 = 3.53, \quad p_t = 221 \text{ kPa}, \quad T_t = 27° \text{ C} \tag{8.29}$$

As a final comment, had this situation actually occurred, the research engineer would have been making use of her understanding of the basics of compressible fluid dynamics to obviate the necessity of additional measurements. Unless the wind tunnel is poorly designed with, for example, strong viscous effects or three dimensionality, the theoretical predictions will very closely match the true conditions in the test section.

8.5 Normal Shock II

Statement of the Problem: *An airplane is equipped with a Pitot-static tube in order to measure its speed. If the tube indicates that the static pressure is p = 13.5 atm and the total pressure is p_t = 20.1 atm, at what Mach number is the airplane moving?*

Solution: As indicated in Figure 8.2, there are two possible solutions to this problem, one corresponding to subsonic motion and the other to the supersonic case. The difference is the presence of a normal shock wave standing in front of the Pitot-static tube when the motion is supersonic.

Figure 8.2: *Pitot-static tubes in subsonic and supersonic streams.*

Since we are given the static and total pressure, we can compute the local Mach number from the isentropic relation, i.e., from

$$M = \sqrt{\frac{2}{\gamma - 1}\left[\left(\frac{p_t}{p}\right)^{(\gamma-1)/\gamma} - 1\right]} = 0.7760 \tag{8.30}$$

Now, since this is the Mach number just ahead of the Pitot-static tube, it is the freestream Mach number in the subsonic case. However, if the airplane is moving supersonically, this is the Mach number behind the shock, which we denote by M_2. Then, from the normal-shock tables, we find

$$M_2 = 0.7760 \quad \Longrightarrow \quad M_1 = 1.32 \tag{8.31}$$

Therefore, from the given information, the airplane is moving either at Mach 0.776 or at Mach 1.32.

8.6 Laval Nozzle I

Statement of the Problem: *You have just finished assembling your "Laval Nozzle at Home Kit" and you are ready to try it out. Due to external constraints, the nozzle-exit Mach number can be no greater than 1/2. The ratio of the throat area, A_t, to the exit area, A_e, is 1/4. Also, a normal shock wave lies at a point where $A = 2A_t$. Will you be able to satisfy the external constraints?*

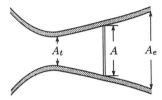

Figure 8.3: *Schematic of a "Laval Nozzle at Home Kit."*

Solution: To solve this problem, we observe that the flow is isentropic both ahead of the shock and behind the shock. We must use the normal-shock tables to determine how conditions in these two isentropic regions are related.

First, we know that the shock occurs at a point where the area ratio is $A/A_t = 2$. Since the flow is isentropic ahead of the shock, necessarily the reference area is $A^* = A_t$, so that $A/A^* = 2$. We can use the isentropic-flow tables of Appendix C to determine the Mach number corresponding to this area ratio. This is, of course, the Mach number ahead of the shock, M_1. Since the values lie in the same rows, we can also obtain the Mach number after the shock, M_2, from the normal-shock part of the same tables. We find

A/A^*	M_1	M_2	Source
1.9698	2.180	0.5498	Table
2.0000	2.197	0.5475	Interpolation
2.0050	2.200	0.5471	Table

Now, we focus on the region behind the shock. When the Mach number is $M_2 = 0.5475$, the isentropic-flow tables tell us that

M	A/A^*	Source
0.5400	1.2703	Table
0.5475	1.2591	Interpolation
0.5600	1.2403	Table

This reflects the fact that A^* changes across the shock, and is thus no longer equal to the actual throat area. Therefore, since we are given $A_e/A_t = 4$, necessarily

$$\frac{A_e}{A^*} = \frac{A_e/A_t}{A/A_t}\frac{A}{A^*} = \frac{4}{2}(1.2591) = 2.5182 \tag{8.32}$$

Finally, we must determine the Mach number corresponding to this area ratio. Since the flow behind the shock is subsonic, and the flow expands between the shock and the exit, it must remain subsonic. The isentropic-flow tables yield the following.

A_e/A^*	M_e	Source
2.7076	0.220	Table
2.5182	0.238	Interpolation
2.4956	0.240	Table

Therefore, the Mach number at the exit of this nozzle is

$$M_e = 0.238 \tag{8.33}$$

Since this value is less than 1/2, your "Laval Nozzle at Home Kit" will operate properly.

8.7 Laval Nozzle II

Statement of the Problem: *A gas with a specific-heat ratio $\gamma = 1.28$ flows through a nozzle and achieves supersonic conditions. A normal shock stands in the nozzle at a point where the cross-sectional area of the shock is A (see Figure 8.4). The area at the throat is A_t, while the nozzle-exit area is A_e. The static pressure ahead of the shock is $p_1 = 16.25$ psi and the Mach number behind the shock is $M_2 = 0.7$.*

(a) *Determine the Mach number ahead of the shock, M_1, and the pressure behind the shock, p_2.*

(b) *Find the total pressure ahead of the shock, p_t, and the static pressure at the throat, p^*.*

(c) *Compute A_e/A_t and A/A_t for a nozzle-exit Mach number of $M_e = 1/2$.*

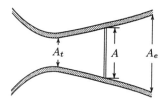

Figure 8.4: *Supersonic flow through a nozzle.*

Solution: Because $\gamma \neq 1.4$, we cannot use the isentropic-flow and normal-shock tables. Rather, we must use the appropriate algebraic relations as needed.

8.7(a): We are given the pressure ahead of the shock and the Mach number behind the shock. First, we note that the Mach numbers are related by

$$M_2^2 = \frac{1 + \frac{\gamma-1}{2}M_1^2}{\gamma M_1^2 - \frac{\gamma-1}{2}} \tag{8.34}$$

However, a straightforward algebraic exercise shows that this equation is valid with the subscripts interchanged, wherefore

$$M_1^2 = \frac{1 + \frac{\gamma-1}{2}M_2^2}{\gamma M_2^2 - \frac{\gamma-1}{2}} \tag{8.35}$$

This is more convenient as we can solve directly for M_1. Using the given value of $M_2 = 0.7$, there follows

$$M_1^2 = \frac{1 + \left(\frac{1.28-1}{2}\right)(0.7)^2}{1.28(0.7)^2 - \left(\frac{1.28-1}{2}\right)} = 2.1933 \implies M_1 = 1.4810 \tag{8.36}$$

8.7. LAVAL NOZZLE II

Turning to the pressure, the appropriate normal-shock relation is

$$\frac{p_2}{p_1} = 1 + \frac{2\gamma}{\gamma+1}\left(M_1^2 - 1\right) = 1 + \frac{2(1.28)}{1.28+1}\left[(1.4810)^2 - 1\right] = 2.3399 \qquad (8.37)$$

Hence, the static pressure behind the shock is

$$p_2 = \frac{p_2}{p_1} p_1 = 2.3399(16.25 \text{ psi}) = 38.02 \text{ psi} \qquad (8.38)$$

8.7(b): Now, the flow ahead of the shock is isentropic. Hence, the total and static pressures are related by

$$p_t = p_1\left[1 + \frac{\gamma-1}{2} M_1^2\right]^{\gamma/(\gamma-1)} \qquad (8.39)$$

Substituting the given value of $p_1 = 16.25$ psi and the computed value of $M_1 = 1.4810$ from Part (a), we find

$$p_t = (16.25 \text{ psi})\left[1 + \frac{1.28-1}{2}(1.4810)^2\right]^{1.28/(1.28-1)} = 55.27 \text{ psi} \qquad (8.40)$$

Because supersonic conditions have been achieved in the nozzle, the flow is sonic at the throat, i.e., $M^* = 1$. Hence, the static pressure at the throat is

$$p^* = \frac{p_t}{\left[1 + \frac{\gamma-1}{2}(M^*)^2\right]^{\gamma/(\gamma-1)}} = \frac{55.27 \text{ psi}}{\left[1 + \frac{1.28-1}{2}\right]^{1.28/(1.28-1)}} = 30.36 \text{ psi} \qquad (8.41)$$

8.7(c): Since the flow is isentropic in the region behind the shock, the area ratios of interest are given by

$$\frac{A}{A^*} = \frac{1}{M_2}\left\{\frac{2}{\gamma+1}\left[1 + \frac{\gamma-1}{2} M_2^2\right]\right\}^{\frac{\gamma+1}{2(\gamma-1)}} \qquad (8.42)$$

$$\frac{A_e}{A^*} = \frac{1}{M_e}\left\{\frac{2}{\gamma+1}\left[1 + \frac{\gamma-1}{2} M_e^2\right]\right\}^{\frac{\gamma+1}{2(\gamma-1)}} \qquad (8.43)$$

We are given $M_2 = 0.7$ and $M_e = 0.5$. Therefore,

$$\frac{A}{A^*} = \frac{1}{0.7}\left\{\frac{2}{1.28+1}\left[1 + \left(\frac{1.28-1}{2}\right)(0.7)^2\right]\right\}^{\frac{1.28+1}{2(1.28-1)}} = 1.098 \qquad (8.44)$$

$$\frac{A_e}{A^*} = \frac{1}{0.5}\left\{\frac{2}{1.28+1}\left[1 + \left(\frac{1.28-1}{2}\right)(0.5)^2\right]\right\}^{\frac{1.28+1}{2(1.28-1)}} = 1.350 \qquad (8.45)$$

However, because of the presence of the shock, $A^* \neq A_t$. Rather, it corresponds to the nozzle cross-sectional area for which the flow would accelerate to sonic conditions. Since the flow is isentropic ahead of the shock, we know that

$$\frac{A}{A_t} = \frac{1}{M_1}\left\{\frac{2}{\gamma+1}\left[1 + \frac{\gamma-1}{2} M_1^2\right]\right\}^{\frac{\gamma+1}{2(\gamma-1)}} \qquad (8.46)$$

Substituting the computed value of $M_1 = 1.4810$ from Part (a), we have

$$\frac{A}{A_t} = \frac{1}{1.4810}\left\{\frac{2}{1.28+1}\left[1+\left(\frac{1.28-1}{2}\right)(1.4810)^2\right]\right\}^{\frac{1.28+1}{2(1.28-1)}} = 1.177 \qquad (8.47)$$

Finally, the ratio of A_e to A_t is

$$\frac{A_e}{A_t} = \frac{A_e/A^*}{A/A^*}\frac{A}{A_t} = \frac{1.350}{1.097}(1.178) = 1.447 \qquad (8.48)$$

In summary, the area ratios are

$$\frac{A}{A_t} = 1.177 \quad \text{and} \quad \frac{A_e}{A_t} = 1.447 \qquad (8.49)$$

Chapter 9

Turbomachinery

Pumps and turbines are the two most basic types of turbomachines, and practical designs of all sizes exist that satisfy a variety of needs in industry and in the home. While the flow through a modern turbomachine is extremely complicated (i.e., three dimensional, viscous and unsteady), useful insight can be obtained from the control volume method and dimensional analysis.

The basic equations that are used to solve simple turbomachine problems follow from a straightforward control-volume analysis, and are called the *Euler turbomachine equations* and *Bernoulli's equation* specialized for motion in a rotating coordinate frame. The former is used to relate the torque and the velocity components, while the latter is used in calculating pressure changes.

Additionally, there are a number of dimensionless parameters that help characterize the performance of a given turbomachine, such as the specific speed and the efficiency. This chapter includes a collection of problems that address each of these aspects of turbomachinery.

Section 9.1 Pump Analysis: This is a pump problem. The goal is to solve for the impeller radius that yields maximum pressure change. The problem is a bit subtle to the extent that its solution requires solving for the pressure change in terms of the radius, and then differentiating to solve for the maximum value.

Section 9.2 Performance Curves: In this problem, which involves a pump, the head and torque are specified analytically as functions of volume-flow rate for a portion of the pump's performance curves. The goal is to determine the type of pump by computing and examining the specific speed.

Section 9.3 Power: For given speed and geometric properties, the solution to this problem uses the Euler turbomachine equations in order to determine the power that is required to run a fan.

Section 9.4 Turbine Analysis: The problem in this section involves determining the type of turbine, its efficiency and selected dimensionless coefficients that characterize the turbine's performance. As in Section 9.2, the specific speed is used to identify the type of turbine under consideration.

Section 9.5 Pelton Wheel: This problem analyzes a special type of turbine known as the Pelton wheel. The solution involves developing a wheel design based on desired efficiency and torque requirements.

9.1 Pump Analysis

Statement of the Problem: *Consider a centrifugal water pump whose impeller geometry (see Figure 9.1) is as follows.*

$$r_2 = 10 \text{ cm}, \quad b_1 = 3 \text{ cm}, \quad b_2 = 5 \text{ cm}, \quad \beta_1 = 15°, \quad \beta_2 = 10°$$

The quantity r_2 is the outlet impeller radius, b_1 and b_2 are the inlet and outlet blade widths, and β_1 and β_2 are the inlet and outlet blade angles. The latter two quantities are thus the angles between the impeller blades and the relative velocities at the inlet and outlet. Also, the impeller rotates at an angular velocity of Ω = 300 rpm, and the inlet absolute velocity is purely radial. Neglecting gravitational effects, determine the value of the inlet impeller radius, r_1, for which the pressure rise is a maximum.

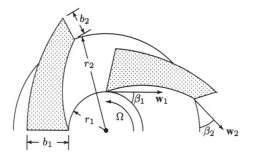

Figure 9.1: *Closeup view of impeller geometry.*

Solution: In order to solve this problem, we make use of Bernoulli's equation in a rotating coordinate system. Neglecting gravitational effects, the equation is

$$p_1 + \frac{1}{2}\rho \mathbf{w}_1 \cdot \mathbf{w}_1 - \frac{1}{2}\rho \Omega^2 r_1^2 = p_2 + \frac{1}{2}\rho \mathbf{w}_2 \cdot \mathbf{w}_2 - \frac{1}{2}\rho \Omega^2 r_2^2 \qquad (9.1)$$

where $\mathbf{w}_1$ and $\mathbf{w}_2$ are the velocities at the inlet and outlet as observed in the hub-fixed, rotating coordinate frame. From elementary pump theory, we know that

$$\mathbf{w}_1 = \frac{\Omega r_1}{\cot \alpha_1 + \cot \beta_1} [\mathbf{e}_r - \cot \beta_1 \, \mathbf{e}_\theta] \qquad (9.2)$$

$$\mathbf{w}_2 = \frac{\Omega r_1}{\cot \alpha_1 + \cot \beta_1} \left(\frac{r_1}{r_2}\right)\left(\frac{b_1}{b_2}\right) [\mathbf{e}_r - \cot \beta_2 \, \mathbf{e}_\theta] \qquad (9.3)$$

where $\mathbf{e}_r$ and $\mathbf{e}_\theta$ are unit vectors in the radial and circumferential directions, respectively. Also, α_1 is the angle between the absolute inlet velocity and the impeller. Hence, taking the dot product of each velocity vector with itself gives

$$\mathbf{w}_1 \cdot \mathbf{w}_1 = \frac{\Omega^2 r_1^2}{(\cot \alpha_1 + \cot \beta_1)^2} \left[1 + \cot^2 \beta_1\right] \qquad (9.4)$$

$$\mathbf{w}_2 \cdot \mathbf{w}_2 = \frac{\Omega^2 r_1^2}{(\cot \alpha_1 + \cot \beta_1)^2} \left(\frac{r_1}{r_2}\right)^2 \left(\frac{b_1}{b_2}\right)^2 \left[1 + \cot^2 \beta_2\right] \qquad (9.5)$$

We can simplify a bit since we are given that the inlet absolute velocity is purely radial, which means $\alpha_1 = \pi/2$. Hence,

$$\cot \alpha_1 = 0 \qquad (9.6)$$

9.2. PERFORMANCE CURVES

Also, we know that
$$1 + \cot^2 \beta_i = \csc^2 \beta_i = \frac{1}{\sin^2 \beta_i} \tag{9.7}$$

Therefore, Equations (9.4) and (9.5) simplify to
$$\mathbf{w}_1 \cdot \mathbf{w}_1 = \frac{\Omega^2 r_1^2}{\cos^2 \beta_1} \quad \text{and} \quad \mathbf{w}_2 \cdot \mathbf{w}_2 = \frac{\Omega^2 r_1^2 \tan^2 \beta_1}{\sin^2 \beta_2} \left(\frac{r_1}{r_2}\right)^2 \left(\frac{b_1}{b_2}\right)^2 \tag{9.8}$$

Combining Equations (9.1) and (9.8), and defining $\Delta p \equiv p_2 - p_1$ yields

$$\begin{aligned}
\Delta p &= \frac{1}{2}\rho\Omega^2 r_1^2 \left[\frac{1}{\cos^2 \beta_1} - \frac{\tan^2 \beta_1}{\sin^2 \beta_2}\left(\frac{r_1}{r_2}\right)^2\left(\frac{b_1}{b_2}\right)^2\right] - \frac{1}{2}\rho\Omega^2 (r_1^2 - r_2^2) \\
&= \frac{1}{2}\rho\Omega^2 \left[\left(\frac{1}{\cos^2 \beta_1} - 1\right) r_1^2 + r_2^2 - \frac{\tan^2 \beta_1}{\sin^2 \beta_2}\left(\frac{b_1}{b_2}\right)^2 \frac{r_1^4}{r_2^2}\right] \\
&= \frac{1}{2}\rho\Omega^2 \left[r_1^2 \tan^2 \beta_1 + r_2^2 - \frac{r_1^4 \tan^2 \beta_1}{r_2^2 \sin^2 \beta_2}\left(\frac{b_1}{b_2}\right)^2\right]
\end{aligned} \tag{9.9}$$

Our primary objective is to determine the value of r_1 that yields the maximum value of Δp. To do so, we seek the value of r_1 for which $d\Delta p/dr_1 = 0$ and $d^2\Delta p/dr_1^2 < 0$. Since r_1 appears only in even powers of r_1, it is easier to work with r_1^2, so that we seek the conditions under which $d\Delta p/d(r_1^2) = 0$ and $d^2\Delta p/d(r_1^2)^2 < 0$. Thus,

$$\frac{d\Delta p}{d(r_1^2)} = \frac{1}{2}\rho\Omega^2 \tan^2 \beta_1 \left[1 - \frac{2}{\sin^2 \beta_2}\left(\frac{b_1}{b_2}\right)^2\left(\frac{r_1}{r_2}\right)^2\right] \tag{9.10}$$

$$\frac{d^2\Delta p}{d(r_1^2)^2} = -\frac{\rho\Omega^2 \tan^2 \beta_1}{r_2^2 \sin^2 \beta_2}\left(\frac{b_1}{b_2}\right)^2 \tag{9.11}$$

Clearly, the second derivative of Δp is always negative, so that we indeed have a maximum value when the first derivative vanishes. Hence,

$$\left(\frac{r_1}{r_2}\right)^2 = \frac{1}{2}\sin^2 \beta_2 \left(\frac{b_2}{b_1}\right)^2 \implies r_1 = \frac{r_2 \sin \beta_2}{\sqrt{2}}\left(\frac{b_2}{b_1}\right) \tag{9.12}$$

Using the given values, the inlet radius for which Δp is a maximum is

$$r_1 = \frac{(10 \text{ cm}) \sin 10°}{\sqrt{2}}\left(\frac{5 \text{ cm}}{3 \text{ cm}}\right) = 2.05 \text{ cm} \tag{9.13}$$

9.2 Performance Curves

Statement of the Problem: *The performance curves for a turbomachine are shown in Figure 9.2. For the practical operating range of the machine, the head, h_p, and power, $\Omega\tau$, can be approximated by*

$$h_p \approx \frac{1}{26}(5125 - Q) \quad \text{and} \quad \Omega\tau \approx \frac{1}{137}(Q + 6974)$$

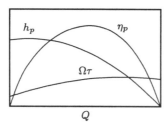

Figure 9.2: *Performance curves for a turbomachine.*

with h_p in m, $\Omega\tau$ in MW (Megawatts) and Q in m³/sec. Determine the type of turbomachine (i.e., is it a fan, blower or compressor) if the torque is $\tau = 2.75 \cdot 10^5$ N·m when the working fluid is air.

Solution: To determine the type of turbomachine, we must compute the specific speed, N_s, defined by

$$N_s \equiv \frac{\Omega \left(Q^*\right)^{1/2}}{\left(gh_p^*\right)^{3/4}} \qquad (9.14)$$

where Q^* and h_p^* are the volume-flow rate and head at the best-efficiency point, respectively. Since the turbomachine is a fan, blower or compressor, the efficiency is

$$\eta_p = \frac{\rho Q g h_p}{\Omega \tau} \qquad (9.15)$$

Now, using the given expressions for h_p and $\Omega\tau$, the efficiency of the machine is

$$\eta_p = \frac{137}{26} \rho g \left(\frac{5125Q - Q^2}{Q + 6974} \right) \qquad (9.16)$$

We seek the value of Q at which η_p is a maximum, which corresponds to the point where $d\eta_p/dQ = 0$. Given the shape of the curve for η_p, it is clear that $d^2\eta_p/dQ^2 < 0$, so there is no need to compute the second derivative. Differentiating η_p yields

$$\frac{d\eta_p}{dQ} = \frac{137}{26} \rho g \left[\frac{(5125 - 2Q)(Q + 6974) - (5125Q - Q^2)}{(Q + 6974)^2} \right] \qquad (9.17)$$

Setting $d\eta_p/dQ = 0$, there follows

$$(5125 - 2Q)(Q + 6974) - (5125Q - Q^2) = 0 \implies Q^2 + 13948Q - 35741750 = 0 \quad (9.18)$$

Solving, the two roots are

$$Q = 2212 \quad \text{and} \quad Q = -16160 \qquad (9.19)$$

So, rejecting the negative root, which is clearly nonphysical, the best-efficiency point occurs when $Q = 2212$ m³/sec, i.e., we have shown that

$$Q^* = 2212 \text{ m}^3/\text{sec} \qquad (9.20)$$

9.3. POWER

Hence, the head at the best-efficiency point is

$$h_p^* = \frac{1}{26}(5125 - 2212) \text{ m} = 112 \text{ m} \tag{9.21}$$

and, using the torque given in the problem statement, the rotation rate is given by

$$\frac{\Omega \tau}{10^6 \text{ (N·m/sec)}/MW} = Q + 6974 \implies \Omega = \frac{(2212 + 6974)}{137(0.275)} \text{ sec}^{-1} = 243.8 \text{ sec}^{-1} \tag{9.22}$$

Therefore, the specific speed is

$$N_s = \frac{(243.8 \text{ sec}^{-1})\left(2212 \text{ m}^3/\text{sec}\right)^{1/2}}{\left[\left(9.801 \text{ m/sec}^2\right)(112 \text{ m})\right]^{3/4}} \approx 60 \tag{9.23}$$

Therefore, since $N_s > 3$, this turbomachine is a very powerful **compressor**.

9.3 Power

Statement of the Problem: *You have just finished painting your living room and you are using a radial fan to dry the walls. Curious about its operation, you find the owner's manual and discover that it includes details of the impeller geometry (see Figure 9.3), which are as follows.*

$$r_1 = 7 \text{ in}, \quad r_2 = 15 \text{ in}, \quad b_1 = 6 \text{ in}, \quad b_2 = 2 \text{ in}, \quad \beta_1 = 10°, \quad \beta_2 = 40°$$

The quantities r_1 and r_2 are the inlet and outlet impeller radii, b_1 and b_2 are the inlet and outlet blade widths, and β_1 and β_2 are the inlet and outlet blade angles. The latter two quantities are thus the angles between the impeller blades and the relative velocities at the inlet and outlet. Also, the impeller rotates at an angular velocity of Ω = 2450 rpm, the inlet absolute velocity is purely radial and the density of air is ρ = 0.00234 slug/ft³. Determine the power required to run the fan and the volume-flow rate.

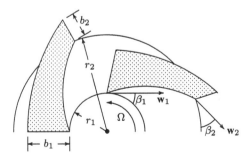

Figure 9.3: *Radial-fan impeller geometry.*

Solution: To determine the power required to run the fan, $\dot{W}_p$, we use one of the Euler turbomachine equations, viz.,

$$\dot{W}_p = \rho \Omega Q \left[r_2 V_{\theta 2} - r_1 V_{\theta 1}\right] \tag{9.24}$$

where $V_{\theta 1}$ and $V_{\theta 2}$ are the circumferential components of the absolute velocity at the impeller inlet and outlet, respectively. We must appeal to elementary pump theory to determine the various quantities in Equation (9.24).

Volume-flow rate: The equation for Q is

$$Q = \frac{2\pi b_1 \Omega r_1^2}{\cot \alpha_1 + \cot \beta_1} \tag{9.25}$$

First, we must convert Ω from rpm to radians per second. Since there are 2π radians in one revolution, we multiply by 2π. Therefore,

$$\Omega = \left(2\pi \, \frac{\text{radians}}{\text{rev}}\right)\left(2450 \, \frac{\text{rev}}{\text{min}}\right)\left(\frac{1}{60} \, \frac{\text{min}}{\text{sec}}\right) = 257 \, \frac{\text{radians}}{\text{sec}} \tag{9.26}$$

Also, since the inlet flow is purely radial, we have $\alpha_1 = 90°$. From Equation (9.25), the volume-flow rate is (replacing radians/sec by sec^{-1} since a radian is dimensionless):

$$Q = \frac{2\pi b_1 \Omega r_1^2}{\cot \alpha_1 + \cot \beta_1} = \frac{2\pi (0.5 \text{ ft}) \left(257 \text{ sec}^{-1}\right) (7/12 \text{ ft})^2}{\cot 90° + \cot 10°} = 48.4 \, \frac{\text{ft}^3}{\text{sec}} \tag{9.27}$$

Absolute circumferential velocities: At the impeller inlet and outlet, we have

$$V_{\theta 1} = \frac{\Omega r_1 \cot \alpha_1}{\cot \alpha_1 + \cot \beta_1} \tag{9.28}$$

$$V_{\theta 2} = \Omega r_2 \left[1 - \left(\frac{r_1}{r_2}\right)^2 \left(\frac{b_1}{b_2}\right) \frac{\cot \beta_2}{\cot \alpha_1 + \cot \beta_1}\right] \tag{9.29}$$

Clearly, since $\alpha_1 = 90°$, we see immediately that

$$V_{\theta 1} = 0 \tag{9.30}$$

Substituting the known values into the equation for $V_{\theta 2}$, Equation (9.29), gives

$$V_{\theta 2} = \left(257 \text{ sec}^{-1}\right)(1.25 \text{ ft}) \left[1 - \left(\frac{7 \text{ in}}{15 \text{ in}}\right)^2 \left(\frac{6 \text{ in}}{2 \text{ in}}\right) \frac{\cot 40°}{\cot 90° + \cot 10°}\right] = 277 \, \frac{\text{ft}}{\text{sec}} \tag{9.31}$$

Therefore, combining Equations (9.24), (9.26), (9.27), (9.30) and (9.31), the power required to run the fan is (observing that 1 lb = 1 slug·ft/sec^2):

$$\begin{aligned} \dot{W}_p &= \left(0.00234 \, \frac{\text{slug}}{\text{ft}^3}\right)\left(257 \text{ sec}^{-1}\right)\left(48.4 \, \frac{\text{ft}^3}{\text{sec}}\right)\left[(1.25 \text{ ft})\left(277 \, \frac{\text{ft}}{\text{sec}}\right) - 0\right] \\ &= 10078 \, \frac{\text{ft} \cdot \text{lb}}{\text{sec}} \end{aligned} \tag{9.32}$$

Finally, since 1 hp = 550 ft·lb/sec, the power is

$$\dot{W}_p = 18.3 \text{ hp} \tag{9.33}$$

9.4 Turbine Analysis

Statement of the Problem: *At its best-efficiency point, the volume-flow rate of a 6 ft radius common water turbine is $Q^* = 400$ ft³/sec. Also, it has a power output of $\Omega\tau = 6500$ hp, a best-efficiency-point head of $h_p^* = 615$ ft, and a torque of $\tau = 500{,}000$ ft·lbs. The angular-rotation rate is $\Omega = 250$ rpm and the density of water is $\rho = 1.94$ slug/ft³.*

(a) What type of turbine is it?

(b) Determine the efficiency of the turbine.

(c) Calculate the capacity and head coefficients.

Solution: The first thing we must do is convert Ω from rpm to radians per second. Since there are 2π radians in one revolution, we multiply by 2π. Therefore,

$$\Omega = \left(2\pi \, \frac{\text{radians}}{\text{rev}}\right)\left(250 \, \frac{\text{rev}}{\text{min}}\right)\left(\frac{1}{60} \, \frac{\text{min}}{\text{sec}}\right) = 26.18 \, \frac{\text{radians}}{\text{sec}} \quad (9.34)$$

In the following computations, we regard radians/sec as sec^{-1} since the radian is a dimensionless quantity.

9.4(a): To determine the type of turbine, we compute the specific speed, N_s, defined by

$$N_s = \frac{\Omega (Q^*)^{1/2}}{(gh_p^*)^{3/4}} \quad (9.35)$$

Using the given values and Equation (9.34), we find

$$N_s = \frac{(26.18 \text{ sec}^{-1})(400 \text{ ft}^3/\text{sec})^{1/2}}{\left[(32.174 \text{ ft}/\text{sec}^2)(615 \text{ ft})\right]^{3/4}} = 0.31 \quad (9.36)$$

Since the specific speed lies in the range $0.25 < N_s < 2.10$, it is a **Francis turbine**.

9.4(b): By definition, the efficiency of the turbine (at its best-efficiency point) is

$$\eta_t = \frac{\Omega\tau}{\rho Q^* g h_p^*} = \frac{(26.18 \text{ sec}^{-1})(500000 \text{ ft} \cdot \text{lb})}{(1.94 \text{ slug}/\text{ft}^3)(400 \text{ ft}^3/\text{sec})(32.174 \text{ ft}/\text{sec}^2)(615 \text{ ft})} = 0.85 \quad (9.37)$$

Thus, the turbine is 85% efficient.

9.4(c): The capacity and head coefficients for the turbine at its best-efficiency point are

$$C_{Q^*} = \frac{Q^*}{\Omega D^3} \quad \text{and} \quad C_{H^*} = \frac{gh_p^*}{\Omega^2 D^2} \quad (9.38)$$

Since the radius is 6 ft, the diameter is $D = 12$ ft. Thus,

$$C_{Q^*} = \frac{400 \text{ ft}^3/\text{sec}}{(26.18 \text{ sec}^{-1})(12 \text{ ft})^3} = 0.0088 \quad (9.39)$$

$$C_{H^*} = \frac{(32.174 \text{ ft}/\text{sec}^2)(615 \text{ ft})}{(26.18 \text{ sec}^{-1})^2 (12 \text{ ft})^2} = 0.2005 \quad (9.40)$$

9.5 Pelton Wheel

Statement of the Problem: *Imagine that you have been employed as a consultant to design a single-jet Pelton wheel that is 90% efficient with a jet velocity, V_j = 90 m/sec. You must use a motor that runs with a torque of $\Omega \tau = 10^7$ N·m. The jet is driven by a reservoir with a head of h_p = 325 m. If you arrive at a design with a specific speed of N_s = 0.11, determine the angular-rotation rate, Ω, the volume-flow rate, Q, and the radius of the wheel, r. Assume this is the best efficiency point and that the density of water is ρ = 998 kg/m³.*

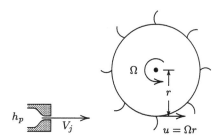

Figure 9.4: *Schematic of a Pelton wheel.*

Solution: We need two equations to solve for the volume-flow rate, Q^*, and the angular-rotation rate, Ω. Since part of our design objective is to attain a prescribed efficiency, η_t, we can use its definition as one of the equations. We are also given the specific speed, N_s, so that its definition can serve as a second equation.

Using the definition of efficiency at the best-efficiency point,

$$\eta_t = \frac{\Omega \tau}{\rho Q^* g h_p^*} \quad \Longrightarrow \quad Q^* = \frac{\tau}{\rho g h_p^* \eta_t} \Omega \qquad (9.41)$$

which involves both unknowns, Q^* and Ω. Next, from the definition of specific speed, i.e.,

$$N_s = \frac{\Omega (Q^*)^{1/2}}{\left(g h_p^*\right)^{3/4}} \quad \Longrightarrow \quad \Omega = \frac{N_s \left(g h_p^*\right)^{3/4}}{(Q^*)^{1/2}} \qquad (9.42)$$

Equation (9.42) also involves both Q^* and Ω. So, combining Equations (9.41) and (9.42), we can solve directly for Q^* as follows.

$$Q^* = \frac{\tau}{\rho g h_p^* \eta_t} \frac{N_s \left(g h_p^*\right)^{3/4}}{(Q^*)^{1/2}} \quad \Longrightarrow \quad (Q^*)^{3/2} = \frac{N_s \tau}{\rho \left(g h_p^*\right)^{1/4} \eta_t} \qquad (9.43)$$

which can be rewritten as

$$Q^* = \left(\frac{N_s \tau}{\rho \eta_t}\right)^{2/3} \left(g h_p^*\right)^{-1/6} \qquad (9.44)$$

Substituting the given values, the volume-flow rate for the Pelton wheel is (observing that 1 N = 1 kg·m/sec²):

$$Q^* = \left[\frac{(0.11)\left(10^7 \text{ N} \cdot \text{m}\right)}{\left(998 \text{ kg/m}^3\right)(0.90)}\right]^{2/3} \left[\left(9.801 \text{ m/sec}^2\right)(325 \text{ m})\right]^{-1/6} = 29.84 \frac{\text{m}^3}{\text{sec}} \qquad (9.45)$$

9.5. PELTON WHEEL

Finally, using Equation (9.42), the corresponding value of the angular-rotation rate is

$$\Omega = \frac{N_s \left(gh_p^*\right)^{3/4}}{(Q^*)^{1/2}} = \frac{(0.11)\left[\left(9.801 \text{ m/sec}^2\right)(325 \text{ m})\right]^{3/4}}{(29.84 \text{ m}^3/\text{sec})^{1/2}} = 8.54 \text{ sec}^{-1} \quad (9.46)$$

To determine the radius of the Pelton wheel, we make use of the fact that the torque, τ, is given by

$$\tau = 2\rho Q r \left(V_j - \Omega r\right) \quad \Longrightarrow \quad 2\rho Q \Omega r^2 - 2\rho Q V_j r + \tau = 0 \quad (9.47)$$

The solution to this quadratic for r is

$$r = \frac{2\rho Q V_j \pm \sqrt{4\rho^2 Q^2 V_j^2 - 8\rho Q \Omega \tau}}{4\rho Q \Omega} = \frac{V_j}{2\Omega} \pm \sqrt{\left(\frac{V_j}{2\Omega}\right)^2 - \frac{\tau}{2\rho Q \Omega}} \quad (9.48)$$

Since the Pelton wheel operates at its best-efficiency point, $Q = Q^*$, we have

$$r = \frac{V_j}{2\Omega} \pm \sqrt{\left(\frac{V_j}{2\Omega}\right)^2 - \frac{\tau}{2\rho Q^* \Omega}} \quad (9.49)$$

Since τ, ρ, Q^* and Ω are all positive, the quantity under the square root is less than $V_j/(2\Omega)$. Hence, there are two possible solutions. Substituting the known values from above,

$$\begin{aligned} r &= \frac{90 \text{ m/sec}}{2\left(8.54 \text{ sec}^{-1}\right)} \pm \sqrt{\left[\frac{90 \text{ m/sec}}{2\left(8.54 \text{ sec}^{-1}\right)}\right]^2 - \frac{10^7 \text{ kg} \cdot \text{m}^2/\text{sec}^2}{2\left(998 \text{ kg/m}^3\right)\left(29.84 \text{ m}^3/\text{sec}\right)\left(8.54 \text{ sec}^{-1}\right)}} \\ &= 5.27 \text{ m} \pm 2.85 \text{ m} \end{aligned} \quad (9.50)$$

Therefore, the two possible wheel radii are:

$$r = 2.42 \text{ m} \quad \text{and} \quad r = 8.12 \text{ m} \quad (9.51)$$

The smaller wheel will be less expensive to fabricate, and would thus be the most likely recommendation of a responsible consultant.

Chapter 10

Vorticity and Viscosity

Chapter 10 investigates the relationship between vorticity and viscosity in a *real*, or viscous, fluid. If a truly inviscid fluid existed, it would be impossible to develop a lifting force on an airplane wing sufficient to lift off the runway in the manner we are accustomed to. This is true because the lifting force depends upon the presence of vorticity in the flow, whose origin is intimately linked to viscous effects. Although viscous effects are normally confined to thin *boundary layers* close to solid boundaries, they have a large global impact on the forces that develop on an object, including not only lift, but drag as well. This chapter addresses some aspects of the relation between viscosity and vorticity, and the attendant forces through the following problems.

Section 10.1 The Vorticity Equation: The problem in this section examines the effect of a non-conservative body force on the differential equation for the vorticity.

Section 10.2 Circulation: A straightforward calculation of the circulation for a prescribed velocity field that is typical of a viscous flow. Note that the integrals are done in the positive sense, with the sign of each portion of the contour integral determined by the differential arclength vector, $d\mathbf{s}$.

Section 10.3 Vorticity in a Boundary Layer: This section's problem compares the magnitude of the vorticity close to a surface in a turbulent flow to its surface value. The calculation is quite realistic, and underscores the fact that vorticity is normally concentrated in very small regions of a flow.

Section 10.4 Drag: The solution to this problem gives the terminal velocity of a falling object for two different orientations, each corresponding to a unique frontal area and drag coefficient. It makes use of the drag data of Appendix D.

Section 10.5 Lift: This problem involves determining the glide angle of a powerless aircraft as a function of its lift-to-drag ratio.

10.1 The Vorticity Equation

Statement of the Problem: *In a coordinate system rotating with constant angular velocity Ω, the Euler equation is*

$$\rho \frac{d\mathbf{u}}{dt} = -\nabla p^* + \rho \mathbf{f}_{Coriolis}, \qquad \mathbf{f}_{Coriolis} = -2\mathbf{\Omega} \times \mathbf{u}$$

where p^* is the reduced pressure, which includes effects of centrifugal acceleration. Assuming the flow is incompressible, verify that the Coriolis force, $\mathbf{f}_{Coriolis}$, is non-conservative, and derive the differential equation for the vorticity, $\boldsymbol{\omega} = \nabla \times \mathbf{u}$, when $\mathbf{\Omega} = \Omega \mathbf{k}$.

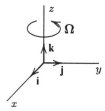

Figure 10.1: *Rotating coordinate system.*

Solution: By definition, a conservative force, $\mathbf{f}$, is one that can be written as the gradient of a scalar potential function, $\mathcal{V}$, i.e., $\mathbf{f} = -\nabla \mathcal{V}$. This implies that $\nabla \times \mathbf{f} = \mathbf{0}$ since the curl of the gradient of any scalar is zero. A force is non-conservative if $\nabla \times \mathbf{f} \neq \mathbf{0}$. So, we must take the curl of the Coriolis force and show that it is nonzero to demonstrate that it is non-conservative. Thus, using the standard vector expression for the curl of the cross product of two vectors,

$$\begin{aligned}
\nabla \times \mathbf{f}_{Coriolis} &= \nabla \times [-2\mathbf{\Omega} \times \mathbf{u}] = -2\nabla \times [\mathbf{\Omega} \times \mathbf{u}] \\
&= -2 \left[\underbrace{(\mathbf{u} \cdot \nabla)\mathbf{\Omega}}_{Term\ 1} + \underbrace{\mathbf{\Omega}(\nabla \cdot \mathbf{u})}_{Term\ 2} - \underbrace{(\mathbf{\Omega} \cdot \nabla)\mathbf{u}}_{Term\ 3} - \underbrace{\mathbf{u}(\nabla \cdot \mathbf{\Omega})}_{Term\ 4} \right] = 2(\mathbf{\Omega} \cdot \nabla)\mathbf{u} \quad (10.1)
\end{aligned}$$

Note that Terms 1 and 4 are zero because $\mathbf{\Omega}$ does not vary with x, y and z. Also, Term 2 vanishes because $\nabla \cdot \mathbf{u} = 0$ for incompressible flow. Finally, for $\mathbf{\Omega} = \Omega \mathbf{k}$, necessarily we have $\mathbf{\Omega} \cdot \nabla = \Omega \partial/\partial z$. Hence, Equation (10.1) becomes

$$\nabla \times \mathbf{f}_{Coriolis} = 2\Omega \frac{\partial \mathbf{u}}{\partial z} \tag{10.2}$$

Therefore, unless $\mathbf{u}$ does not vary with z, the Coriolis force is non-conservative. The easiest way to proceed is to make use of the following vector identity.

$$\mathbf{u} \cdot \nabla \mathbf{u} = \nabla\left(\frac{1}{2}\mathbf{u} \cdot \mathbf{u}\right) - \mathbf{u} \times (\nabla \times \mathbf{u}) \tag{10.3}$$

Then, since $\boldsymbol{\omega} = \nabla \times \mathbf{u}$, the acceleration term in Euler's equation is

$$\frac{d\mathbf{u}}{dt} = \frac{\partial \mathbf{u}}{\partial t} + \nabla\left(\frac{1}{2}\mathbf{u} \cdot \mathbf{u}\right) - \mathbf{u} \times \boldsymbol{\omega} \tag{10.4}$$

Therefore, substituting this result into Euler's equation yields

$$\frac{\partial \mathbf{u}}{\partial t} + \nabla\left(\frac{1}{2}\mathbf{u} \cdot \mathbf{u}\right) - \mathbf{u} \times \boldsymbol{\omega} = -\frac{1}{\rho}\nabla p^* + \mathbf{f}_{Coriolis} \tag{10.5}$$

10.2. CIRCULATION

We derive a differential equation for ω by taking the curl of Euler's equation. Proceeding term by term, we find the following. The unsteady term becomes

$$\nabla \times \frac{\partial \mathbf{u}}{\partial t} = \frac{\partial}{\partial t}(\nabla \times \mathbf{u}) = \frac{\partial \omega}{\partial t} \tag{10.6}$$

This is true because spatial and temporal derivatives can be interchanged, i.e., time and space are independent variables. Also, the two terms involving the gradient of a scalar give

$$\nabla \times \nabla \left(\frac{1}{2}\mathbf{u}\cdot\mathbf{u}\right) = 0 \quad \text{and} \quad \nabla \times \left(-\frac{1}{\rho}\nabla p^*\right) = -\frac{1}{\rho}\nabla \times \nabla p^* = 0 \tag{10.7}$$

which is true because the curl of the gradient of any scalar function is zero. Turning to the cross product of the velocity and the vorticity, the so-called *vortex force*, we have

$$\nabla \times (\mathbf{u} \times \omega) = (\omega \cdot \nabla)\mathbf{u} + \underbrace{\mathbf{u}(\nabla \cdot \omega)}_{=0} - (\mathbf{u} \cdot \nabla)\omega - \underbrace{\omega(\nabla \cdot \mathbf{u})}_{=0} = \omega \cdot \nabla \mathbf{u} - \mathbf{u} \cdot \nabla \omega \tag{10.8}$$

Clearly, the term proportional to $\nabla \cdot \mathbf{u}$ vanishes because the flow is incompressible, and since the divergence of the curl of any vector is zero, we can say $\nabla \cdot \omega = \nabla \cdot (\nabla \times \mathbf{u}) = 0$. Finally, we have already taken the curl of the Coriolis force, and the result is given in Equation (10.2). Therefore, combining Equations (10.2), (10.5), (10.6), (10.7) and (10.8), we find

$$\frac{\partial \omega}{\partial t} - \omega \cdot \nabla \mathbf{u} + \mathbf{u} \cdot \nabla \omega = 2\Omega \frac{\partial \mathbf{u}}{\partial z} \tag{10.9}$$

Rearranging terms, the final form of the vorticity equation is

$$\frac{d\omega}{dt} = \omega \cdot \nabla \mathbf{u} + 2\Omega \frac{\partial \mathbf{u}}{\partial z} \tag{10.10}$$

10.2 Circulation

Statement of the Problem: *Consider viscous flow close to a solid boundary for with*

$$\mathbf{u} = U\left[\frac{y}{h} + K\left(\frac{y}{h}\right)^2\right]\mathbf{i}$$

where y is distance normal to the boundary, h is the thickness of the viscous layer and is independent of x, U is a constant reference velocity and K is a dimensionless constant.

(a) *Using the closed contour shown in Figure 10.2, compute the circulation in terms of K, U and Δx.*

(b) *Is there a value of K for which the circulation is zero?*

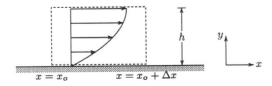

Figure 10.2: *Viscous flow near a solid boundary—velocity profile shown is for $K = -\frac{1}{2}$.*

Solution: By definition, the circulation is

$$\Gamma = \oint_C \mathbf{u} \cdot d\mathbf{s} \tag{10.11}$$

where we use the standard mathematical convention that integration is positive in the counterclockwise direction.

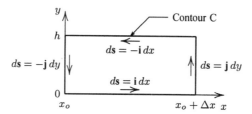

Figure 10.3: *Rectangular contour with differential direction vectors.*

10.2(a): In general, to evaluate a line integral, we treat the integral as the sum of conventional integrals on each segment of the contour, with the sign determined by the differential distance vector, $d\mathbf{s}$. Hence, referring to Figure 10.3, we see that

$$\begin{aligned}\Gamma &= \int_{x_o}^{x_o+\Delta x} \mathbf{u}(x,0) \cdot (\mathbf{i}\, dx) + \int_0^h \mathbf{u}(x_o+\Delta x, y) \cdot (\mathbf{j}\, dy) \\ &+ \int_{x_o}^{x_o+\Delta x} \mathbf{u}(x,h) \cdot (-\mathbf{i}\, dx) + \int_0^h \mathbf{u}(x_o, y) \cdot (-\mathbf{j}\, dy) \\ &= \int_{x_o}^{x_o+\Delta x} [u(x,0) - u(x,h)]\, dx + \int_0^h [v(x_o+\Delta x, y) - v(x_o, y)]\, dy \end{aligned} \tag{10.12}$$

Thus, for the given velocity vector, we have

$$u(x,0) = 0, \quad u(x,h) = (K+1)U, \quad v(x_o+\Delta x, y) = 0, \quad v(x_o, y) = 0 \tag{10.13}$$

Therefore, the circulation is

$$\Gamma = \int_{x_o}^{x_o+\Delta x} [-(K+1)U]\, dx = -(K+1)U\Delta x \tag{10.14}$$

10.2(b): Obviously, the circulation vanishes when

$$K = -1 \qquad \text{(Zero circulation condition)} \tag{10.15}$$

10.3 Vorticity in a Boundary Layer

Statement of the Problem: *For two-dimensional turbulent flow over a solid boundary, the velocity close to the surface is approximately*

$$\mathbf{u} \approx u_\tau \left[\frac{1}{\kappa}\ell n \frac{u_\tau y}{\nu} + C\right]\mathbf{i}, \qquad y \geq 30 \frac{\nu}{u_\tau}$$

10.3. VORTICITY IN A BOUNDARY LAYER

where **i** is a unit vector parallel to the surface, $u_\tau = \sqrt{\tau_w/\rho}$ is the friction velocity, $\tau_w = \mu(\partial u/\partial y)_{y=0}$, y is distance normal to the surface and ν is kinematic viscosity. Also, $\kappa = 0.41$ and $C = 5.0$. Determine the vorticity at $y = 30\nu/u_\tau$ and compute its ratio to the vorticity at $y = 0$.

Solution: In general, the vorticity for a two-dimensional flow is

$$\omega = \left(\frac{\partial v}{\partial x} - \frac{\partial u}{\partial y}\right)\mathbf{k} \qquad (10.16)$$

Since $v = 0$, if we write $\omega = \omega\,\mathbf{k}$, then

$$\omega = -\frac{\partial u}{\partial y} \qquad (10.17)$$

So, for the given velocity profile, we have

$$\omega = -\frac{u_\tau}{\kappa y} \qquad (10.18)$$

At $y = 30\nu/u_\tau$, the vorticity is

$$\omega = -\frac{u_\tau^2}{30\kappa\nu} = -\frac{\tau_w}{30\kappa\rho\nu} = -\frac{1}{30\kappa}\frac{\tau_w}{\mu} \qquad (10.19)$$

where we use the definition of the friction velocity ($u_\tau \equiv \sqrt{\tau_w/\rho}$) and the fact that kinematic and molecular viscosities are related by $\mu = \rho\nu$. By definition, the shear stress at the wall is given by

$$\tau_w = \mu\left(\frac{\partial u}{\partial y}\right)_{y=0} \implies \frac{\tau_w}{\mu} = \left(\frac{\partial u}{\partial y}\right)_{y=0} \qquad (10.20)$$

Therefore, combining Equations (10.19) and (10.20), the vorticity is

$$\omega = -\frac{1}{30\kappa}\left(\frac{\partial u}{\partial y}\right)_{y=0} \qquad (10.21)$$

Finally, at the surface, for any velocity profile, the vorticity is related to the velocity gradient by

$$\omega_{y=0} = -\left(\frac{\partial u}{\partial y}\right)_{y=0} \qquad (10.22)$$

Dividing left and right sides of Equation (10.21) by the corresponding side of Equation (10.22), we find

$$\frac{\omega}{\omega_{y=0}} = \frac{1}{30\kappa} = 0.08 \qquad (10.23)$$

Hence, the vorticity has fallen to 8% of its surface value at $y = 30\nu/u_\tau$.

To appreciate the implication of this result, consider the following. The magnitude of $30\nu/u_\tau$ on an automobile traveling at 65 mph is less than the diameter of the head of a pin. Thus, the vorticity in flows of practical engineering interest is confined to very small regions of the flow.

10.4 Drag

Statement of the Problem: *Due to fatigue from temperatures in excess of 100° F, a construction worker drops a standard 15 inch long 1 by 2 wooden board of nominal cross section $\frac{3}{4}$ inch by $\frac{3}{2}$ inch from the roof of a very tall building. The wood is pine and has a density of $\rho_o = 1$ slug/ft³. Compute the terminal velocity of the board as it falls with a frontal area as shown in each of the two configurations shown in Figure 10.4. Assume the density of air is $\rho = 0.00220$ slug/ft³ and that the building is sufficiently high that the board reaches its terminal velocity before striking the ground.*

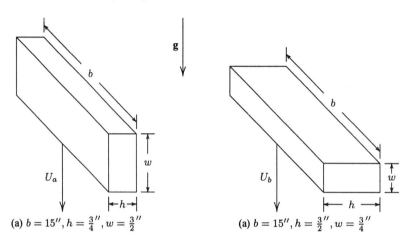

Figure 10.4: *Board falling from a building.*

Solution: In general, the terminal velocity is achieved when the board's weight equals its aerodynamic drag. Since the volume of the board is $V = bhw$, the weight is $\rho_o g bhw$. The drag force is

$$F_D = \frac{1}{2}\rho U^2 A C_D \tag{10.24}$$

where A is the *frontal area* and C_D is the drag coefficient. From Figure D.2, the frontal area for a rectangular cross section is $A = bh$, and the drag coefficient depends upon b/h. So, balancing forces gives

$$\rho_o g bhw = \frac{1}{2}\rho U^2 bh C_D \quad \Longrightarrow \quad U^2 = \frac{2}{C_D}\frac{\rho_o}{\rho}gw \tag{10.25}$$

Therefore, the terminal velocity is

$$U = \sqrt{\frac{2}{C_D}\frac{\rho_o}{\rho}gw} \tag{10.26}$$

10.4(a): If the board falls with its $\frac{3}{4}$ in side facing downward as in Figure 10.4(a), we have $b/h = 20$ and $w = \frac{3}{2}$ inch $= \frac{1}{8}$ ft. Hence, from Figure D.2, the drag coefficient is

$$C_D = 1.50 \tag{10.27}$$

10.5. LIFT

Therefore, the terminal velocity is

$$U = \sqrt{\left(\frac{2}{1.50}\right)\left(\frac{1 \text{ slug/ft}^3}{0.0022 \text{ slug/ft}^3}\right)\left(32.174 \frac{\text{ft}}{\text{sec}^2}\right)\left(\frac{1}{8} \text{ ft}\right)} = 49.4 \frac{\text{ft}}{\text{sec}} \quad (10.28)$$

10.4(b): If the board falls with its $\frac{3}{2}$ in side facing downward as in Figure 10.4(b), we have $b/h = 10$ and $w = \frac{3}{4}$ in $= \frac{1}{16}$ ft. Again referring to Figure D.2, the drag coefficient is

$$C_D = 1.30 \quad (10.29)$$

Therefore, the terminal velocity is

$$U = \sqrt{\left(\frac{2}{1.30}\right)\left(\frac{1 \text{ slug/ft}^3}{0.0022 \text{ slug/ft}^3}\right)\left(32.174 \frac{\text{ft}}{\text{sec}^2}\right)\left(\frac{1}{16} \text{ ft}\right)} = 37.5 \frac{\text{ft}}{\text{sec}} \quad (10.30)$$

10.5 Lift

Statement of the Problem: *The lift coefficient of an unpowered aircraft is 28.5 times its drag coefficient. Assuming the aircraft moves at constant speed with lift, drag and weight in equilibrium, at what angle to the horizontal is it inclined?*

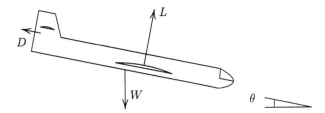

Figure 10.5: *Forces acting on the aircraft.*

Solution: Balancing the lift, L, drag, D, and weight, W, in the vertical and horizontal directions yields

$$\begin{aligned} W &= L\cos\theta + D\sin\theta & \text{(Vertical)} \\ 0 &= L\sin\theta - D\cos\theta & \text{(Horizontal)} \end{aligned} \quad (10.31)$$

So, from the balance of forces in the horizontal direction, we have

$$\tan\theta = \frac{D}{L} = \frac{C_D}{C_L} \quad (10.32)$$

Therefore, since we are given $C_L/C_D = 28.5$, the glide angle is

$$\theta = \tan^{-1}\left(\frac{1}{28.5}\right) = 2.0° \quad (10.33)$$

Chapter 11

Potential Flow

The material in Chapter 11 represents a change in focus from the global approach of the control volume method, which is based on the integral form of the conservation laws, to a detailed view that requires solution of the differential equations of motion at each point in a given flow. In potential-flow theory, we focus on the special case of incompressible, irrotational flow. In this special case, the equations of motion reduce to Laplace's equation for the velocity potential, ϕ. Additionally, for two-dimensional flow we introduce the streamfunction, ψ, which also satisfies Laplace's equation. The pressure and velocity are related by Bernoulli's equation, which is where the inherent nonlinearity of the momentum equation appears in potential-flow theory.

Solutions to Laplace's equation, which is a linear equation, can be built up by superposition of fundamental solutions such as *Uniform Flow*, the *Potential Vortex*, the *Source*, the *Sink* and the *Doublet*. Problems are included in this chapter that involve calculating the streamfunction, velocity potential, velocity components and pressure using these fundamental solutions. The chapter also includes problems that exercise the well-known solution for flow past a circular cylinder with and without rotation, and for an accelerating point source. The chapter concludes with problems on linearized airfoil theory and some basic aspects of *Computational Fluid Dynamics (CFD)*.

Note that one common theme of the problems in this chapter is the use of both rectangular Cartesian coordinates and cylindrical polar coordinates. This is characteristic of potential-flow theory, and originates from the fact that the fundamental solutions are most naturally formulated in cylindrical coordinates, while many practical problems are more naturally developed in terms of Cartesian coordinates. Hence, to become proficient in the application of potential-flow theory, we must be sufficiently familiar with both coordinate systems to use them interchangeably.

Section 11.1 Basic Concepts—Cartesian Coordinates: This problem involves calculation of the velocity components and stagnation points from a given streamfunction. The streamfunction is given in Cartesian coordinates.

Section 11.2 Basic Concepts—Cylindrical Coordinates: As in the problem of Section 11.1, we again compute the velocity components and stagnation points for a given streamfunction. However, this time the operations are done in cylindrical coordinates, rather than in Cartesian coordinates.

Section 11.3 Computing ψ from a Given ϕ: This is the first of two problems dealing with the relationship between the velocity potential and the streamfunction. We derive the streamfunction from a given velocity potential that is prescribed in terms of Cartesian coordinates.

Section 11.4 Computing ϕ from a Given ψ: This is the second of two problems dealing with the relationship between the velocity potential and the streamfunction. This time, we derive the velocity potential from a given streamfunction using cylindrical coordinates.

Section 11.5 Bernoulli's Equation—Cylindrical Coordinates: The problem in this section starts with the velocity potential specified in cylindrical coordinates. We first calculate the velocity components and then use Bernoulli's equation to compute the pressure.

Section 11.6 Superposition of Fundamental Solutions I: The problem in this section involves superposition of fundamental solutions to construct the flow induced by the so-called *spiral vortex*.

Section 11.7 Superposition of Fundamental Solutions II: The problem in this section involves superposition of fundamental solutions to construct flow past a lifting body emitting a stream of fluid that can be thought of as the exhaust of a jet engine.

Section 11.8 Flow Past a Rotating Cylinder I: This is the first of two cylinder-flow problems. Here, the cylinder rotates at different speeds. The solution involves computing the critical speed and the location of stagnation points.

Section 11.9 Flow Past a Rotating Cylinder II: This is the second of two cylinder-flow problems. The solution involves determining the direction of rotation of a lifting cylinder, and the circulation corresponding to a given lift coefficient.

Section 11.10 Unsteady Potential Flow: All the problems analyzed up to this point have been for steady flows. This problem addresses an unsteady potential flow, using the generalized Bernoulli equation to compute pressure.

Section 11.11 Linearized Airfoil Theory: This is an airfoil problem. It uses the principles of linearized airfoil theory to compute the camber line.

Section 11.12 CFD—Richardson Extrapolation: This problem explores the use of central differencing and Richardson extrapolation. These are important aspects of numerical solution methods for potential-flow problems.

11.1 Basic Concepts—Cartesian Coordinates

Statement of the Problem: *Consider the streamfunction given by*

$$\psi(x,y) = Ax^2 + Bxy - Ay^2$$

where A and B are constants and $B^2 \neq 4A^2$.

(a) Compute the velocity components and locate all stagnation points.

(b) In what part of the xy plane are the x and y velocity components equal?

Solution: The primary information we need to solve this problem is how the velocity is related to ψ, and how stagnation points are defined. In terms of the streamfunction, the velocity components are

$$u = \frac{\partial \psi}{\partial y} \quad \text{and} \quad v = -\frac{\partial \psi}{\partial x} \qquad (11.1)$$

11.2. BASIC CONCEPTS—CYLINDRICAL COORDINATES

Also, the stagnation points are the points where both u and v are zero.

11.1(a): Hence, computing the appropriate derivatives according to Equation (11.1), the velocity components are

$$u(x,y) = Bx - 2Ay \quad \text{and} \quad v(x,y) = -2Ax - By \qquad (11.2)$$

Now, $u = 0$ when $Bx - 2Ay = 0$ and $v = 0$ when $2Ax + By = 0$. We can write these two equations as follows.

$$\begin{bmatrix} B & -2A \\ 2A & B \end{bmatrix} \begin{Bmatrix} x \\ y \end{Bmatrix} = \begin{Bmatrix} 0 \\ 0 \end{Bmatrix} \qquad (11.3)$$

But, the determinant of this matrix is $B^2 + 4A^2$, which is zero only if $A = B = 0$. Therefore, only the trivial solution exists so that there is one stagnation point in this flow and it lies at

$$x = y = 0 \qquad \text{(Stagnation point location)} \qquad (11.4)$$

11.1(b): The condition $u = v$ is satisfied when

$$Bx - 2Ay = -2Ax - By \quad \Longrightarrow \quad (2A + B)x = (2A - B)y \qquad (11.5)$$

Now, since we are given $B^2 \neq 4A^2$, necessarily $2A \neq \pm B$, which means $(2A \pm B)$ is nonzero. Therefore, $u = v$ along the straight line defined by

$$y = \left(\frac{2A + B}{2A - B}\right) x \qquad (11.6)$$

11.2 Basic Concepts—Cylindrical Coordinates

Statement of the Problem: *Consider the streamfunction given by*

$$\psi(r, \theta) = Qr^3 \sin 3\theta$$

where Q is a constant.

(a) Compute the velocity components and locate all stagnation points.

(b) If $\psi = 4$ when $x = 2$ and $y = 2$, what is the value of Q?

(c) Develop an equation of the form $x = f(\psi, y)$ for the streamlines in this flow.

Solution: The primary information we need to solve this problem is how the velocity is related to ψ, and how stagnation points are defined. In terms of the streamfunction, the velocity components are

$$u_r = \frac{1}{r} \frac{\partial \psi}{\partial \theta} \quad \text{and} \quad u_\theta = -\frac{\partial \psi}{\partial r} \qquad (11.7)$$

Also, the stagnation points are the points where both u_r and u_θ are zero.

11.2(a): Differentiating the streamfunction as indicated in Equation (11.7), the velocity components are

$$u_r(r, \theta) = 3Qr^2 \cos 3\theta \quad \text{and} \quad u_\theta(r, \theta) = -3Qr^2 \sin 3\theta \qquad (11.8)$$

Now, $u_r = 0$ when $r^2 \cos 3\theta = 0$ and $u_\theta = 0$ when $r^2 \sin 3\theta = 0$. It is impossible to have both $\sin 3\theta$ and $\cos 3\theta$ equal to zero for the same value of θ. Since both velocity components vanish when $r = 0$, we conclude that there is a single stagnation point at

$$r = 0 \qquad \text{(Stagnation point location)} \qquad (11.9)$$

11.2(b): The analysis can be simplified by rewriting the streamfunction in terms of Cartesian coordinates. First, we note that

$$\sin 3\theta = 3\sin\theta - 4\sin^3\theta \qquad (11.10)$$

so that the streamfunction can be rewritten as

$$\psi(r,\theta) = Q\left(3r^3 \sin\theta - 4r^3 \sin^3\theta\right) \qquad (11.11)$$

Next, we take account of the following relations between Cartesian and cylindrical coordinates.

$$r^2 = x^2 + y^2 \quad \text{and} \quad r\sin\theta = y \qquad (11.12)$$

Thus, the streamfunction becomes

$$\psi(x,y) = Q\left[3\left(x^2+y^2\right)y - 4y^3\right] = Qy\left(3x^2 - y^2\right) \qquad (11.13)$$

So, using the fact that $\psi(2,2) = 4$, we can solve for Q as follows.

$$Q = \frac{\psi(2,2)}{2\left(3\cdot 2^2 - 2^2\right)} = \frac{4}{16} = \frac{1}{4} \qquad (11.14)$$

11.2(c): It is again convenient to work with Cartesian coordinates. Therefore, combining Equations (11.13) and (11.14), we begin with

$$\psi(x,y) = \frac{y}{4}\left(3x^2 - y^2\right) \qquad (11.15)$$

Since a streamline corresponds to a constant value of ψ, we have

$$\frac{4\psi}{y} = 3x^2 - y^2 \qquad (11.16)$$

Then, solving for x as a function of ψ and y yields the following equation for the streamlines.

$$x = \pm\sqrt{\frac{1}{3}\left(y^2 + \frac{4\psi}{y}\right)} \qquad (11.17)$$

11.3 Computing ψ from a Given ϕ

Statement of the Problem: *The velocity potential for a two-dimensional flow is*

$$\phi(x,y) = U\left[x + \lambda e^{-x/\lambda}\sin(y/\lambda)\right], \qquad x \geq 0$$

where U and λ are constant velocity and length scales, respectively. Determine the corresponding streamfunction.

11.4. COMPUTING ϕ FROM A GIVEN ψ

Solution: The solution strategy is to first compute the velocity components from the velocity potential. Then, introducing the streamfunction, we obtain two equations relating the derivatives of the streamfunction and the velocity components. Integration of these equations yields the streamfunction. Hence, we begin by writing

$$u = \frac{\partial \phi}{\partial x} = U\left[1 - e^{-x/\lambda}\sin(y/\lambda)\right] \qquad (11.18)$$

$$v = \frac{\partial \phi}{\partial y} = Ue^{-x/\lambda}\cos(y/\lambda) \qquad (11.19)$$

By definition, the velocity components are related to the streamfunction as follows.

$$u = \frac{\partial \psi}{\partial y} \quad \text{and} \quad v = -\frac{\partial \psi}{\partial x} \qquad (11.20)$$

Combining Equation (11.18) with the first of Equations (11.20), we have

$$\frac{\partial \psi}{\partial y} = U\left[1 - e^{-x/\lambda}\sin(y/\lambda)\right] \qquad (11.21)$$

Integrating over y, the velocity potential is

$$\psi(x,y) = U\left[y + \lambda e^{-x/\lambda}\cos(y/\lambda)\right] + f(x) \qquad (11.22)$$

where $f(x)$ is a function of integration. Now, we differentiate with respect to x to evaluate v according to the second of Equations (11.20), wherefore,

$$v = Ue^{-x/\lambda}\cos(y/\lambda) - \frac{df}{dx} \qquad (11.23)$$

Comparing Equations (11.19) and (11.23), clearly

$$\frac{df}{dx} = 0 \quad \Longrightarrow \quad f(x) = \text{constant} \qquad (11.24)$$

With no loss of generality, we can set the constant equal to zero. Therefore, the streamfunction for this flow is

$$\psi(x,y) = U\left[y + \lambda e^{-x/\lambda}\cos(y/\lambda)\right] \qquad (11.25)$$

11.4 Computing ϕ from a Given ψ

Statement of the Problem: *The streamfunction for a two-dimensional flow is*

$$\psi(r,\theta) = Qr^n\left(\sin n\theta + \cos n\theta\right)$$

where Q and n are constants. Determine the corresponding velocity potential.

Solution: The solution strategy is to first compute the velocity components from the streamfunction. Then, introducing the velocity potential, we obtain two equations relating the derivatives of the velocity potential and the velocity components. Integration of these equations yields the velocity potential. Hence, we begin by writing

$$u_r = \frac{1}{r}\frac{\partial \psi}{\partial \theta} = nQr^{n-1}\left(\cos n\theta - \sin n\theta\right) \qquad (11.26)$$

$$u_\theta = -\frac{\partial \psi}{\partial r} = -nQr^{n-1}\left(\sin n\theta + \cos n\theta\right) \qquad (11.27)$$

By definition, the velocity components are related to the velocity potential as follows.

$$u_r = \frac{\partial \phi}{\partial r} \quad \text{and} \quad u_\theta = \frac{1}{r}\frac{\partial \phi}{\partial \theta} \qquad (11.28)$$

Combining Equation (11.26) with the first of Equations (11.28), we have

$$\frac{\partial \phi}{\partial r} = nQr^{n-1}(\cos n\theta - \sin n\theta) \qquad (11.29)$$

Integrating over r, the velocity potential is

$$\phi(r, \theta) = Qr^n(\cos n\theta - \sin n\theta) + f(\theta) \qquad (11.30)$$

where $f(\theta)$ is a function of integration. Now, we differentiate with respect to θ and divide by r to evaluate u_θ according to the second of Equations (11.28), wherefore,

$$u_\theta = -nQr^{n-1}(\sin n\theta + \cos n\theta) + \frac{1}{r}\frac{df}{d\theta} \qquad (11.31)$$

Comparing Equations (11.27) and (11.31), clearly

$$\frac{df}{d\theta} = 0 \quad \Longrightarrow \quad f(\theta) = \text{constant} \qquad (11.32)$$

With no loss of generality, we can set the constant equal to zero. Therefore, the velocity potential for this flow is

$$\phi(r, \theta) = Qr^n(\cos n\theta - \sin n\theta) \qquad (11.33)$$

11.5 Bernoulli's Equation

Statement of the Problem: *Determine the pressure induced by a potential vortex in an otherwise motionless vat of mercury at 20° C, assuming $p \to p_\infty$ as $r \to \infty$. If the farfield pressure is $p_\infty = 100$ kPa and the vortex strength is $\Gamma = 10$ m²/sec, what is the minimum value of r for which the solution is physically meaningful?*

Solution: For a potential vortex, we know that

$$\phi(r, \theta) = \frac{\Gamma}{2\pi}\theta \qquad (11.34)$$

Hence, the velocity components are

$$u_r = \frac{\partial \phi}{\partial r} = 0 \quad \text{and} \quad u_\theta = \frac{1}{r}\frac{\partial \phi}{\partial \theta} = \frac{\Gamma}{2\pi r} \qquad (11.35)$$

So, using these velocities in Bernoulli's equation,

$$\frac{p}{\rho} + \frac{1}{2}(u_r^2 + u_\theta^2) = \frac{p}{\rho} + \frac{1}{2}\left(\frac{\Gamma}{2\pi r}\right)^2 = \frac{p_\infty}{\rho} \qquad (11.36)$$

Therefore, the pressure is given by

$$p = p_\infty - \frac{1}{2}\rho\left(\frac{\Gamma}{2\pi r}\right)^2 \qquad (11.37)$$

11.6. SUPERPOSITION OF FUNDAMENTAL SOLUTIONS I

Clearly, the solution is meaningless when $p < 0$, which can occur as r approaches zero. Hence, the minimum value of r corresponds to the point where $p = 0$, viz.,

$$0 = p_\infty - \frac{1}{2}\rho\left(\frac{\Gamma}{2\pi r_{min}}\right)^2 \implies r_{min} = \frac{\Gamma}{2\pi}\sqrt{\frac{\rho}{2p_\infty}} \qquad (11.38)$$

For mercury at 20° C, Table A.2 shows that its density is $\rho = 13550$ kg/m^3. Also, the pressure as $r \to \infty$ is $p_\infty = 100$ kPa $= 10^5$ kg/(m·sec^2). Thus, for the given values,

$$r_{min} = \frac{10 \text{ m}^2/\text{sec}}{2\pi}\sqrt{\frac{13550 \text{ kg/m}^3}{2\,[10^5 \text{ kg}/(\text{m}\cdot\text{sec}^2)]}} = 0.41 \text{ m} \qquad (11.39)$$

11.6 Superposition of Fundamental Solutions I

Statement of the Problem: *Determine the equipotentials for a spiral vortex, which is the superposition of a point source of strength Q and a potential vortex of strength Γ. Plot a streamline for $\Gamma = -\frac{1}{2}Q$.*

Solution: For the superposition of a potential vortex and a point source, the velocity potential is

$$\phi(r,\theta) = \frac{Q}{2\pi}\ell n\, r + \frac{\Gamma}{2\pi}\theta \qquad (11.40)$$

So, regarding ϕ as a constant, the equipotentials are given by

$$\frac{Q}{2\pi}\ell n\, r = \phi - \frac{\Gamma}{2\pi}\theta \implies \ell n\, r = \frac{2\pi\phi}{Q} - \frac{\Gamma\theta}{Q} \qquad (11.41)$$

Therefore, the equation of an equipotential is

$$r = r_o e^{-\Gamma\theta/Q}, \qquad r_o \equiv e^{2\pi\phi/Q} \qquad (11.42)$$

For $Q = -2\Gamma$, we have

$$\frac{r}{r_o} = e^{\theta/2}, \qquad r_o \equiv e^{-\pi\phi/\Gamma} \qquad (11.43)$$

Figure 11.1 displays this equipotential for $0 \le \theta \le 4\pi$.

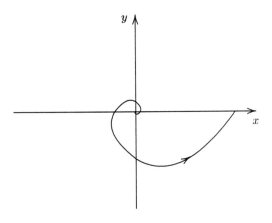

Figure 11.1: *Equipotential for a spiral vortex.*

11.7 Superposition of Fundamental Solutions II

Statement of the Problem: *If we superimpose a uniform flow of velocity U with a doublet of strength D, a point vortex of strength Γ and a point source of strength Q, the farfield solution corresponds to a two-dimensional lifting body that emits a stream of fluid. The emitted stream could represent the exhaust of a jet engine. State the streamfunction and velocity potential for this flow, noting that the lift should be upward as indicated in Figure 11.2.*

(a) *If there is a stagnation point at $r = R$ and $\theta = 210°$, how are the vortex, source and doublet strengths related?*

(b) *Suppose now that the body is a circular cylinder with a doublet strength $\mathcal{D} = 2\pi U R^2$. Including an additive constant in the streamfunction, explain why $\psi(R, \theta) = 0$ no longer corresponds to the cylinder surface. Also, verify that $\psi(R, \pi) - \psi(R, 0) = \frac{1}{2}Q$.*

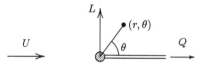

Figure 11.2: *Flow past a lifting body with a jet engine; L = lift.*

Solution: The sign of the vortex term must correspond to clockwise vorticity in order to have positive lift, i.e., according to the *Kutta-Joukowski lift theorem*, the lift is $L = -\rho U \Gamma$. Hence, we choose $\Gamma = -\Gamma_a$, where $\Gamma_a > 0$ is the aerodynamic circulation. So, using superposition, the streamfunction and velocity potential for this flow are

$$\psi(r,\theta) = \underbrace{U r \sin\theta}_{Uniform\ flow} + \underbrace{\frac{\Gamma_a}{2\pi}\ell n r}_{Vortex} + \underbrace{\frac{Q}{2\pi}\theta}_{Source} - \underbrace{\frac{\mathcal{D}}{2\pi r}\sin\theta}_{Doublet} \quad (11.44)$$

$$\phi(r,\theta) = \underbrace{U r \cos\theta}_{Uniform\ flow} - \underbrace{\frac{\Gamma_a}{2\pi}\theta}_{Vortex} + \underbrace{\frac{Q}{2\pi}\ell n r}_{Source} + \underbrace{\frac{\mathcal{D}}{2\pi r}\cos\theta}_{Doublet} \quad (11.45)$$

11.7(a): Working with the streamfunction, the velocity components for this flow are

$$u_r = \frac{1}{r}\frac{\partial \psi}{\partial \theta} = U\cos\theta + \frac{Q}{2\pi r} - \frac{\mathcal{D}}{2\pi r^2}\cos\theta \quad (11.46)$$

$$u_\theta = -\frac{\partial \psi}{\partial r} = -U\sin\theta - \frac{\Gamma_a}{2\pi r} - \frac{\mathcal{D}}{2\pi r^2}\sin\theta \quad (11.47)$$

When $\theta = 210°$, the trigonometric functions are

$$\sin\theta = -\frac{1}{2} \quad \text{and} \quad \cos\theta = -\frac{\sqrt{3}}{2} \quad (11.48)$$

So, for $r = R$ and $\theta = 210°$, the velocity components are

$$u_r = -\frac{\sqrt{3}}{2}U + \frac{Q}{2\pi R} + \frac{\sqrt{3}}{2}\frac{\mathcal{D}}{2\pi R^2} \quad (11.49)$$

$$u_\theta = \frac{1}{2}U - \frac{\Gamma_a}{2\pi R} + \frac{1}{2}\frac{\mathcal{D}}{2\pi R^2} \quad (11.50)$$

11.8. FLOW PAST A ROTATING CYLINDER I

Therefore, since $u_r = u_\theta = 0$ at a stagnation point, we arrive at the following two equations relating U, Q, Γ_a and $\mathcal{D}$.

$$U - \frac{Q}{\sqrt{3}\,\pi R} - \frac{\mathcal{D}}{2\pi R^2} = 0 \qquad (11.51)$$

$$U - \frac{\Gamma_a}{\pi R} + \frac{\mathcal{D}}{2\pi R^2} = 0 \qquad (11.52)$$

Subtracting Equation (11.52) from Equation (11.51), we find

$$-\frac{Q}{\sqrt{3}\,\pi R} + \frac{\Gamma_a}{\pi R} - \frac{\mathcal{D}}{\pi R^2} = 0 \implies Q = \sqrt{3}\,(\Gamma_a - \mathcal{D}/R) \qquad (11.53)$$

11.7(b): Now, substituting $\mathcal{D} = 2\pi U R^2$ into Equation (11.44), the streamfunction becomes

$$\psi(r,\theta) = Ur\left(1 - \frac{R^2}{r^2}\right)\sin\theta + \frac{\Gamma_a}{2\pi}\ell n\, r + \frac{Q}{2\pi}\theta + \text{constant} \qquad (11.54)$$

Now, when $r = R$, we have

$$\psi(R,\theta) = \underbrace{\frac{\Gamma_a}{2\pi}\ell n\, R}_{=\text{constant}} + \frac{Q}{2\pi}\theta + \text{constant} = \frac{Q}{2\pi}\theta + \text{constant} \qquad (11.55)$$

Thus, $\psi(R,\theta)$ varies linearly with θ, so that $\psi = 0$ no longer represents the surface of the cylinder. Also, from Equation (11.55), we find

$$\psi(R,\pi) - \psi(R,0) = \frac{Q}{2\pi}\pi = \frac{Q}{2} \qquad (11.56)$$

This result reflects the fact that fluid is being injected into the flow, half of which flows above the x axis, and half below. The difference between two values of ψ is equal to the volume flux between the two streamlines, which in this flow is exactly $Q/2$.

11.8 Flow Past a Rotating Cylinder I

Statement of the Problem: *A circular cylinder of radius $R = 10$ inches is placed in a wind tunnel with a test-section velocity of $U = 10$ ft/sec. The cylinder rotates in the counterclockwise direction about its longitudinal axis with an angular velocity, Ω, which can be varied. Using the potential-flow solution, determine the angular velocities corresponding to critical flow, subcritical flow with $\Gamma = \frac{1}{2}\Gamma_c$ and supercritical flow with $\Gamma = 2\Gamma_c$. Express your answers in rpm. Also, determine the location of the stagnation points for each case.*

Solution: In general, as depicted in Figure 11.3, the flow is critical when $\Gamma = \Gamma_c \equiv 4\pi RU$. Now, from the potential-flow solution for flow past a rotating cylinder, the circumferential velocity at the cylinder surface is

$$u_\theta(R,\theta) = -2U\sin\theta + \frac{\Gamma}{2\pi R} \qquad (11.57)$$

If the cylinder rotates with an angular velocity of Ω, the velocity due to the rotation at a radius R is $\Delta u_\theta = \Omega R$. Therefore, identifying the second term on the right-hand side of Equation (11.57) as Δu_θ, the angular velocity of the spinning cylinder is

$$\Omega = \frac{\Gamma}{2\pi R^2} \qquad (11.58)$$

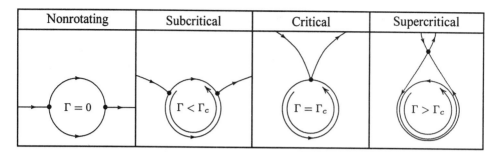

Figure 11.3: *Dividing streamlines for rotating-cylinder flow;* $\Gamma_c = 4\pi RU$ *is the critical circulation;* • *denotes a stagnation point.*

Critical Flow: For this case, there is a single stagnation point at $r = R$ and $\theta = 90°$ (see Figure 11.3). The angular velocity for this case is

$$\Omega_c = \frac{\Gamma_c}{2\pi R^2} = \frac{4\pi RU}{2\pi R^2} = 2\frac{U}{R} \tag{11.59}$$

Substituting the given values, we find

$$\Omega_c = 2\left(\frac{10 \text{ ft/sec}}{10/12 \text{ ft}}\right)\left(\frac{60}{2\pi}\frac{\text{rpm}}{\text{sec}^{-1}}\right) = 229 \text{ rpm} \tag{11.60}$$

Subcritical Flow: For this case with $\Gamma = \frac{1}{2}\Gamma_c$, necessarily $\Omega = \frac{1}{2}\Omega_c$. Therefore, calling this value Ω_{sub}, we have

$$\Omega_{sub} = 115 \text{ rpm} \tag{11.61}$$

From the potential-flow solution, there are two stagnation points (see Figure 11.3) located at

$$r = R, \quad \theta = \sin^{-1}\left(\frac{\Gamma}{4\pi RU}\right) \tag{11.62}$$

Using the given values for this case, since $\Gamma = \frac{1}{2}\Gamma_c = 2\pi RU$, we have

$$\theta = \sin^{-1}\left(\frac{1}{2}\right) = 30° \text{ and } 150° \tag{11.63}$$

Supercritical Flow: For this case with $\Gamma = 2\Gamma_c$, we have $\Omega = 2\Omega_c$. Calling this value Ω_{sup}, we have

$$\Omega_{sup} = 458 \text{ rpm} \tag{11.64}$$

As shown in Figure 11.3, there is a single stagnation point at $\theta = 90°$. The solution for supercritical flow tells us the stagnation point is located at

$$r = \frac{\Gamma}{4\pi U}\left[1 + \sqrt{1 - \left(\frac{4\pi RU}{\Gamma}\right)^2}\right] \tag{11.65}$$

So, since $\Gamma = 2\Gamma_c = 8\pi RU$, we have

$$r = 2R\left[1 + \sqrt{1 - \left(\frac{1}{2}\right)^2}\right] = 3.73R \tag{11.66}$$

11.9 Flow Past a Rotating Cylinder II

Statement of the Problem: *An incompressible fluid of density ρ and velocity U flows past a rotating circular cylinder of diameter D as shown in Figure 11.4. The cylinder develops lift, L, in the direction shown. In what direction is the cylinder rotating? If the lift coefficient is*

$$C_L \equiv \frac{L}{\frac{1}{2}\rho U^2 D} = 2\pi$$

what is the circulation? Remember that Γ is positive for counterclockwise rotation.

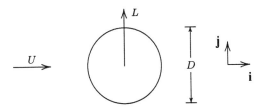

Figure 11.4: *Incompressible flow past a rotating cylinder.*

Solution: We can determine the direction of rotation of the cylinder by appealing to the *Kutta-Joukowski lift theorem*, which tells us that the lift force vector, **L**, is

$$\mathbf{L} = \rho \mathbf{U} \times \mathbf{\Gamma} \qquad (11.67)$$

Now, in terms of the unit vectors **i** and **j** indicated in Figure 11.4, the lift and velocity vectors are

$$\mathbf{L} = L\mathbf{j} \quad \text{and} \quad \mathbf{U} = U\mathbf{i} \qquad (11.68)$$

By inspection, the circulation vector, $\mathbf{\Gamma}$, must point into the page, i.e.,

$$\mathbf{\Gamma} = -\Gamma \mathbf{k} \qquad (11.69)$$

Since positive circulation corresponds to counterclockwise direction (i.e., the direction in which the cylindrical coordinate θ increases), and since the circulation is negative, the cylinder must be rotating in the **clockwise** direction. With the understanding that Γ is negative, the lift on the cylinder is

$$L = -\rho U \Gamma \qquad (11.70)$$

Thus, the lift coefficient becomes

$$C_L \equiv \frac{L}{\frac{1}{2}\rho U^2 D} = -\frac{\rho U \Gamma}{\frac{1}{2}\rho U^2 D} = -\frac{2\Gamma}{UD} \qquad (11.71)$$

Finally, using the fact that $C_L = 2\pi$, we find

$$\Gamma = -\pi U D \qquad (11.72)$$

11.10 Unsteady Potential Flow

Statement of the Problem: *Consider a point source whose strength, $Q(t)$, varies with time. The source is immersed in an infinite fluid of density ρ. The pressure far from the point source (i.e., as $r \to \infty$) is p_∞. Determine the pressure as a function of p_∞, ρ, $Q(t)$ and the coordinates r and θ. Ignore body forces.*

Solution: Since the flow is unsteady, the pressure is given by the generalized Bernoulli equation, viz.,

$$\frac{\partial \phi}{\partial t} + \frac{p}{\rho} + \frac{1}{2}\mathbf{u}\cdot\mathbf{u} = F(t) \qquad (11.73)$$

where $F(t)$ is a function that must be determined from the boundary conditions. Now, the velocity potential for the point source is

$$\phi(r,\theta,t) = \frac{Q(t)}{2\pi}\theta \qquad (11.74)$$

Hence, we have

$$\frac{\partial \phi}{\partial t} = \frac{\dot{Q}(t)}{2\pi}\theta \qquad (11.75)$$

where $\dot{Q} \equiv dQ/dt$. Also, the velocity components are

$$u_r = \frac{\partial \phi}{\partial r} = 0 \quad \text{and} \quad u_\theta = \frac{1}{r}\frac{\partial \phi}{\partial \theta} = \frac{Q(t)}{2\pi r} \qquad (11.76)$$

Combining Equations (11.73), (11.75) and (11.76) yields

$$\frac{\dot{Q}(t)}{2\pi}\theta + \frac{p}{\rho} + \frac{Q^2(t)}{8\pi^2 r^2} = F(t) \qquad (11.77)$$

Finally, to determine the function $F(t)$, we need a suitable reference point. Clearly, since we are given $p \to p_\infty$ as $r \to \infty$, we can use points distant from the source. Inspection of Equation (11.77) shows that for $\theta = 0$, all terms on the left-hand side are known, and are independent of r and θ (their sum can be a function only of t). Hence,

$$0 + \frac{p_\infty}{\rho} + 0 = F(t) \quad \Longrightarrow \quad F(t) = \frac{p_\infty}{\rho} \qquad (11.78)$$

Combining Equations (11.77) and (11.78) and rearranging terms gives the pressure, viz.,

$$p(r,\theta,t) = p_\infty - \frac{\rho \dot{Q}(t)}{2\pi}\theta - \frac{\rho Q^2(t)}{8\pi^2 r^2} \qquad (11.79)$$

11.11 Linearized Airfoil Theory

Statement of the Problem: *Consider a vortex sheet with strength*

$$\gamma(\xi) = \gamma_o\left(\frac{\xi}{c} - \frac{1}{2}\right)\sqrt{\frac{\xi}{c}\left(1 - \frac{\xi}{c}\right)}\,, \qquad \gamma_o = \text{constant}$$

Find the shape of the camber line for an airfoil simulated by superimposing this vortex sheet with a uniform flow of velocity $\mathbf{U} = \mathbf{i}\,U\cos\alpha + \mathbf{j}\,U\sin\alpha$, proceeding as follows.

11.11. LINEARIZED AIRFOIL THEORY

(a) Making the change of variables defined by

$$\xi = \frac{c}{2}(1 + \cos\theta), \qquad x = \frac{c}{2}(1 + \cos\theta')$$

determine the differential equation for the camber line, $C(x)$.

(b) Using the fact that for any integer $n \geq 0$,

$$\int_0^\pi \frac{\cos n\theta\, d\theta}{\cos\theta - \cos\theta'} = \pi \frac{\sin n\theta'}{\sin\theta'}$$

determine the camber line. Set $C(0) = 0$ to eliminate any constant of integration.

Solution: Linear-airfoil theory tells us the equation for the camber line of an airfoil is given by

$$2U \cos\alpha \frac{dC}{dx} - 2U \sin\alpha = -\frac{1}{\pi} \int_0^c \frac{\gamma(\xi)\, d\xi}{x - \xi} \tag{11.80}$$

11.11(a): Making the change of variables given in the problem statement,

$$\xi = \frac{c}{2}(1 + \cos\theta), \qquad x = \frac{c}{2}(1 + \cos\theta') \tag{11.81}$$

we have

$$x - \xi = \frac{c}{2}(\cos\theta' - \cos\theta), \qquad d\xi = -\frac{c}{2}\sin\theta\, d\theta \tag{11.82}$$

Also, the limits of the integral transform according to:

$$\xi = 0 \implies \theta = \pi, \qquad \xi = c \implies \theta = 0 \tag{11.83}$$

The quantity under the square root in the prescribed vortex-sheet strength, $\gamma(\xi)$, becomes

$$(\xi/c)(1 - \xi/c) = \frac{1}{2}(1 + \cos\theta)\frac{1}{2}(1 - \cos\theta) = \frac{1}{4}(1 - \cos^2\theta) = \frac{1}{4}\sin^2\theta \tag{11.84}$$

Also, we have

$$\frac{\xi}{c} - \frac{1}{2} = \frac{1}{2}(1 + \cos\theta) - \frac{1}{2} = \frac{1}{2}\cos\theta \tag{11.85}$$

This means the product of the vortex sheet strength and $d\xi$ is

$$\begin{aligned}
\gamma(\xi)d\xi &= \gamma_o \left(\frac{1}{2}\cos\theta\right)\left(\frac{1}{2}\sin\theta\right)\left(-\frac{c}{2}\sin\theta\, d\theta\right) \\
&= -\frac{\gamma_o c}{8}\sin^2\theta \cos\theta\, d\theta = -\frac{\gamma_o c}{8}(1 - \cos^2\theta)\cos\theta\, d\theta \\
&= -\frac{\gamma_o c}{8}(\cos\theta - \cos^3\theta)\, d\theta
\end{aligned} \tag{11.86}$$

Now, we can make use of the following trigonometric identity.

$$\cos^3\theta = \frac{1}{4}(3\cos\theta + \cos 3\theta) \tag{11.87}$$

Combining Equations (11.86) and (11.87) gives

$$\gamma(\xi)d\xi = -\frac{\gamma_o c}{8}\left(\cos\theta - \frac{3}{4}\cos\theta - \frac{1}{4}\cos 3\theta\right)d\theta = -\frac{\gamma_o c}{32}(\cos\theta - \cos 3\theta)\, d\theta \tag{11.88}$$

Finally, using Equation (11.82), there follows

$$\frac{\gamma(\xi)d\xi}{x-\xi} = -\frac{\gamma_o c}{32}\frac{2}{c}\frac{(\cos\theta - \cos 3\theta)}{(\cos\theta' - \cos\theta)}d\theta = \frac{\gamma_o}{16}\frac{(\cos\theta - \cos 3\theta)}{(\cos\theta - \cos\theta')}d\theta \qquad (11.89)$$

Substituting Equation (11.89) into Equation (11.80) yields

$$2U\cos\alpha\frac{dC}{dx} - 2U\sin\alpha = -\frac{\gamma_o}{16\pi}\int_\pi^0 \frac{(\cos\theta - \cos 3\theta)d\theta}{(\cos\theta - \cos\theta')} \qquad (11.90)$$

So, dividing through by $2U\cos\alpha$, the camber function, $C(x)$, satisfies the following differential equation.

$$\frac{dC}{dx} = \tan\alpha + \frac{\gamma_o}{32\pi U\cos\alpha}\int_0^\pi \frac{(\cos\theta - \cos 3\theta)d\theta}{(\cos\theta - \cos\theta')} \qquad (11.91)$$

11.11(b): Noting that

$$\int_0^\pi \frac{\cos n\theta\, d\theta}{\cos\theta - \cos\theta'} = \pi\frac{\sin n\theta'}{\sin\theta'} \qquad (11.92)$$

for any integer $n \geq 0$, there follows

$$\begin{aligned}\frac{dC}{dx} &= \tan\alpha + \frac{\gamma_o}{32\pi U\cos\alpha}\left[\pi\frac{\sin\theta'}{\sin\theta'} - \pi\frac{\sin 3\theta'}{\sin\theta'}\right] \\ &= \tan\alpha + \frac{\gamma_o}{32U\cos\alpha}\left(1 - \frac{\sin 3\theta'}{\sin\theta'}\right)\end{aligned} \qquad (11.93)$$

We can now make use of another trigonometric identity, viz.,

$$\sin 3\theta' = 3\sin\theta' - 4\sin^3\theta' \qquad (11.94)$$

Therefore, the differential equation for $C(x)$ becomes

$$\begin{aligned}\frac{dC}{dx} &= \tan\alpha + \frac{\gamma_o}{32U\cos\alpha}\left[1 - 3 + 4\sin^2\theta'\right] \\ &= \tan\alpha + \frac{\gamma_o}{16U\cos\alpha}\left[2\sin^2\theta' - 1\right] \\ &= \tan\alpha + \frac{\gamma_o}{16U\cos\alpha}\left[1 - 2\cos^2\theta'\right]\end{aligned} \qquad (11.95)$$

But, the inverse of the transformation of Part (a) tells us

$$x = \frac{c}{2}(1 + \cos\theta') \implies \cos\theta' = 2\frac{x}{c} - 1 \qquad (11.96)$$

Thus, the differential equation for $C(x)$ simplifies to

$$\frac{dC}{dx} = \tan\alpha + \frac{\gamma_o}{16U\cos\alpha}\left[1 - 8\left(\frac{x}{c} - \frac{1}{2}\right)^2\right] \qquad (11.97)$$

Integrating over x yields

$$C(x) = x\tan\alpha + \frac{\gamma_o}{16U\cos\alpha}\left[x - \frac{8}{3}c\left(\frac{x}{c} - \frac{1}{2}\right)^3\right] + C_o \qquad (11.98)$$

11.12. CFD—RICHARDSON EXTRAPOLATION

where C_o is a constant of integration. To satisfy the given boundary condition, we have

$$C(0) = 0 \implies C_o = -\frac{\gamma_o c}{48 U \cos \alpha} \qquad (11.99)$$

Therefore, the equation of the camber line is

$$C(x) = x \tan \alpha + \frac{\gamma_o c}{48 U \cos \alpha} \left[3\frac{x}{c} - 1 - 8\left(\frac{x}{c} - \frac{1}{2}\right)^3 \right] \qquad (11.100)$$

11.12 CFD—Richardson Extrapolation

Statement of the Problem: *Consider the function $\phi(y) = y^6 - 3y^2 + 4$. Use the second-order accurate finite-difference approximation to compute $d^2\phi/dy^2$ at $y = 1$ for $\Delta y = 0.10, 0.20, 0.25$ and 0.50. Compare the numerical results with the exact value in tabular form, and use Richardson extrapolation to estimate the error for $\Delta y = 0.10$ and $\Delta y = 0.25$.*

Solution: First, note that the exact second derivative of $\phi(y) = y^6 - 3y^2 + 4$ is

$$\frac{d\phi}{dy} = 30y^4 - 6 \qquad (11.101)$$

Hence, when $y = 1$, the exact second derivative is

$$\phi''(1) = 24 \qquad (11.102)$$

Now, using the standard second-order accurate central-difference approximation, the numerical approximation to the second derivative at $y = 1$ is

$$\frac{d^2\phi}{dy^2} \approx \frac{\phi_{j+1} - 2\phi_{i,j} + \phi_{j-1}}{(\Delta y)^2} \qquad (11.103)$$

For the given function, we have

$$\begin{aligned}\phi''(1) &= \frac{(1+\Delta y)^6 - 3(1+\Delta y)^2 + 4 - 2(1^6 - 3 \cdot 1^2 + 4) + (1-\Delta y)^6 - 3(1-\Delta y)^2 + 4}{(\Delta y)^2} \\ &= \frac{(1+\Delta y)^6 - 3(1+\Delta y)^2 + (1-\Delta y)^6 - 3(1-\Delta y)^2 + 4}{(\Delta y)^2}\end{aligned} \qquad (11.104)$$

The table below compares the discretized value of $\phi''(1)$ for $\Delta y = 0.10, 0.20, 0.25$ and 0.50.

Δy	$\phi''(1)$
0	24.0000
0.10	24.3002
0.20	25.2032
0.25	25.8828
0.50	31.6250

The estimated errors for the $\Delta y = 0.10$ and $\Delta y = 0.25$ results according to Richardson extrapolation are

$$E(0.10) \approx \frac{1}{3}(\phi''_{0.10} - \phi''_{0.20}) = -0.301 \qquad (11.105)$$

$$E(0.25) \approx \frac{1}{3}(\phi''_{0.25} - \phi''_{0.50}) = -1.914 \qquad (11.106)$$

Observe that the sum of the numerical value for $\Delta y = 0.10$ and the Richardson-extrapolation error for ϕ'' yields 23.9992, which is within 0.003% of the exact value. A similar calculation for $\Delta y = 0.25$ yields 23.9688, which is within 0.13% of the exact value.

Chapter 12

Viscous Effects

Chapter 12 addresses development of the Navier-Stokes equation, including analysis of the strain-rate tensor and the viscous-stress tensor. While Chapter 4 addresses kinematics, it focuses on differences between the Lagrangian and Eulerian descriptions and the Reynolds Transport Theorem. Here, we take a closer look at kinematics, which shows that general motion and deformation of a fluid particle can be described in terms of the vorticity and the strain-rate tensor.

Also, preceding chapters deal with inviscid fluids for which only normal stresses (pressure) act on a given surface. For a viscous fluid, we must deal with oblique stresses as represented by the combined effects of both pressure and viscous stresses. Thus, we must also formulate the stress tensor. For Newtonian fluids, the viscous-stress tensor is proportional to the strain-rate tensor, which is the Stokes hypothesis. Combining this relation with Navier's dynamical equation of motion yields the famous Navier-Stokes equation. This chapter includes the following problems.

Section 12.1 Kinematics: A problem that focuses on basic kinematical considerations, including calculation of the strain-rate tensor and the vorticity vector.

Section 12.2 Strain-Rate Tensor: The solution to this problem gives the strain-rate tensor for a point source and a potential vortex.

Section 12.3 Viscous-Stress Tensor: The problem in this section involves the viscous-stress tensor and its divergence, which appears in the Navier-Stokes equation.

Section 12.4 Navier-Stokes Equation: An exercise that determines whether or not a given velocity field satisfies the Navier-Stokes equation, including determination of the pressure. The analysis is done in cylindrical coordinates.

Section 12.5 CFD—Stability Analysis: The problem in this section examines the stability of a finite-difference algorithm using von Neumann stability analysis.

12.1 Kinematics

Statement of the Problem: *Consider the following velocity vector.*

$$\mathbf{u} = \frac{U}{L^2}[2Kyz\,\mathbf{i} - 3xz\,\mathbf{j} - 4Cxy\,\mathbf{k}]$$

where U and L are constant velocity and length scales, respectively. Also, C and K are dimensionless constants. Determine the values of C and K that correspond to irrotational flow. Using these values, compute the strain-rate tensor.

Solution: In order for the flow to be irrotational, the vorticity must vanish. Thus, letting $\boldsymbol{\omega} = \nabla \times \mathbf{u}$ denote the vorticity, we proceed as follows in order to determine the conditions on C and K.

$$\begin{aligned}\boldsymbol{\omega} &= \frac{U}{L^2}\begin{vmatrix}\mathbf{i} & \mathbf{j} & \mathbf{k}\\ \frac{\partial}{\partial x} & \frac{\partial}{\partial y} & \frac{\partial}{\partial z}\\ 2Kyz & -3xz & -4Cxy\end{vmatrix} = \mathbf{i}\frac{U}{L^2}\left[\frac{\partial}{\partial y}(-4Cxy) - \frac{\partial}{\partial z}(-3xz)\right]\\ &\quad + \mathbf{j}\frac{U}{L^2}\left[\frac{\partial}{\partial z}(2Kyz) - \frac{\partial}{\partial x}(-4Cxy)\right] + \mathbf{k}\frac{U}{L^2}\left[\frac{\partial}{\partial x}(-3xz) - \frac{\partial}{\partial y}(2Kyz)\right]\\ &= \frac{U}{L^2}[(3-4C)x\,\mathbf{i} + 2(K+2C)y\,\mathbf{j} - (3+2K)z\,\mathbf{k}]\end{aligned} \qquad (12.1)$$

In order for the vorticity to be zero, all three components must be zero, which will be true only if

$$3 - 4C = 0, \quad K + 2C = 0, \quad 3 + 2K = 0 \qquad (12.2)$$

In general this system of equations is overdetermined as we have three equations for two unknowns. However, the equations are linearly dependent (half the sum of the first and third equations equals the second), and the solution is

$$C = \frac{3}{4} \quad \text{and} \quad K = -\frac{3}{2} \qquad (12.3)$$

In general, the strain-rate tensor, $[\mathbf{S}]$, is represented by the following 3 by 3 matrix.

$$[\mathbf{S}] = \begin{bmatrix} \frac{\partial u}{\partial x} - \nabla\cdot\mathbf{u} & \frac{1}{2}\left(\frac{\partial v}{\partial x} + \frac{\partial u}{\partial y}\right) & \frac{1}{2}\left(\frac{\partial w}{\partial x} + \frac{\partial u}{\partial z}\right)\\ \frac{1}{2}\left(\frac{\partial u}{\partial y} + \frac{\partial v}{\partial x}\right) & \frac{\partial v}{\partial y} - \nabla\cdot\mathbf{u} & \frac{1}{2}\left(\frac{\partial w}{\partial y} + \frac{\partial v}{\partial z}\right)\\ \frac{1}{2}\left(\frac{\partial u}{\partial z} + \frac{\partial w}{\partial x}\right) & \frac{1}{2}\left(\frac{\partial v}{\partial z} + \frac{\partial w}{\partial y}\right) & \frac{\partial w}{\partial z} - \nabla\cdot\mathbf{u} \end{bmatrix} \qquad (12.4)$$

Now, for the flow at hand, using the values of C and K computed above, the velocity components are

$$u = -\frac{3Uyz}{L^2}, \quad v = -\frac{3Uxz}{L^2}, \quad w = -\frac{3Uxy}{L^2} \qquad (12.5)$$

12.2. STRAIN-RATE TENSOR

Therefore, the velocity derivatives are as follows.

$$\frac{\partial u}{\partial x} = 0, \qquad \frac{\partial v}{\partial x} = -\frac{3Uz}{L^2}, \qquad \frac{\partial w}{\partial x} = -\frac{3Uy}{L^2}$$

$$\frac{\partial u}{\partial y} = -\frac{3Uz}{L^2}, \qquad \frac{\partial v}{\partial y} = 0, \qquad \frac{\partial w}{\partial y} = -\frac{3Ux}{L^2} \qquad (12.6)$$

$$\frac{\partial u}{\partial z} = -\frac{3Uy}{L^2}, \qquad \frac{\partial v}{\partial z} = -\frac{3Ux}{L^2}, \qquad \frac{\partial w}{\partial z} = 0$$

Also, the divergence of the velocity is

$$\nabla \cdot \mathbf{u} = \frac{\partial u}{\partial x} + \frac{\partial v}{\partial y} + \frac{\partial w}{\partial z} = 0 \qquad (12.7)$$

Substituting Equations (12.6) and (12.7) into Equation (12.4) gives

$$[\mathbf{S}] = -\frac{3U}{L^2} \begin{bmatrix} 0 & z & y \\ z & 0 & x \\ y & x & 0 \end{bmatrix} \qquad (12.8)$$

12.2 Strain-Rate Tensor

Statement of the Problem: *Determine the strain-rate tensor for the following incompressible flows:*

(a) *Point source, $u_r = Q/(2\pi r), u_\theta = 0, w = 0$;*

(b) *Potential vortex, $u_r = 0, u_\theta = \Gamma/(2\pi r), w = 0$.*

where Q and Γ are constant source and vortex strength, respectively..

Solution: In cylindrical coordinates, the strain-rate tensor is

$$[\mathbf{S}] = \begin{bmatrix} \frac{\partial u_r}{\partial r} & \frac{1}{2}\left(\frac{\partial u_\theta}{\partial r} - \frac{u_\theta}{r} + \frac{1}{r}\frac{\partial u_r}{\partial \theta}\right) & \frac{1}{2}\left(\frac{\partial w}{\partial r} + \frac{\partial u_r}{\partial z}\right) \\ \frac{1}{2}\left(\frac{1}{r}\frac{\partial u_r}{\partial \theta} + \frac{\partial u_\theta}{\partial r} - \frac{u_\theta}{r}\right) & \frac{1}{r}\frac{\partial u_\theta}{\partial \theta} + \frac{u_r}{r} & \frac{1}{2}\left(\frac{1}{r}\frac{\partial w}{\partial \theta} + \frac{\partial u_\theta}{\partial z}\right) \\ \frac{1}{2}\left(\frac{\partial u_r}{\partial z} + \frac{\partial w}{\partial r}\right) & \frac{1}{2}\left(\frac{\partial u_\theta}{\partial z} + \frac{1}{r}\frac{\partial w}{\partial \theta}\right) & \frac{\partial w}{\partial z} \end{bmatrix} \qquad (12.9)$$

Since both cases of interest here are planar with $w = 0$, the strain-rate tensor simplifies immediately to the following.

$$[\mathbf{S}] = \begin{bmatrix} \frac{\partial u_r}{\partial r} & \frac{1}{2}\left(\frac{\partial u_\theta}{\partial r} - \frac{u_\theta}{r} + \frac{1}{r}\frac{\partial u_r}{\partial \theta}\right) & 0 \\ \frac{1}{2}\left(\frac{1}{r}\frac{\partial u_r}{\partial \theta} + \frac{\partial u_\theta}{\partial r} - \frac{u_\theta}{r}\right) & \frac{1}{r}\frac{\partial u_\theta}{\partial \theta} + \frac{u_r}{r} & 0 \\ 0 & 0 & 0 \end{bmatrix} \qquad (12.10)$$

12.2(a): For the point source, the only nonzero velocity derivative is

$$\frac{\partial u_r}{\partial r} = -\frac{Q}{2\pi r^2} \qquad (12.11)$$

Hence, the only two nonzero strain-rate components are

$$S_{rr} = \frac{\partial u_r}{\partial r} = -\frac{Q}{2\pi r^2} \quad \text{and} \quad S_{\theta\theta} = \frac{u_r}{r} = \frac{Q}{2\pi r^2} \qquad (12.12)$$

Therefore, the strain-rate tensor for the point source is

$$[\mathbf{S}] = \begin{bmatrix} -\frac{Q}{2\pi r^2} & 0 & 0 \\ 0 & \frac{Q}{2\pi r^2} & 0 \\ 0 & 0 & 0 \end{bmatrix} = -\frac{Q}{2\pi r^2} \begin{bmatrix} 1 & 0 & 0 \\ 0 & -1 & 0 \\ 0 & 0 & 0 \end{bmatrix} \qquad (12.13)$$

12.2(b): For the potential vortex, the only nonzero velocity derivative is

$$\frac{\partial u_\theta}{\partial r} = -\frac{\Gamma}{2\pi r^2} \qquad (12.14)$$

Hence, the only two nonzero strain-rate components are

$$S_{r\theta} = S_{\theta r} = \frac{1}{2}\left(\frac{\partial u_\theta}{\partial r} - \frac{U_\theta}{r}\right) = -\frac{\Gamma}{2\pi r^2} \qquad (12.15)$$

Therefore, the strain-rate tensor for the potential vortex is

$$[\mathbf{S}] = \begin{bmatrix} 0 & -\frac{\Gamma}{2\pi r^2} & 0 \\ -\frac{\Gamma}{2\pi r^2} & 0 & 0 \\ 0 & 0 & 0 \end{bmatrix} = -\frac{\Gamma}{2\pi r^2} \begin{bmatrix} 0 & 1 & 0 \\ 1 & 0 & 0 \\ 0 & 0 & 0 \end{bmatrix} \qquad (12.16)$$

12.3 Viscous-Stress Tensor

Statement of the Problem: *Consider an incompressible flow for which*

$$\mathbf{u} = \frac{U}{L^3}\left(3xy^2\mathbf{i} + 5xz^2\mathbf{j} - 3y^2z\,\mathbf{k}\right)$$

where U and L are constant velocity and length scales, respectively. Determine the viscous-stress tensor, $[\tau]$, and its divergence, $\nabla \cdot [\tau]$.

Solution: In general, for a monatomic gas, the viscous-stress tensor is

$$[\tau] = 2\mu[\mathbf{S}] - \frac{2}{3}\mu(\nabla \cdot \mathbf{u})[\delta] \qquad (12.17)$$

Since the flow is incompressible, the term proportional to $\nabla \cdot \mathbf{u}$ is zero (inspection of the given velocity vector confirms that this is indeed true). Hence, in matrix form, the viscous-stress tensor is

$$\begin{bmatrix} \tau_{xx} & \tau_{xy} & \tau_{xz} \\ \tau_{yx} & \tau_{yy} & \tau_{yz} \\ \tau_{zx} & \tau_{zy} & \tau_{zz} \end{bmatrix} = 2\mu \begin{bmatrix} \frac{\partial u}{\partial x} & \frac{1}{2}\left(\frac{\partial v}{\partial x} + \frac{\partial u}{\partial y}\right) & \frac{1}{2}\left(\frac{\partial w}{\partial x} + \frac{\partial u}{\partial z}\right) \\ \frac{1}{2}\left(\frac{\partial u}{\partial y} + \frac{\partial v}{\partial x}\right) & \frac{\partial v}{\partial y} & \frac{1}{2}\left(\frac{\partial w}{\partial y} + \frac{\partial v}{\partial z}\right) \\ \frac{1}{2}\left(\frac{\partial u}{\partial z} + \frac{\partial w}{\partial x}\right) & \frac{1}{2}\left(\frac{\partial v}{\partial z} + \frac{\partial w}{\partial y}\right) & \frac{\partial w}{\partial z} \end{bmatrix} \qquad (12.18)$$

12.3. VISCOUS-STRESS TENSOR

For the given velocity components, the velocity derivatives are as follows.

$$\frac{\partial u}{\partial x} = \frac{3Uy^2}{L^3}, \quad \frac{\partial v}{\partial x} = \frac{5Uz^2}{L^3}, \quad \frac{\partial w}{\partial x} = 0$$

$$\frac{\partial u}{\partial y} = \frac{6Uxy}{L^3}, \quad \frac{\partial v}{\partial y} = 0, \quad \frac{\partial w}{\partial y} = -\frac{6Uyz}{L^3} \quad (12.19)$$

$$\frac{\partial u}{\partial z} = 0, \quad \frac{\partial v}{\partial z} = \frac{10Uxz}{L^3}, \quad \frac{\partial w}{\partial z} = -\frac{3Uy^2}{L^3}$$

Thus, the viscous-stress tensor components are

$$\tau_{xx} = 2\mu \frac{\partial u}{\partial x} = 6\frac{\mu U y^2}{L^3}, \quad \tau_{yy} = 2\mu \frac{\partial v}{\partial y} = 0, \quad \tau_{zz} = 2\mu \frac{\partial w}{\partial z} = -6\frac{\mu U y^2}{L^3} \quad (12.20)$$

$$\tau_{xy} = \tau_{yx} = \mu \left(\frac{\partial u}{\partial y} + \frac{\partial v}{\partial x} \right) = \frac{\mu U}{L^3} \left(6xy + 5z^2 \right) \quad (12.21)$$

$$\tau_{xz} = \tau_{zx} = \mu \left(\frac{\partial u}{\partial z} + \frac{\partial w}{\partial x} \right) = 0 \quad (12.22)$$

$$\tau_{yz} = \tau_{zy} = \mu \left(\frac{\partial v}{\partial z} + \frac{\partial w}{\partial y} \right) = \frac{\mu U}{L^3} (10xz - 6yz) \quad (12.23)$$

Therefore, the viscous-stress tensor is

$$[\tau] = \frac{\mu U}{L^3} \begin{bmatrix} 6y^2 & (6xy + 5z^2) & 0 \\ (6xy + 5z^2) & 0 & (10xz - 6yz) \\ 0 & (10xz - 6yz) & -6y^2 \end{bmatrix} \quad (12.24)$$

Taking the divergence of $[\tau]$, we have

$$\nabla \cdot [\tau] = \frac{\mu U}{L^3} \begin{bmatrix} \frac{\partial}{\partial x} & \frac{\partial}{\partial y} & \frac{\partial}{\partial z} \end{bmatrix} \begin{bmatrix} 6y^2 & (6xy + 5z^2) & 0 \\ (6xy + 5z^2) & 0 & (10xz - 6yz) \\ 0 & (10xz - 6yz) & -6y^2 \end{bmatrix}$$

$$= \frac{\mu U}{L^3} [6x \quad (6y + 10x - 6y) \quad (-6y)]$$

$$= \frac{\mu U}{L^3} [6x \quad 10x \quad -6y] \quad (12.25)$$

So, in vector form, the divergence of the viscous-stress tensor, which is the viscous force per unit volume acting at a given point in this flow, is

$$\nabla \cdot [\tau] = \frac{2\mu U}{L^3} \left(3x \mathbf{i} + 5x \mathbf{j} - 3y \mathbf{k} \right) \quad (12.26)$$

12.4 Navier-Stokes Equation

Statement of the Problem: *Show that the incompressible flowfield for potential flow into a 90° corner, i.e.,*

$$\psi(r,\theta) = \psi_o r^2 \sin 2\theta$$

with constant ψ_o is an exact solution to the Navier-Stokes (with no body force) and continuity equations in cylindrical coordinates. Compute the pressure, $p(r,\theta)$, in terms of density, ρ, as well as ψ_o, r and θ. What value can p assume at the origin, $r = 0$, and in the farfield, $r \to \infty$? Why is this unsuitable as an "exact" solution for a viscous flow?

Solution: First, note that the velocity components for this flow are

$$u_r = \frac{1}{r}\frac{\partial \psi}{\partial \theta} = 2\psi_o r \cos 2\theta, \quad u_\theta = -\frac{\partial \psi}{\partial r} = -2\psi_o r \sin 2\theta, \quad w = 0 \quad (12.27)$$

We begin with the continuity equation in cylindrical coordinates, and make use of the fact that $w = 0$, whence

$$\frac{1}{r}\frac{\partial}{\partial r}(r u_r) + \frac{1}{r}\frac{\partial u_\theta}{\partial \theta} + \frac{\partial w}{\partial z} = 0 \quad (12.28)$$

So, substituting the velocity components from Equation (12.27) into Equation (12.28),

$$4\psi_o r \cos 2\theta - 4\psi_o r \cos 2\theta + 0 = 0 \quad (12.29)$$

Therefore, the continuity equation is satisfied.

For steady flow in the absence of body forces, the radial, circumferential and axial components of the Navier-Stokes equation, respectively, in cylindrical coordinates are:

$$u_r \frac{\partial u_r}{\partial r} + \frac{u_\theta}{r}\frac{\partial u_r}{\partial \theta} + w\frac{\partial u_r}{\partial z} - \frac{u_\theta^2}{r} = -\frac{1}{\rho}\frac{\partial p}{\partial r}$$
$$+ \nu \left(\frac{\partial^2 u_r}{\partial r^2} + \frac{1}{r^2}\frac{\partial^2 u_r}{\partial \theta^2} + \frac{\partial^2 u_r}{\partial z^2} + \frac{1}{r}\frac{\partial u_r}{\partial r} - \frac{2}{r^2}\frac{\partial u_\theta}{\partial \theta} - \frac{u_r}{r^2} \right) \quad (12.30)$$

$$u_r \frac{\partial u_\theta}{\partial r} + \frac{u_\theta}{r}\frac{\partial u_\theta}{\partial \theta} + w\frac{\partial u_\theta}{\partial z} + \frac{u_r u_\theta}{r} = -\frac{1}{\rho r}\frac{\partial p}{\partial \theta}$$
$$+ \nu \left(\frac{\partial^2 u_\theta}{\partial r^2} + \frac{1}{r^2}\frac{\partial^2 u_\theta}{\partial \theta^2} + \frac{\partial^2 u_\theta}{\partial z^2} + \frac{1}{r}\frac{\partial u_\theta}{\partial r} + \frac{2}{r^2}\frac{\partial u_r}{\partial \theta} - \frac{u_\theta}{r^2} \right) \quad (12.31)$$

$$u_r \frac{\partial w}{\partial r} + \frac{u_\theta}{r}\frac{\partial w}{\partial \theta} + w\frac{\partial w}{\partial z} = -\frac{1}{\rho}\frac{\partial p}{\partial z} + \nu \left(\frac{\partial^2 w}{\partial r^2} + \frac{1}{r^2}\frac{\partial^2 w}{\partial \theta^2} + \frac{\partial^2 w}{\partial z^2} + \frac{1}{r}\frac{\partial w}{\partial r} \right) \quad (12.32)$$

Using the fact that $w = 0$ and that u_r and u_θ are functions only of r and θ, the three components of the Navier-Stokes equation simplify to the following.

$$u_r \frac{\partial u_r}{\partial r} + \frac{u_\theta}{r}\frac{\partial u_r}{\partial \theta} - \frac{u_\theta^2}{r} = -\frac{1}{\rho}\frac{\partial p}{\partial r} + \nu \left(\frac{\partial^2 u_r}{\partial r^2} + \frac{1}{r^2}\frac{\partial^2 u_r}{\partial \theta^2} + \frac{1}{r}\frac{\partial u_r}{\partial r} - \frac{2}{r^2}\frac{\partial u_\theta}{\partial \theta} - \frac{u_r}{r^2} \right) \quad (12.33)$$

$$u_r \frac{\partial u_\theta}{\partial r} + \frac{u_\theta}{r}\frac{\partial u_\theta}{\partial \theta} + \frac{u_r u_\theta}{r} = -\frac{1}{\rho r}\frac{\partial p}{\partial \theta} + \nu \left(\frac{\partial^2 u_\theta}{\partial r^2} + \frac{1}{r^2}\frac{\partial^2 u_\theta}{\partial \theta^2} + \frac{1}{r}\frac{\partial u_\theta}{\partial r} + \frac{2}{r^2}\frac{\partial u_r}{\partial \theta} - \frac{u_\theta}{r^2} \right) \quad (12.34)$$

$$0 = -\frac{1}{\rho}\frac{\partial p}{\partial z} \quad (12.35)$$

12.4. NAVIER-STOKES EQUATION

Working first with the convective acceleration terms, noting from Equations (12.27) that $u_r = 2\psi_o r \cos 2\theta$ and $u_\theta = -2\psi_o r \sin 2\theta$, the sum of the acceleration terms in the radial-momentum equation, Equation (12.33), is

$$u_r \frac{\partial u_r}{\partial r} + \frac{u_\theta}{r}\frac{\partial u_r}{\partial \theta} - \frac{u_\theta^2}{r} = 4\psi_o^2 r \cos^2 2\theta + 8\psi_o^2 r \sin^2 2\theta$$
$$-4\psi_o^2 r \sin^2 2\theta = 4\psi_o^2 r \quad (12.36)$$

Similarly, for the circumferential-momentum equation, Equation (12.34), again appealing to Equations (12.27), we have

$$u_r \frac{\partial u_\theta}{\partial r} + \frac{u_\theta}{r}\frac{\partial u_\theta}{\partial \theta} + \frac{u_r u_\theta}{r} = -4\psi_o^2 r \sin 2\theta \cos 2\theta + 8\psi_o^2 r \sin 2\theta \cos 2\theta$$
$$-4\psi_o^2 r \sin 2\theta \cos 2\theta = 0 \quad (12.37)$$

Turning now to the radial and circumferential viscous terms, we find

$$\nu\left(\frac{\partial^2 u_r}{\partial r^2} + \frac{1}{r^2}\frac{\partial^2 u_r}{\partial \theta^2} + \frac{1}{r}\frac{\partial u_r}{\partial r} - \frac{2}{r^2}\frac{\partial u_\theta}{\partial \theta} - \frac{u_r}{r^2}\right) = \nu(0 - 8 + 2 + 8 - 2)\frac{\psi_o \cos 2\theta}{r} = 0 \quad (12.38)$$

$$\nu\left(\frac{\partial^2 u_\theta}{\partial r^2} + \frac{1}{r^2}\frac{\partial^2 u_\theta}{\partial \theta^2} + \frac{1}{r}\frac{\partial u_\theta}{\partial r} + \frac{2}{r^2}\frac{\partial u_r}{\partial \theta} - \frac{u_\theta}{r^2}\right) = \nu(0 + 8 - 2 - 8 + 2)\frac{\psi_o \sin 2\theta}{r} = 0 \quad (12.39)$$

The fact that the viscous terms vanish for a potential-flow solution should be expected because the streamfunction satisfies Laplace's equation, i.e., $\nabla^2 \psi = 0$. But, the velocity is given by $\mathbf{u} = \nabla \times (\psi \mathbf{k})$. Hence, there follows $\nabla^2 \mathbf{u} = \nabla \times (\nabla^2 \psi \, \mathbf{k}) = \mathbf{0}$.

Thus, using Equations (12.36), (12.37), (12.38) and (12.39), we can simplify the radial and circumferential components of the Navier-Stokes equation, viz., Equations (12.33) and (12.34). What remains is

$$4\psi_o^2 r = -\frac{1}{\rho}\frac{\partial p}{\partial r} \quad (12.40)$$

$$0 = -\frac{1}{\rho r}\frac{\partial p}{\partial \theta} \quad (12.41)$$

$$0 = -\frac{1}{\rho}\frac{\partial p}{\partial z} \quad (12.42)$$

Now, Equations (12.41) and (12.42) show that the pressure is independent of θ and z, so that

$$\frac{\partial p}{\partial \theta} = \frac{\partial p}{\partial z} = 0 \implies p = p(r) \quad (12.43)$$

i.e., the pressure depends only upon radial distance, r. Finally, integrating Equation (12.40) over r yields

$$p(r) = -2\rho\psi_o^2 r^2 + \text{constant} \quad (12.44)$$

Then, since this is flow toward a stagnation point, we can select $p(0) = p_o$, where p_o is the stagnation pressure. Therefore, the pressure is

$$p(r) = p_o - 2\rho\psi_o^2 r^2 \quad (12.45)$$

So, this solution indeed satisfies the exact continuity and Navier-Stokes equations. Also, the pressure assumes a finite value at the origin, $r = 0$. However, this cannot truly be the solution to the equations of motion everywhere as it is impossible to have $p \to -\infty$, which this solution predicts as $r \to \infty$.

12.5 CFD—Stability Analysis

Statement of the Problem: *Using von Neumann stability analysis, determine G and any condition required for stability of the following method applied to the inviscid Burgers' equation, $u_t + U u_x = 0$:*

$$u_j^{n+1} = u_j^n - \frac{U \Delta t}{2 \Delta x} \left[\frac{3}{4} \left(u_{j+1}^{n+1} - u_{j-1}^{n+1} \right) + \frac{1}{4} \left(u_{j+1}^n - u_{j-1}^n \right) \right], \qquad U > 0$$

Solution: We begin with the following finite-difference scheme:

$$u_j^{n+1} = u_j^n - \frac{U \Delta t}{2 \Delta x} \left[\frac{3}{4} \left(u_{j+1}^{n+1} - u_{j-1}^{n+1} \right) + \frac{1}{4} \left(u_{j+1}^n - u_{j-1}^n \right) \right] \qquad (12.46)$$

To implement von Neumann stability analysis, we assume

$$u_j^n = G^n e^{ij\theta}, \qquad \theta = \kappa \Delta x \qquad (12.47)$$

Substituting into the difference equation, we find

$$G^{n+1} e^{ij\theta} = G^n e^{ij\theta} - \frac{U \Delta t}{8 \Delta x} \left[3 G^{n+1} + G^n \right] \left[e^{i(j+1)\theta} - e^{i(j-1)\theta} \right] \qquad (12.48)$$

Dividing through by $G^n e^{ij\theta}$, we have

$$G = 1 - \frac{U \Delta t}{8 \Delta x} (3G + 1) \underbrace{\left(e^{i\theta} - e^{-i\theta} \right)}_{=2i \sin \theta} = 1 - i \frac{U \Delta t}{4 \Delta x} (3G + 1) \sin \theta \qquad (12.49)$$

Rearranging terms a bit, we have

$$\left[1 + i \frac{3 U \Delta t}{4 \Delta x} \sin \theta \right] G = \left[1 - i \frac{U \Delta t}{4 \Delta x} \sin \theta \right] \qquad (12.50)$$

Solving for G, we find

$$G = \frac{1 - i \lambda \sin \theta}{1 + 3 i \lambda \sin \theta}, \qquad \lambda \equiv \frac{U \Delta t}{4 \Delta x} \qquad (12.51)$$

Since G is complex, we compute $|G|^2$ as follows.

$$|G|^2 = \frac{|1 - i \lambda \sin \theta|^2}{|1 + 3 i \lambda \sin \theta|^2} = \frac{1 + \lambda^2 \sin^2 \theta}{1 + 9 \lambda^2 \sin^2 \theta} \qquad (12.52)$$

For stability, we must have $|G|^2 \leq 1$. Thus,

$$1 + \lambda^2 \sin^2 \theta \leq 1 + 9 \lambda^2 \sin^2 \theta \quad \Longrightarrow \quad 1 \leq 9 \qquad (12.53)$$

which is obviously true. Therefore, this finite-difference scheme is **unconditionally stable**.

Chapter 13

Navier-Stokes Solutions

Chapter 13 concentrates on exact solutions to the continuity and Navier-Stokes equations. In general, exact solutions can be obtained only for very simple geometries that are usually infinite in extent. These solutions are nevertheless instructive, often revealing subtle aspects of fluid motion found in more-complicated geometries. Closed-form solutions are most-commonly found for so-called *parallel flows*, which are flows for which the convective acceleration, $\mathbf{u} \cdot \nabla \mathbf{u}$, is zero. For incompressible, constant-property flows, all of the Navier-Stokes equation's nonlinearity lies in the convective acceleration term, so that the equation is linear for parallel flows. A few exact closed-form solutions have been found for non-parallel flows, usually in terms of a similarity solution. With the advent of powerful computers, numerical solutions can be routinely done in order to obtain Navier-Stokes solutions for nontrivial problems. This chapter touches on each of these issues in the following problems.

Section 13.1 Couette Flow: A straightforward analysis based on the exact solution for Couette flow with constant pressure.

Section 13.2 Rotating Channel Flow: The problem in this section involves generating an exact Navier-Stokes solution for fully-developed flow in a rotating channel. The channel is infinite in length, so that the convective acceleration terms are exactly zero.

Section 13.3 Trailing-Vortex Flow: This problem uses a given similarity solution for an approximate model of a trailing vortex to deduce properties of the solution based on its self-similar evolution.

Section 13.4 Stokes' Flow with Suction: This problem develops a time-dependent solution for an infinite flat plate that is impulsively accelerated to a constant velocity in its own plane, similar to the classical flow known as *Stokes' First Problem*. In addition, the flow includes time-dependent mass removal at the surface. The solution requires determining how the suction velocity must vary with time in order to maintain self-similar motion.

Section 13.5 CFD—Thomas' Algorithm: An exercise in formulating and solving a finite-difference problem using Thomas' algorithm.

13.1 Couette Flow

Statement of the Problem: *The exact solution for fully-developed Couette flow with constant pressure is $u(y) = Uy/h$. Compute the mass flux, $\dot{m}$, momentum flux, $\dot{P}$ and kinetic-energy flux, $\dot{E}$, and compare $\dot{E}$ with $\dot{P}^2/(2\dot{m})$.*

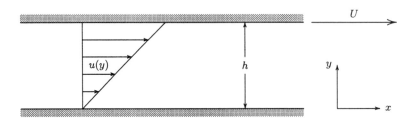

Figure 13.1: *Constant-pressure Couette flow.*

Solution: For the given velocity profile, the differential mass flux is $d\dot{m} = \rho u \, dy$, so that the total mass flux, $\dot{m}$, is

$$\dot{m} = \int_0^h \rho u(y) dy = \rho U \int_0^h \frac{y}{h} dy = \rho U h \int_0^1 \eta \, d\eta \qquad (\eta \equiv y/h)$$

$$= \rho U h \left[\frac{1}{2}\eta^2\right]_{\eta=0}^{\eta=1} = \frac{1}{2}\rho U h \qquad (13.1)$$

Similarly, the differential momentum flux is $d\dot{P} = \rho u^2 dy$, and $\dot{P}$ is

$$\dot{P} = \int_0^h \rho u^2(y) dy = \rho U^2 \int_0^h \left(\frac{y}{h}\right)^2 dy = \rho U^2 h \int_0^1 \eta^2 d\eta$$

$$= \rho U^2 h \left[\frac{1}{3}\eta^3\right]_{\eta=0}^{\eta=1} = \frac{1}{3}\rho U^2 h \qquad (13.2)$$

Finally, the differential kinetic-energy flux is $d\dot{E} = \frac{1}{2}\rho u^3 dy$, wherefore

$$\dot{E} = \int_0^h \frac{1}{2}\rho u^3(y) dy = \frac{1}{2}\rho U^3 \int_0^h \left(\frac{y}{h}\right)^3 dy = \frac{1}{2}\rho U^3 h \int_0^1 \eta^3 d\eta$$

$$= \frac{1}{2}\rho U^3 h \left[\frac{1}{4}\eta^4\right]_{\eta=0}^{\eta=1} = \frac{1}{8}\rho U^3 h \qquad (13.3)$$

Now, in classical mechanics, the kinetic energy of a point mass is equal to the square of its momentum divided by twice its mass. From Equations (13.1) through (13.3), we find

$$\dot{E} - \frac{\dot{P}^2}{2\dot{m}} = \frac{1}{8}\rho U^3 h - \frac{(\frac{1}{3}\rho U^2 h)^2}{2(\frac{1}{2}\rho U h)} = \frac{1}{72}\rho U^3 h = \frac{1}{9}\dot{E} \qquad (13.4)$$

Thus $\dot{E}$ and $\dot{P}^2/(2\dot{m})$ differ by a little more than 10%. This is unsurprising since the integrals represent averages. There is no a priori reason why the square of one averaged quantity divided by another averaged quantity $[\dot{P}^2/(2\dot{m})]$ should be equal to the average of the corresponding integrand [i.e, $d\dot{E} = \frac{1}{2}\rho u^3 dy = (d\dot{P})^2/(2d\dot{m})$ does not imply that $\dot{E} = \int (d\dot{P})^2/(2d\dot{m})$].

13.2 Rotating Channel Flow

Statement of the Problem: *Consider fully-developed flow in a very long channel of height h that rotates as shown in Figure 13.2. The flow is incompressible and steady. Solve for the velocity, $u(y)$, and the pressure, $p(x,y)$. Assume $\partial p^*/\partial x = -\rho P$ where the quantity $p^* \equiv p - \frac{1}{2}\Omega^2(x^2+y^2)$ is the reduced pressure and P is a constant. Also, assume the pressure is equal to p_o when $x = y = 0$.*

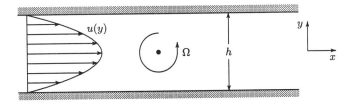

Figure 13.2: *Idealized rotating channel-flow geometry for fully-developed flow.*

Solution: Since this flow is two dimensional and incompressible, the continuity equation is

$$\frac{\partial u}{\partial x} + \frac{\partial v}{\partial y} = 0 \tag{13.5}$$

Because the flow is specified to be fully developed, by definition, the flow is independent of x. Therefore, continuity simplifies to

$$\frac{\partial v}{\partial y} = 0 \quad \Longrightarrow \quad v(y) = \text{constant} \tag{13.6}$$

Since v must vanish at the channel walls, we have

$$v(y) = 0 \tag{13.7}$$

Also, note that because (a) the flow is independent of x and (b) $v = 0$, the convective acceleration is zero, i.e.,

$$\mathbf{u}\cdot\nabla\mathbf{u} = u\frac{\partial \mathbf{u}}{\partial x} + v\frac{\partial \mathbf{u}}{\partial y} = 0 \tag{13.8}$$

For steady flow in a rotating coordinate frame with $\boldsymbol{\Omega} = \Omega\,\mathbf{k}$, the Navier-Stokes equation is

$$\mathbf{u}\cdot\nabla\mathbf{u} + 2\boldsymbol{\Omega}\times\mathbf{u} = -\frac{1}{\rho}\nabla p^* + \nu\nabla^2\mathbf{u}, \qquad p^* \equiv p - \frac{1}{2}\Omega^2\left(x^2+y^2\right) \tag{13.9}$$

Now, the Coriolis acceleration is

$$\boldsymbol{\Omega}\times\mathbf{u} = \begin{vmatrix} \mathbf{i} & \mathbf{j} & \mathbf{k} \\ 0 & 0 & \Omega \\ u & v & 0 \end{vmatrix} = -\Omega v\,\mathbf{i} + \Omega u\,\mathbf{j} \tag{13.10}$$

Combining Equations (13.8), (13.9) and (13.10), using the fact that $\mathbf{u} = \mathbf{u}(y)$, and expressing the results in component form, we find

$$-2\Omega v = -\frac{1}{\rho}\frac{\partial p^*}{\partial x} + \nu\frac{d^2 u}{dy^2} \tag{13.11}$$

$$2\Omega u = -\frac{1}{\rho}\frac{\partial p^*}{\partial y} + \nu\frac{d^2 v}{dy^2} \tag{13.12}$$

Using the fact that $v = 0$ [Equation (13.7)] and that we are given $\partial p^*/\partial x = -\rho P$, we conclude that

$$\nu \frac{d^2 u}{dy^2} = -P \quad \text{and} \quad \frac{\partial p^*}{\partial y} = -2\rho\Omega u \qquad (13.13)$$

We must solve these equations subject to the no-slip boundary conditions at both channel walls and to the given condition that $p^*(0,0) = p_o$, wherefore

$$u(\pm h/2) = 0, \quad p^*(0,0) = p_o \qquad (13.14)$$

Integrating the first of Equations (13.13) twice gives

$$u(y) = A + By - \frac{P}{2\nu} y^2 \qquad (13.15)$$

Then, applying no-slip yields

$$u(\pm h/2) = A \pm B\frac{h}{2} - \frac{Ph^2}{8\nu} = 0 \qquad (13.16)$$

Hence, we obtain

$$A = \frac{Ph^2}{8\nu} \quad \text{and} \quad B = 0 \qquad (13.17)$$

Therefore, the velocity is

$$u(y) = \frac{Ph^2}{8\nu}\left[1 - 4\left(\frac{y}{h}\right)^2\right] \qquad (13.18)$$

Before proceeding, it is instructive to observe that Equation (13.18) is identical to the velocity profile for a non-rotating channel. This reflects the fact that the x component of the Coriolis acceleration is zero, so that the velocity satisfies the same equation and boundary conditions, independent of the rotation rate.

Turning to the reduced pressure, substituting Equation (13.18) into the second of Equations (13.13) yields

$$\frac{\partial p^*}{\partial y} = -2\rho\Omega \frac{Ph^2}{8\nu}\left[1 - 4\left(\frac{y}{h}\right)^2\right] = -\frac{\rho\Omega P}{4\nu}\left[h^2 - 4y^2\right] \qquad (13.19)$$

Integrating this equation over y tells us that

$$p^*(x,y) = -\frac{\rho\Omega P}{4\nu}\left[h^2 y - \frac{4}{3}y^3\right] + g(x) \qquad (13.20)$$

where $g(x)$ is a function of integration. Next, differentiating Equation (13.20) with respect to x, we have

$$\frac{\partial p^*}{\partial x} = \frac{dg}{dx} = -\rho P \quad \Longrightarrow \quad g(x) = -\rho P x + \text{constant} \qquad (13.21)$$

So, combining Equations (13.20) and (13.21) to eliminate the integration function, $g(x)$, the reduced pressure is

$$p^*(x,y) = p_o - \rho P x - \frac{\rho P \Omega h^3}{4\nu}\left[\frac{y}{h} - \frac{4}{3}\left(\frac{y}{h}\right)^3\right] \qquad (13.22)$$

Note that, in writing Equation (13.22), we have imposed the given condition, $p^*(0,0) = p_o$ to evaluate the constant in the function $g(x)$.

13.3. TRAILING-VORTEX FLOW

Finally, as noted in Equation (13.9), the physical and reduced pressures are related by $p = p^* + \frac{1}{2}\rho\Omega^2(x^2 + y^2)$, wherefore

$$p(x,y) = p_o + \frac{1}{2}\rho\Omega^2\left(x^2 + y^2\right) - \rho P x - \frac{\rho P \Omega h^3}{4\nu}\left[\frac{y}{h} - \frac{4}{3}\left(\frac{y}{h}\right)^3\right] \quad (13.23)$$

Unlike the velocity, the pressure is affected by the rotation. In addition to the centrifugal acceleration effect, as represented by the term proportional to Ω^2 on the right-hand side of Equation (13.23), the last term follows from the nonzero Coriolis acceleration appearing in the y component of the Navier-Stokes equation.

13.3 Trailing-Vortex Flow

Statement of the Problem: *An approximate model of the trailing vortex left by an airplane that has just taken off is given by the following exact solution to the Navier-Stokes equation:*

$$\boldsymbol{\omega}(r,t) = \frac{\Gamma}{\pi\nu t}e^{-\eta^2}\mathbf{k}; \qquad \eta \equiv \frac{r}{2\sqrt{\nu t}}$$

where Γ is circulation, ν is kinematic viscosity, t is time and r is radial distance. If the vorticity is $\boldsymbol{\omega} = \omega_o\mathbf{k}$ and $\eta = 2$ when $t = t_o$ and $r = r_o$, at what radial distance is $\boldsymbol{\omega} = \omega_o\mathbf{k}$ when $t = 4t_o$?

Solution: From the given information at $t = t_o$, i.e., $\boldsymbol{\omega} = \omega_o\mathbf{k}$ when $\eta = 2$, there follows

$$\omega_o\mathbf{k} = \frac{\Gamma}{\pi\nu t_o}e^{-4}\mathbf{k} \quad (13.24)$$

Now, when $t = 4t_o$, observing that $\eta^2 = r^2/(4\nu t) = r^2/(16\nu t_o)$, we have

$$\boldsymbol{\omega}(r, 4t_o) = \frac{\Gamma}{4\pi\nu t_o}e^{-r^2/(16\nu t_o)}\mathbf{k} \quad (13.25)$$

Hence, to determine the value of r at which $\boldsymbol{\omega}(r, 4t_o) = \omega_o\mathbf{k}$, we combine Equations (13.24) and (13.25) to obtain

$$\frac{\Gamma}{\pi\nu t_o}e^{-4} = \frac{\Gamma}{4\pi\nu t_o}e^{-r^2/(16\nu t_o)} \quad\Longrightarrow\quad 4e^{-4} = e^{-r^2/(16\nu t_o)} \quad (13.26)$$

Taking the natural log of Equation (13.26) yields

$$\ln 4 - 4 = -\frac{r^2}{16\nu t_o} \quad\Longrightarrow\quad r^2 = 16(4 - \ln 4)\nu t_o \quad (13.27)$$

But, from the given information, we know that when $\eta = 2$,

$$\eta^2 = \frac{r_o^2}{4\nu t_o} \quad\Longrightarrow\quad r_o^2 = 16\nu t_o \quad (13.28)$$

Therefore, combining Equations (13.27) and (13.28) and solving for r gives

$$r = r_o\sqrt{4 - \ln 4} = 1.62 r_o \quad (13.29)$$

13.4 Stokes' Flow with Suction

Statement of the Problem: *Consider an infinite flat plate that is impulsively accelerated in its own plane at time $t = 0$, and translates at constant velocity, U, for $t > 0$. The plate is porous and fluid is removed with $v(0,t) = -V(t)$ for $t > 0$. Assume the fluid is incompressible and that the pressure is constant. Determine what $V(t)$ must be if the viscous layer grows in a self-similar manner so that*

$$u(y,t) = UF(\eta), \qquad \eta \equiv \frac{y}{2\sqrt{\nu t}}$$

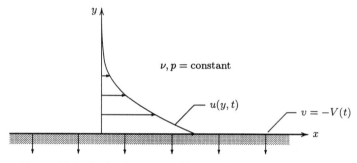

Figure 13.3: *Stokes' First Problem with suction.*

Solution: Because the flow is infinite in extent, we expect to have $\partial u/\partial x = 0$. So, the continuity equation becomes

$$\frac{\partial u}{\partial x} + \frac{\partial v}{\partial y} = 0 \implies \frac{\partial v}{\partial y} = 0 \tag{13.30}$$

Integrating over y and noting that $v = -V(t)$ at $y = 0$, necessarily

$$v(y,t) = -V(t) \tag{13.31}$$

So, the x component of the Navier-Stokes equation is as follows.

$$\frac{\partial u}{\partial t} + \underbrace{u\frac{\partial u}{\partial x}}_{=0} + \underbrace{v\frac{\partial u}{\partial y}}_{=-V\partial u/\partial y} = -\underbrace{\frac{1}{\rho}\frac{\partial p}{\partial x}}_{=0} + \nu\underbrace{\frac{\partial^2 u}{\partial x^2}}_{=0} + \nu\frac{\partial^2 u}{\partial y^2} \tag{13.32}$$

Therefore, the equation governing the motion is

$$\frac{\partial u}{\partial t} - V(t)\frac{\partial u}{\partial y} = \nu\frac{\partial^2 u}{\partial y^2} \tag{13.33}$$

In order to formulate the similarity solution, we must verify that the equation and boundary conditions can be expressed entirely in terms of the similarity variable, η. The first step is to transform Equation (13.33). Using the chain rule, the first derivatives of u are

$$\frac{\partial u}{\partial t} = \frac{du}{d\eta}\frac{\partial \eta}{\partial t} = U\frac{dF}{d\eta}\left(-\frac{1}{2}\frac{y}{2\nu^{1/2}t^{3/2}}\right) = -\frac{U}{2t}\eta\frac{dF}{d\eta} \tag{13.34}$$

$$\frac{\partial u}{\partial y} = \frac{du}{d\eta}\frac{\partial \eta}{\partial y} = U\frac{dF}{d\eta}\left(\frac{1}{2\sqrt{\nu t}}\right) = \frac{U}{2\sqrt{\nu t}}\frac{dF}{d\eta} \tag{13.35}$$

13.5. CFD—THOMAS' ALGORITHM

Hence, the diffusion term in Equation (13.33) becomes

$$\nu \frac{\partial^2 u}{\partial y^2} = \nu \frac{\partial}{\partial y}\left(\frac{U}{2\sqrt{\nu t}}\frac{dF}{d\eta}\right) = \frac{\nu U}{2\sqrt{\nu t}}\frac{\partial}{\partial y}\left(\frac{dF}{d\eta}\right)$$

$$= \frac{\nu U}{2\sqrt{\nu t}}\frac{d^2 F}{d\eta^2}\frac{\partial \eta}{\partial y} = \frac{\nu U}{2\sqrt{\nu t}}\left(\frac{1}{2\sqrt{\nu t}}\right)\frac{d^2 F}{d\eta^2}$$

$$= \frac{U}{4t}\frac{d^2 F}{d\eta^2} \qquad (13.36)$$

Substituting Equations (13.34), (13.35) and (13.36) into Equation (13.33), we have

$$-\frac{U}{2t}\eta\frac{dF}{d\eta} - \frac{U}{2\sqrt{\nu t}}V(t)\frac{dF}{d\eta} = \frac{U}{4t}\frac{d^2 F}{d\eta^2} \qquad (13.37)$$

So, the transformed Navier-Stokes equation is

$$\frac{d^2 F}{d\eta^2} + 2\left[\eta - V(t)\sqrt{\frac{t}{\nu}}\right]\frac{dF}{d\eta} = 0 \qquad (13.38)$$

Focusing on the boundary conditions, the velocity must be equal to the plate velocity at the surface, i.e., $u(0,t) = U$. Far above the plate, the velocity goes to zero so that $u(y,t) \to 0$ as $y \to \infty$. Clearly, these boundary conditions transform to

$$F(0) = 1 \quad \text{and} \quad F(\eta) \to 0 \text{ as } \eta \to \infty \qquad (13.39)$$

Clearly, the boundary conditions involve only η. In order to have a similarity solution, the coefficient of $dF/d\eta$ in Equation (13.38) must be independent of t. Therefore, the condition for existence of a similarity solution is

$$V(t) = C\sqrt{\frac{\nu}{t}}, \qquad C = \text{constant} \qquad (13.40)$$

13.5 CFD—Thomas' Algorithm

Statement of the Problem: *We want to solve the following problem using a finite-difference formulation.*

$$\frac{d^2\phi}{dy^2} - \phi = 0, \qquad \phi(0) = 0, \quad \phi(1) = 1$$

To develop and validate the solution, proceed as follows.

(a) Generate the exact solution analytically.

(b) Using central-difference approximations, discretize the differential equation and develop a matrix equation of the form $[A]\{\phi\} = \{b\}$. Use equally-spaced points with a stepsize of $\Delta y = 0.25$.

(c) Solve the matrix equation developed in Part (b) using Thomas' algorithm. Compare the results with the exact solution.

Solution: Before proceeding, note that a conventional finite-difference solution would normally use much smaller values of Δy. The purpose of this problem is to help develop a complete understanding of the discretization process and details of how Thomas' algorithm works. This end is best served with an example that can be reasonably solved without the aid of a computer.

13.5(a): To solve the differential equation, assume a solution of the form $\phi(y) = e^{\lambda y}$. Then,

$$\frac{d^2\phi}{dy^2} - \phi = \left(\lambda^2 - 1\right) e^{\lambda y} = 0 \quad \Longrightarrow \quad \lambda = \pm 1 \tag{13.41}$$

Therefore, the general solution is

$$\phi(y) = A e^y + B e^{-y} \tag{13.42}$$

where A and B are constants. Applying the boundary conditions yields

$$\phi(0) = A + B = 0 \quad \text{and} \quad \phi(1) = Ae + Be^{-1} = 1 \tag{13.43}$$

Solving for A and B, we find

$$A = \frac{1}{e - e^{-1}} \quad \text{and} \quad B = -\frac{1}{e - e^{-1}} \tag{13.44}$$

Therefore, the exact solution for $\phi(y)$ is

$$\phi(y) = \frac{e^y - e^{-y}}{e - e^{-1}} \tag{13.45}$$

13.5(b): Using central-difference approximations, the discretized form of the differential equation is

$$\frac{\phi_{j+1} - 2\phi_j + \phi_{j-1}}{(\Delta y)^2} - \phi_j = 0 \quad \Longrightarrow \quad \phi_{j+1} - \left[2 + (\Delta y)^2\right] \phi_j + \phi_{j-1} = 0 \tag{13.46}$$

Now, we denote the grid points by y_j so that, since $\Delta y = 0.25$,

$$y_1 = 0, \quad y_2 = 0.25, \quad y_3 = 0.50, \quad y_4 = 0.75, \quad y_5 = 1 \tag{13.47}$$

Two equations come from the boundary conditions, viz.,

$$\phi_1 = \phi(0) = 0 \quad \text{and} \quad \phi_5 = \phi(1) = 1 \tag{13.48}$$

Three more equations come from the interior points. Noting that $2 + (\Delta y)^2 = 33/16$, the three equations are

$$\phi_3 - (33/16)\phi_2 + \phi_1 = 0 \tag{13.49}$$
$$\phi_4 - (33/16)\phi_3 + \phi_2 = 0 \tag{13.50}$$
$$\phi_5 - (33/16)\phi_4 + \phi_3 = 0 \tag{13.51}$$

Hence, in matrix form, we have

$$\begin{bmatrix} 1 & 0 & 0 & 0 & 0 \\ 1 & -33/16 & 1 & 0 & 0 \\ 0 & 1 & -33/16 & 1 & 0 \\ 0 & 0 & 1 & -33/16 & 1 \\ 0 & 0 & 0 & 0 & 1 \end{bmatrix} \begin{Bmatrix} \phi_1 \\ \phi_2 \\ \phi_3 \\ \phi_4 \\ \phi_5 \end{Bmatrix} = \begin{Bmatrix} 0 \\ 0 \\ 0 \\ 0 \\ 1 \end{Bmatrix} \tag{13.52}$$

13.5. CFD—THOMAS' ALGORITHM

13.5(c): Denoting the diagonal elements of the matrix by B_j, off-diagonal elements below and above the diagonal by A_j and C_j, respectively, and right-hand side vector elements by D_j, then for the following matrix equation, we can construct the following table defining the coefficients.

j	A_j	B_j	C_j	D_j
1	-	1	0	0
2	1	-33/16	1	0
3	1	-33/16	1	0
4	1	-33/16	1	0
5	0	1	-	1

To solve using Thomas' algorithm, we have

$$\phi_j = P_j + Q_j \phi_{j+1} \quad \text{where} \quad P_j = \frac{D_j - A_j P_{j-1}}{B_j + A_j Q_{j-1}}, \quad Q_j = \frac{-C_j}{B_j + A_j Q_{j-1}} \quad (13.53)$$

We begin by noting that the equation defined by the first row of the matrix is, by design, $\phi_1 = 0$. This is the boundary condition, and we must select

$$P_1 = 0 \quad \text{and} \quad Q_1 = 0$$

Now, proceeding with Thomas' algorithm, the operations are as follows.

$$P_2 = \frac{D_2 - A_2 P_1}{B_2 + A_2 Q_1} = \frac{0 - (1)(0)}{(-33/16) + (1)(0)} = 0 \quad (13.54)$$

$$Q_2 = \frac{-C_2}{B_2 + A_2 Q_1} = \frac{-1}{(-33/16) + (1)(0)} = \frac{16}{33} = 0.484848 \quad (13.55)$$

$$P_3 = \frac{D_3 - A_3 P_2}{B_3 + A_3 Q_2} = \frac{0 - (1)(0)}{(-33/16) + (1)(16/33)} = 0 \quad (13.56)$$

$$Q_3 = \frac{-C_3}{B_3 + A_3 Q_2} = \frac{-1}{(-33/16) + (1)(16/33)} = \frac{528}{833} = 0.633854 \quad (13.57)$$

$$P_4 = \frac{D_4 - A_4 P_3}{B_4 + A_4 Q_3} = \frac{0 - (1)(0)}{(-33/16) + (1)(528/833)} = 0 \quad (13.58)$$

$$Q_4 = \frac{-C_4}{B_4 + A_4 Q_3} = \frac{-1}{(-33/16) + (1)(528/833)} = \frac{13328}{19041} = 0.699963 \quad (13.59)$$

Therefore, the coefficients P_j and Q_j are as summarized in the following table.

j	P_j	Q_j
1	0	0.000000
2	0	0.484848
3	0	0.633854
4	0	0.699963

Again by design, the equation defined by the last row of the matrix is $\phi_5 = 1$, which is the other boundary condition for this problem. Thus,

$$\phi_5 = 1 \tag{13.60}$$
$$\phi_4 = P_4 + Q_4\phi_5 = 0 + 0.699963(1) = 0.699963 \tag{13.61}$$
$$\phi_3 = P_3 + Q_3\phi_4 = 0 + 0.633854(0.699963) = 0.443674 \tag{13.62}$$
$$\phi_2 = P_2 + Q_2\phi_3 = 0 + 0.484848(0.443674) = 0.215114 \tag{13.63}$$

Therefore, recalling that $\phi_1 = 0$, the solution vector is

$$\{\phi\} = \begin{Bmatrix} 0 \\ 0.215114 \\ 0.443674 \\ 0.699963 \\ 1 \end{Bmatrix} \tag{13.64}$$

Finally, from the exact solution of Part (a) [see Equation (13.45)], gives

$$\{\phi_{exact}\} = \begin{Bmatrix} 0 \\ 0.214952 \\ 0.443409 \\ 0.699724 \\ 1 \end{Bmatrix} \tag{13.65}$$

Comparison of Equations (13.64) and (13.65) shows that the largest percentage error is 0.08% at $y = 0.25$.

Chapter 14

Boundary Layers

Chapter 14 deals with boundary layers for both laminar and turbulent flow. Exact solutions to these equations have been obtained, featuring self-similar velocity profiles, for laminar boundary layers with and without pressure gradient. The former is known as the Falkner-Skan solution and the latter as the Blasius solution. Turbulent boundary layers are generally analyzed using approximate models for the turbulent stresses. More general boundary layers (i.e., boundary layers for which similarity solutions do not exist) are routinely computed using numerical methods.

The classical momentum integral equation plays a useful role in assessing accuracy of both experimental and computational results. Serving as a relation amongst so-called *integral parameters* such as momentum thickness, displacement thickness, shape factor and skin friction, it can used as a consistency check for boundary-layer data.

The Falkner-Skan and Blasius solutions provide exact integral parameters for laminar boundary layers. For turbulent flow, integral parameters based on correlation of measurements often provide reasonable estimates of boundary-layer properties. This chapter's problems address these aspects of boundary-layer theory.

Section 14.1 Boundary-Layer Equations I: Manipulation of the two-dimensional boundary-layer equations that yields a differential equation for the vorticity.

Section 14.2 Boundary-Layer Equations II: The problem in this section involves finding the solution to the boundary-layer equations that is compatible with specified normal velocity and pressure.

Section 14.3 Momentum Integral Equation I: This problem uses the momentum integral equation to assess the accuracy of wind-tunnel measurements for a laminar flat-plate boundary layer. Working with a correlation of measured velocities, the problem involves computing skin friction, momentum thickness, displacement thickness and boundary-layer thickness and comparing to the Blasius solution.

Section 14.4 Momentum Integral Equation II: In this application of the momentum integral equation, the velocity profile and boundary-layer thickness are given. The solution involves finding the corresponding freestream velocity.

Section 14.5 Blasius Solution: The problem in this section makes use of the Blasius solution to determine the viscosity of a fluid.

Section 14.6 Transitional Boundary Layer: This problem focuses on a transitional boundary layer. Ignoring the width of the transition region, the solution uses the Blasius skin friction and a turbulent-flow skin-friction correlation to determine the force on a flat plate.

Section 14.7 Turbulent Boundary Layers: A straightforward analysis of a turbulent boundary layer. The solution involves estimation of *Coles' wake-strength parameter* for boundary layers with favorable, zero and adverse pressure gradients.

Section 14.8 Law of the Wall I: This problem is the first of two that deals with the classical law of the wall for turbulent flow over smooth and rough surfaces. Given measured velocity and skin friction, the problem's solution requires determining if a surface is smooth or rough.

Section 14.9 Law of the Wall II: This is the second of two problems that examines the difference between turbulent flow over smooth and rough surfaces. Given measured velocity at a known distance from the surface, we determine the difference between smooth-wall and rough-wall skin friction for a given roughness height.

Section 14.10 CFD—Truncation Error: A straightforward computation that illustrates how the order of accuracy of a given finite-difference approximation is determined.

14.1 Boundary-Layer Equations I

Statement of the Problem: *Beginning with the two-dimensional boundary-layer equations, and observing that the vorticity is $\omega \approx \partial u/\partial y$ in a boundary layer, derive a differential equation for the vorticity.*

Solution: The equations of motion for a two-dimensional incompressible boundary layer are

$$\frac{\partial u}{\partial x} + \frac{\partial v}{\partial y} = 0 \qquad (14.1)$$

$$u\frac{\partial u}{\partial x} + v\frac{\partial u}{\partial y} = -\frac{1}{\rho}\frac{\partial p}{\partial x} + \nu\frac{\partial^2 u}{\partial y^2} \qquad (14.2)$$

$$0 = -\frac{1}{\rho}\frac{\partial p}{\partial y} \qquad (14.3)$$

We begin by differentiating each term in Equation (14.2) with respect to y, which yields

$$\frac{\partial u}{\partial y}\frac{\partial u}{\partial x} + u\frac{\partial^2 u}{\partial x \partial y} + \frac{\partial v}{\partial y}\frac{\partial u}{\partial y} + v\frac{\partial^2 u}{\partial y^2} = -\frac{1}{\rho}\frac{\partial^2 p}{\partial x \partial y} + \nu\frac{\partial^3 u}{\partial y^3} \qquad (14.4)$$

Differentiating Equation (14.3) with respect to x shows that $\partial^2 p/\partial x \partial y = 0$. Regrouping terms, we have

$$\frac{\partial u}{\partial y}\left(\frac{\partial u}{\partial x} + \frac{\partial v}{\partial y}\right) + u\frac{\partial}{\partial x}\left(\frac{\partial u}{\partial y}\right) + v\frac{\partial}{\partial y}\left(\frac{\partial u}{\partial y}\right) = \nu\frac{\partial^2}{\partial y^2}\left(\frac{\partial u}{\partial y}\right) \qquad (14.5)$$

Appealing to the continuity equation, Equation (14.1), clearly the first term on the left-hand side of Equation (14.5) is zero. Hence, since the vorticity, ω, in a boundary layer is approximately equal to $\partial u/\partial y$, the desired equation for ω is as follows.

$$u\frac{\partial \omega}{\partial x} + v\frac{\partial \omega}{\partial y} = \nu\frac{\partial^2 \omega}{\partial y^2} \qquad (14.6)$$

14.2 Boundary-Layer Equations II

Statement of the Problem: *Consider the laminar boundary layer that develops on a flat plate aligned with the freestream flow direction. The flow is incompressible, the freestream flow speed is U and the pressure is constant in the flow direction, i.e., $\partial p/\partial x = 0$. The vertical velocity component is constant and equal to v_w. Determine the horizontal velocity component, $u(x,y)$. Is there any restriction on the value of v_w?*

Solution: First, note that, in two dimensions, the equations of motion for an incompressible boundary layer are

$$\frac{\partial u}{\partial x} + \frac{\partial v}{\partial y} = 0 \tag{14.7}$$

$$u\frac{\partial u}{\partial x} + v\frac{\partial u}{\partial y} = -\frac{1}{\rho}\frac{\partial p}{\partial x} + \nu\frac{\partial^2 u}{\partial y^2} \tag{14.8}$$

$$0 = -\frac{1}{\rho}\frac{\partial p}{\partial y} \tag{14.9}$$

Since we are given $v = v_w$ where v_w is constant, the continuity equation, Equation (14.7), simplifies to

$$\frac{\partial u}{\partial x} = 0 \implies u = u(y) \tag{14.10}$$

Therefore, since we know from the problem statement that the pressure gradient in the x direction is zero, the horizontal-momentum equation, Equation (14.8), becomes

$$v_w\frac{du}{dy} = \nu\frac{d^2u}{dy^2} \tag{14.11}$$

This equation must be solved subject to the no-slip surface boundary condition and the velocity must approach the freestream velocity far above the plate. Thus, we must solve subject to

$$u(0) = 0 \quad \text{and} \quad u(y) \to U \text{ as } y \to \infty \tag{14.12}$$

We begin by rewriting Equation (14.11) as

$$\frac{\frac{d}{dy}\left(\frac{du}{dy}\right)}{\frac{du}{dy}} = \frac{v_w}{\nu} \implies \ell n\left(\frac{du}{dy}\right) = \frac{v_w y}{\nu} + C_0 \tag{14.13}$$

where C_0 is an integration constant. Exponentiating both sides of Equation (14.13) gives

$$\frac{du}{dy} = C_1 e^{v_w y/\nu} \tag{14.14}$$

where $C_1 = e^{C_0}$. Integrating again over y yields

$$u(y) = C_2 e^{v_w y/\nu} + C_3 \tag{14.15}$$

where $C_2 = C_1 \nu/v_w$ and C_3 is another integration constant. In order to satisfy the no-slip boundary condition, $u(0) = 0$, we must have $C_2 = -C_3$. Therefore,

$$u(y) = C_3 \left(1 - e^{v_w y/\nu}\right) \tag{14.16}$$

Finally, in order for the velocity to approach the freestream value, U, far above the plate, the exponential term must decay. This will be true provided v_w is negative, in which case $C_3 = U$. Consequently, the horizontal velocity component is

$$u(y) = U\left(1 - e^{v_w y/\nu}\right), \qquad v_w < 0 \qquad (14.17)$$

14.3 Momentum Integral Equation I

Statement of the Problem: *An experimenter has made a series of wind-tunnel measurements to calibrate his tunnel. His measurements correspond to what he believes is a constant-pressure, laminar boundary layer. Based on his data, he has developed the following correlation for the velocity profile.*

$$\frac{u}{U} = \frac{3}{2}\left(\frac{y}{\delta}\right) - \frac{2}{5}\left(\frac{y}{\delta}\right)^3 - \frac{1}{10}\left(\frac{y}{\delta}\right)^5$$

(a) *Compute the skin friction, c_f, displacement thickness, δ^*, momentum thickness, θ, and shape factor, H.*

(b) *Using the momentum integral equation, determine the boundary-layer thickness, δ.*

(c) *Quantify the differences between the results of Parts (a) and (b) and the Blasius solution. Comment on whether or not the measurements establish the intended purpose of certifying the accuracy of the measurements.*

Solution: Before proceeding, note that we will compute the skin friction from the given velocity profile. In practice, skin friction, c_f, and velocities are often measured independently. For constant pressure, the momentum integral equation is

$$\frac{d\theta}{dx} = \frac{c_f}{2} \qquad (14.18)$$

where θ is the momentum thickness. The momentum integral equation serves as a consistency check in the sense that the left-hand side of Equation (14.18) depends only upon the measured velocities and the right-hand side depends only upon c_f.

14.3(a): For the given velocity profile, the surface shear stress, τ_w, is

$$\begin{aligned}\tau_w &= \mu\left(\frac{\partial u}{\partial y}\right)_{y=0} = \mu U \frac{\partial}{\partial y}\left[\frac{3}{2}\left(\frac{y}{\delta}\right) - \frac{2}{5}\left(\frac{y}{\delta}\right)^3 - \frac{1}{10}\left(\frac{y}{\delta}\right)^5\right]_{y=0} \\ &= \mu U \left[\frac{3}{2}\frac{1}{\delta} - \frac{6}{5}\frac{y^2}{\delta^3} - \frac{1}{2}\frac{y^4}{\delta^5}\right]_{y=0} = \frac{3}{2}\frac{\mu U}{\delta}\end{aligned} \qquad (14.19)$$

Hence, the skin friction coefficient, c_f, is

$$c_f = \frac{\tau_w}{\frac{1}{2}\rho U^2} = \frac{3\mu U}{\rho U^2 \delta} = 3\frac{\nu}{U\delta} \qquad (14.20)$$

14.3. MOMENTUM INTEGRAL EQUATION I

Also, the displacement thickness, δ^*, is

$$\begin{aligned}
\delta^* &= \int_0^\delta \left[1 - \frac{u}{U}\right] dy \\
&= \int_0^\delta \left[1 - \frac{3}{2}\frac{y}{\delta} + \frac{2}{5}\left(\frac{y}{\delta}\right)^3 + \frac{1}{10}\left(\frac{y}{\delta}\right)^5\right] dy \\
&= \delta \int_0^1 \left[1 - \frac{3}{2}\eta + \frac{2}{5}\eta^3 + \frac{1}{10}\eta^5\right] d\eta \qquad (\eta \equiv y/\delta) \\
&= \delta \left[\eta - \frac{3}{4}\eta^2 + \frac{1}{10}\eta^4 + \frac{1}{60}\eta^6\right]_{\eta=0}^{\eta=1} \\
&= \delta \left[1 - \frac{3}{4} + \frac{1}{10} + \frac{1}{60}\right] = \frac{11}{30}\delta \qquad (14.21)
\end{aligned}$$

Similarly, the momentum thickness, θ, is

$$\begin{aligned}
\theta &= \int_0^\delta \frac{u}{U}\left[1 - \frac{u}{U}\right] dy \\
&= \int_0^\delta \left[\frac{3}{2}\frac{y}{\delta} - \frac{2}{5}\left(\frac{y}{\delta}\right)^3 - \frac{1}{10}\left(\frac{y}{\delta}\right)^5\right]\left[1 - \frac{3}{2}\frac{y}{\delta} + \frac{2}{5}\left(\frac{y}{\delta}\right)^3 + \frac{1}{10}\left(\frac{y}{\delta}\right)^5\right] dy \\
&= \delta \int_0^1 \left[\frac{3}{2}\eta - \frac{2}{5}\eta^3 - \frac{1}{10}\eta^5\right]\left[1 - \frac{3}{2}\eta + \frac{2}{5}\eta^3 + \frac{1}{10}\eta^5\right] d\eta \\
&= \delta \int_0^1 \left[\frac{3}{2}\eta - \frac{9}{4}\eta^2 - \frac{2}{5}\eta^3 + \frac{6}{5}\eta^4 - \frac{1}{10}\eta^5 + \frac{7}{50}\eta^6 - \frac{2}{25}\eta^8 - \frac{1}{100}\eta^{10}\right] d\eta \\
&= \delta \left[\frac{3}{4}\eta^2 - \frac{3}{4}\eta^3 - \frac{1}{10}\eta^4 + \frac{6}{25}\eta^5 - \frac{1}{60}\eta^6 + \frac{1}{50}\eta^7 - \frac{2}{225}\eta^9 - \frac{1}{1100}\eta^{11}\right]_{\eta=0}^{\eta=1} \\
&= \delta \left[\frac{3}{4} - \frac{3}{4} - \frac{1}{10} + \frac{6}{25} - \frac{1}{60} + \frac{1}{50} - \frac{2}{225} - \frac{1}{1100}\right] = \frac{661}{4950}\delta \qquad (14.22)
\end{aligned}$$

Finally, the shape factor, H, is

$$H = \frac{\delta^*}{\theta} = \frac{11\delta/30}{661\delta/4950} = \frac{1815}{661} = 2.75 \qquad (14.23)$$

14.3(b): Substituting the computed c_f and θ from Part (a) into the momentum integral equation, Equation (14.18), we have

$$\frac{661}{4950}\frac{d\delta}{dx} = \frac{3}{2}\frac{\nu}{U\delta} \qquad (14.24)$$

So, multiplying through by $4950\delta/661$, we find

$$\delta \frac{d\delta}{dx} = \frac{7425}{661}\frac{\nu}{U} \quad \Longrightarrow \quad \frac{1}{2}\frac{d\delta^2}{dx} = \frac{7425}{661}\frac{\nu}{U} \qquad (14.25)$$

Hence, the boundary-layer thickness is

$$\delta^2 = \frac{14850}{661}\frac{\nu x}{U} + \text{constant} \qquad (14.26)$$

Since $\delta = 0$ at the leading edge of the plate, $x = 0$, we thus have

$$\delta = \sqrt{\frac{14850}{661}\frac{\nu x}{U}} = 4.74x Re_x^{-1/2} \tag{14.27}$$

14.3(c): From the results of Parts (a) and (b), there follows

$$c_f = 3\frac{\nu}{U\delta} = 3\frac{\nu}{U}\sqrt{\frac{661}{14850}\frac{U}{\nu x}} = \sqrt{\frac{661}{1650 Re_x}} = 0.633 Re_x^{-1/2} \tag{14.28}$$

$$\theta = \frac{661}{4950}\delta = \frac{661}{4950}\sqrt{\frac{14850}{661}\frac{\nu x}{U}} = x\sqrt{\frac{661}{1650 Re_x}} = 0.633 x Re_x^{-1/2} \tag{14.29}$$

$$\delta^* = \frac{11}{30}\delta = \frac{11}{30}\sqrt{\frac{14850}{661}\frac{\nu x}{U}} = x\sqrt{\frac{3993}{1322 Re_x}} = 1.738 x Re_x^{-1/2} \tag{14.30}$$

Comparing Equations (14.23) and (14.28) – (14.30) to the Blasius solution, we have the following results.

Property	Computed	Blasius	Error
$Re_x^{1/2} c_f$	0.633	0.664	-5%
$Re_x^{1/2} \theta/x$	0.633	0.664	-5%
$Re_x^{1/2} \delta^*/x$	1.738	1.721	1%
H	2.75	2.59	6%

These results are sufficiently close to the Blasius solution to validate the accuracy of the wind tunnel instrumentation, freestream conditions, etc.

14.4 Momentum Integral Equation II

Statement of the Problem: *Correlation of measurements shows that the velocity profile and the boundary-layer thickness for a laminar boundary layer can be approximated as follows:*

$$\frac{u}{u_e} = 2\left(\frac{y}{\delta}\right) - \left(\frac{y}{\delta}\right)^2 \quad \text{and} \quad \frac{\delta}{x} = 2\left(\frac{\nu}{U_o x}\right)^{3/5}$$

where ν is kinematic viscosity and U_o is a reference velocity. Using the momentum integral equation, determine the corresponding freestream velocity, $u_e(x)$.

Solution: In general, the momentum integral equation is

$$\frac{d\theta}{dx} + (2+H)\frac{\theta}{u_e}\frac{du_e}{dx} = \frac{c_f}{2} \tag{14.31}$$

where θ is momentum thickness, $H = \delta^*/\theta$ is shape factor, δ^* is displacement thickness and c_f is skin friction. For the given velocity profile, the surface shear stress, τ_w, is

$$\tau_w = \mu\left(\frac{\partial u}{\partial y}\right)_{y=0} = \mu u_e \frac{\partial}{\partial y}\left[2\left(\frac{y}{\delta}\right) - \left(\frac{y}{\delta}\right)^2\right]_{y=0}$$

$$= \mu u_e\left[2\frac{1}{\delta} - 2\frac{y}{\delta^2}\right]_{y=0} = \frac{2\mu u_e}{\delta} \tag{14.32}$$

14.4. MOMENTUM INTEGRAL EQUATION II

Hence, the skin friction coefficient, c_f, is

$$c_f = \frac{\tau_w}{\frac{1}{2}\rho u_e^2} = \frac{4\mu u_e}{\rho u_e^2 \delta} = 4\frac{\nu}{u_e \delta} \tag{14.33}$$

For the given velocity profile, the displacement thickness, δ^*, is

$$\begin{aligned}
\delta^* &= \int_0^\delta \left[1 - \frac{u}{u_e}\right] dy \\
&= \int_0^\delta \left[1 - 2\frac{y}{\delta} + \left(\frac{y}{\delta}\right)^2\right] dy \\
&= \delta \int_0^1 \left[1 - 2\eta + \eta^2\right] d\eta \qquad (\eta \equiv y/\delta) \\
&= \delta \left[\eta - \eta^2 + \frac{1}{3}\eta^3\right]_{\eta=0}^{\eta=1} \\
&= \delta \left[1 - 1 + \frac{1}{3}\right] = \frac{1}{3}\delta \tag{14.34}
\end{aligned}$$

Similarly, the momentum thickness, θ, is

$$\begin{aligned}
\theta &= \int_0^\delta \frac{u}{u_e}\left[1 - \frac{u}{u_e}\right] dy \\
&= \int_0^\delta \left[2\frac{y}{\delta} - \left(\frac{y}{\delta}\right)^2\right]\left[1 - 2\frac{y}{\delta} + \left(\frac{y}{\delta}\right)^2\right] dy \\
&= \delta \int_0^1 \left[2\eta - \eta^2\right]\left[1 - 2\eta + \eta^2\right] d\eta \\
&= \delta \int_0^1 \left[2\eta - 5\eta^2 + 4\eta^3 - \eta^4\right] d\eta \\
&= \delta \left[\eta^2 - \frac{5}{3}\eta^3 + \eta^4 - \frac{1}{5}\eta^5\right]_{\eta=0}^{\eta=1} \\
&= \delta \left[1 - \frac{5}{3} + 1 - \frac{1}{5}\right] = \frac{2}{15}\delta \tag{14.35}
\end{aligned}$$

Therefore, the shape factor, H, is

$$H = \frac{\delta^*}{\theta} = \frac{\delta/3}{2\delta/15} = \frac{5}{2} \tag{14.36}$$

Substituting the computed c_f, θ and H into the momentum integral equation, Equation (14.31), yields

$$\frac{2}{15}\frac{d\delta}{dx} + \left(2 + \frac{5}{2}\right)\frac{2}{15}\frac{\delta}{u_e}\frac{du_e}{dx} = 2\frac{\nu}{u_e \delta} \implies \frac{2}{15}\frac{d\delta}{dx} + \frac{3}{5}\frac{\delta}{u_e}\frac{du_e}{dx} = 2\frac{\nu}{u_e \delta} \tag{14.37}$$

Now, multiplying through by $15 u_e \delta$, we find

$$2 u_e \delta \frac{d\delta}{dx} + 9\delta^2 \frac{du_e}{dx} = 30\nu \implies u_e \frac{d\delta^2}{dx} + 9\delta^2 \frac{du_e}{dx} = 30\nu \tag{14.38}$$

We are given that the boundary-layer thickness is

$$\delta = 2\left(\frac{\nu}{U_o}\right)^{3/5} x^{2/5} \implies \delta^2 = 4\left(\frac{\nu}{U_o}\right)^{6/5} x^{4/5} \tag{14.39}$$

Combining Equations (14.38) and (14.39) gives the following differential equation for u_e.

$$\left(\frac{\nu}{U_o}\right)^{6/5}\left[\frac{16}{5}x^{-1/5}u_e + 36x^{4/5}\frac{du_e}{dx}\right] = 30\nu \tag{14.40}$$

We can solve this first-order, ordinary differential equation by assuming a solution of the form $u_e = Bx^m$. By inspection, the exponent m must be $1/5$, for then all terms in Equation (14.40) are constant. So, substituting $u_e = Bx^{1/5}$ into Equation (14.40) tells us that

$$\left(\frac{\nu}{U_o}\right)^{6/5}\left[\frac{16}{5}B + \frac{36}{5}B\right] = 30\nu \implies \left(\frac{\nu}{U_o}\right)^{6/5}\frac{52}{5}B = 30\nu \tag{14.41}$$

Solving for the constant B, there follows

$$B = \frac{75}{26}\frac{U_o^{6/5}}{\nu^{1/5}} \tag{14.42}$$

Therefore, the freestream velocity for this boundary layer is

$$u_e(x) = \frac{75}{26}U_o\left(\frac{U_o x}{\nu}\right)^{1/5} \tag{14.43}$$

14.5 Blasius Solution

Statement of the Problem: *A liquid is flowing with constant pressure and a freestream velocity of U = 4 ft/sec over a flat plate. The boundary layer thickness is δ = 0.9 inch at a distance x = 1.1 ft from the leading edge of the plate. Assuming the flow is laminar, determine the kinematic viscosity of the fluid. If the liquid is one of those listed in Table A.5 of Appendix A, which is it?*

Solution: The Blasius solution provides the properties of a constant-pressure, laminar boundary layer such as the one in this application. Hence, the boundary-layer thickness is

$$\delta = 5\frac{x}{\sqrt{Re_x}} \tag{14.44}$$

where Re_x is the plate-length Reynolds number given by

$$Re_x = \frac{Ux}{\nu} \tag{14.45}$$

Combining these two equations, we find

$$\delta = 5x\sqrt{\frac{\nu}{Ux}} = 5\sqrt{\frac{\nu x}{U}} \implies \nu = \frac{U\delta^2}{25x} \tag{14.46}$$

14.6. TRANSITIONAL BOUNDARY LAYER

Using the given information, the numerical value of the kinematic viscosity is

$$\nu = \frac{(4 \text{ ft/sec})(0.9/12 \text{ ft})^2}{25(1.1 \text{ ft})} = 8.18 \cdot 10^{-4} \frac{\text{ft}^2}{\text{sec}} \quad (14.47)$$

Reference to Table A.5 shows that the liquid is **SAE 10W-30 oil**.

As a final comment, it is important to verify that the boundary layer is indeed laminar. This will be true provided the Reynolds number is less than about 10^5. Although laminar flow can exist at higher Reynolds numbers, the flow must be laminar at this low a value. So,

$$Re_x = \frac{(4 \text{ ft/sec})(1.1 \text{ ft})}{8.18 \cdot 10^{-4} \text{ ft}^2/\text{sec}} = 5.38 \cdot 10^3 \quad (14.48)$$

This confirms that the boundary layer is laminar.

14.6 Transitional Boundary Layer

Statement of the Problem: *A flat plate that is L = 10 ft long by W = 5 ft wide is immersed in a stream of air moving at U = 60 ft/sec with kinematic viscosity $\nu = 1.62 \cdot 10^{-4}$ ft²/sec and density $\rho = 0.00234$ slug/ft³. The boundary layer undergoes transition from laminar to turbulent flow at a plate-length Reynolds number, Re_{x_t}, of 10^6. Ignoring the transition region, i.e., assuming the width of the transition region is negligibly small, compute the total force on the plate.*

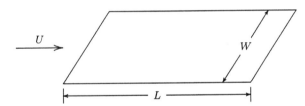

Figure 14.1: *Flow past a flat plate.*

Solution: In order to compute the force on the plate, we observe that there are boundary layers on both sides. Hence, the total force is

$$F = 2W \int_0^L \frac{1}{2} \rho U^2 c_f dx = \rho U^2 W \int_0^L c_f dx \quad (14.49)$$

Now, to determine the skin friction, we use the Blasius skin friction from the leading edge to the transition point. Ignoring the finite, but thin, transition region, we use a correlation of turbulent-flow skin friction. That is, we use

$$c_f = \begin{cases} \dfrac{0.664}{Re_x^{1/2}}, & x \leq x_t \quad \text{(Laminar: Blasius solution)} \\ \dfrac{0.0576}{Re_x^{1/5}}, & x \geq x_t \quad \text{(Turbulent: Correlation)} \end{cases} \quad (14.50)$$

where x_t denotes the distance from the leading edge of the plate at which transition occurs. Combining Equations (14.49) and (14.50), we find

$$\begin{aligned} F &= \rho U^2 W \left[\int_0^{x_t} \frac{0.664 \, dx}{Re_x^{1/2}} + \int_{x_t}^L \frac{0.0576 \, dx}{Re_x^{1/5}} \right] \\ &= \rho U^2 W \left[0.664 \left(\frac{\nu}{U}\right)^{1/2} \int_0^{x_t} \frac{dx}{x^{1/2}} + 0.0576 \left(\frac{\nu}{U}\right)^{1/5} \int_{x_t}^L \frac{dx}{x^{1/5}} \right] \\ &= \rho U^2 W \left[0.664 \left(\frac{\nu}{U}\right)^{1/2} 2 x^{1/2} \Big|_{x=0}^{x=x_t} + 0.0576 \left(\frac{\nu}{U}\right)^{1/5} \frac{5}{4} x^{4/5} \Big|_{x=x_t}^{x=L} \right] \\ &= \rho U^2 W \left[1.328 \left(\frac{\nu}{U}\right)^{1/2} x_t^{1/2} + 0.072 \left(\frac{\nu}{U}\right)^{1/5} \left(L^{4/5} - x_t^{4/5} \right) \right] \end{aligned} \quad (14.51)$$

Finally, we can write the final expression for the force in terms of dimensionless transition length, x_t/L, and Reynolds number based on L, i.e., Re_L, as follows.

$$F = \rho U^2 W L \left[\frac{1.328}{Re_L^{1/2}} \left(\frac{x_t}{L}\right)^{1/2} + \frac{0.072}{Re_L^{1/5}} \left\{ 1 - \left(\frac{x_t}{L}\right)^{4/5} \right\} \right] \quad (14.52)$$

To determine the force from the given values, note first that the Reynolds numbers based on the plate length at which transition occurs, x_t, and on the total plate length, L, are given by

$$Re_{x_t} = \frac{U x_t}{\nu} \quad \text{and} \quad Re_L = \frac{U L}{\nu} \quad (14.53)$$

Therefore, the ratio of x_t to L is

$$\frac{x_t}{L} = \frac{Re_{x_t}}{Re_L} \quad (14.54)$$

Also, the plate-length Reynolds number is

$$Re_L = \frac{U L}{\nu} = \frac{(60 \text{ ft/sec})(10 \text{ ft})}{1.62 \cdot 10^{-4} \text{ ft}^2/\text{sec}} = 3.70 \cdot 10^6 \quad (14.55)$$

So, since we are given $Re_{x_t} = 10^6$, substituting Equation (14.55) into Equation (14.54) yields

$$\frac{x_t}{L} = \frac{10^6}{3.70 \cdot 10^6} = 0.27 \quad (14.56)$$

Another useful constant appearing in the equation for the force on the plate, Equation (14.52), is the dimensional quantity $\rho U^2 W L$. Substituting the given values gives

$$\rho U^2 W L = \left(0.00234 \text{ slug/ft}^3 \right) (60 \text{ ft/sec})^2 (5 \text{ ft})(10 \text{ ft}) = 421.2 \text{ lb} \quad (14.57)$$

Thus, substituting Equations (14.55), (14.56) and (14.57) into Equation (14.52), we have

$$\begin{aligned} F &= (421.2 \text{ lb}) \left[\frac{1.328}{(3.70 \cdot 10^6)^{1/2}} (0.27)^{1/2} + \frac{0.072}{(3.70 \cdot 10^6)^{1/5}} \left\{ 1 - (0.27)^{4/5} \right\} \right] \\ &= \underbrace{0.151 \text{ lb}}_{\text{Laminar part}} + \underbrace{0.956 \text{ lb}}_{\text{Turbulent part}} = 1.107 \text{ lb} \end{aligned} \quad (14.58)$$

14.7 Turbulent Boundary Layers

Statement of the Problem: *Determine Coles' wake-strength parameter, $\tilde{\Pi}$, for the following boundary layers. Assume the surface is perfectly smooth in all cases.*

(a) *Favorable pressure gradient with $c_f = 0.00326$ and $Re_\delta = 5.80 \cdot 10^4$.*

(b) *Zero pressure gradient with $c_f = 0.00246$ and $Re_\delta = 1.47 \cdot 10^5$.*

(c) *Adverse pressure gradient with $c_f = 0.00163$ and $Re_\delta = 1.20 \cdot 10^5$.*

Solution: We begin with Coles' composite law-of-the-wall/law-of-the wake profile, which is:

$$\frac{\bar{u}}{u_\tau} = \frac{1}{\kappa}\ell n \frac{u_\tau y}{\nu} + C + \frac{2\tilde{\Pi}}{\kappa}\sin^2\left(\frac{\pi}{2}\frac{y}{\delta}\right) \qquad (14.59)$$

where $\bar{u}$ is mean velocity, u_τ is friction velocity, $\kappa = 0.41$ is Kármán's constant, y is distance normal to the surface, ν is kinematic viscosity, $C = 5.0$ for perfectly-smooth surfaces, $\tilde{\Pi}$ is Coles' wake-strength parameter and δ is boundary-layer thickness. So, denoting the velocity at the edge of the boundary layer by u_e, we have

$$\frac{u_e}{u_\tau} = \frac{1}{\kappa}\ell n \frac{u_\tau \delta}{\nu} + C + \frac{2\tilde{\Pi}}{\kappa} \qquad (14.60)$$

Solving for the wake-strength parameter yields

$$\tilde{\Pi} = \frac{\kappa}{2}\left[\frac{u_e}{u_\tau} - \frac{1}{\kappa}\ell n \frac{u_\tau \delta}{\nu} - C\right] \qquad (14.61)$$

We can write this in terms of skin friction, c_f, and Reynolds number based on boundary-layer thickness, Re_δ, by noting that $u_\tau = \sqrt{\tau_w/\rho}$ and $c_f = \tau_w/(\tfrac{1}{2}\rho u_e^2)$ so that

$$\frac{u_\tau}{u_e} = \sqrt{\frac{c_f}{2}} \quad \text{and} \quad \frac{u_\tau \delta}{\nu} = \frac{u_e \delta}{\nu}\frac{u_\tau}{u_e} = Re_\delta \sqrt{\frac{c_f}{2}} \qquad (14.62)$$

Substituting Equations (14.62) into Equation (14.61) gives $\tilde{\Pi}$ as a function of c_f and Re_δ:

$$\tilde{\Pi} = \frac{\kappa}{2}\left[\sqrt{\frac{2}{c_f}} - \frac{1}{\kappa}\ell n\left(Re_\delta \sqrt{\frac{c_f}{2}}\right) - C\right] \qquad (14.63)$$

14.7(a): Using the given values of $c_f = 0.00326$ and $Re_\delta = 5.80 \cdot 10^4$ yields:

$$\tilde{\Pi} = \frac{0.41}{2}\left[\sqrt{\frac{2}{0.00326}} - \frac{1}{0.41}\ell n\left(5.80 \cdot 10^4 \sqrt{\frac{0.00326}{2}}\right) - 5\right] = 0.173 \qquad (14.64)$$

14.7(b): Using the given values of $c_f = 0.00246$ and $Re_\delta = 1.47 \cdot 10^5$, we have:

$$\tilde{\Pi} = \frac{0.41}{2}\left[\sqrt{\frac{2}{0.00246}} - \frac{1}{0.41}\ell n\left(1.47 \cdot 10^5 \sqrt{\frac{0.00246}{2}}\right) - 5\right] = 0.546 \qquad (14.65)$$

14.7(c): Using the given values of $c_f = 0.00163$ and $Re_\delta = 1.20 \cdot 10^5$, there follows:

$$\tilde{\Pi} = \frac{0.41}{2}\left[\sqrt{\frac{2}{0.00163}} - \frac{1}{0.41}\ell n\left(1.20 \cdot 10^5 \sqrt{\frac{0.00163}{2}}\right) - 5\right] = 2.086 \qquad (14.66)$$

14.8 Law of the Wall I

Statement of the Problem: *Measurements show that for an incompressible turbulent boundary layer, the velocity is $\bar{u}$ = 15.85 m/sec at a distance y = 6 mm above a solid surface. Also, the measured skin friction is c_f = 0.003. The freestream velocity is u_e = 33 m/sec and the kinematic viscosity is $\nu = 1.5 \cdot 10^{-5}$ m^2/sec. Is the surface smooth or rough? If it's rough, what is the average surface-roughness height, k_s?*

Solution: The solution strategy is as follows. We know that close to a solid boundary, the law of the wall holds, i.e.,

$$\frac{\bar{u}}{u_\tau} = \frac{1}{\kappa}\ln\frac{u_\tau y}{\nu} + C \tag{14.67}$$

where $\bar{u}$ is mean velocity, u_τ is friction velocity, $\kappa = 0.41$ is Kármán's constant and C is a constant that depends upon the surface roughness. For perfectly-smooth surfaces, measurements show that $C = 5.0$. By contrast, for very rough surfaces, the law of the wall assumes an alternate form, viz.,

$$\frac{\bar{u}}{u_\tau} = \frac{1}{\kappa}\ln\frac{y}{k_s} + 8.5, \quad k_s^+ \equiv \frac{u_\tau k_s}{\nu} > 50 \tag{14.68}$$

The information given is sufficient to compute C. Should the computed value of C be significantly less than 5.0, we can manipulate Equation (14.68) to determine k_s.

First, note that the friction velocity and skin friction are given by

$$u_\tau = \sqrt{\frac{\tau_w}{\rho}} \quad \text{and} \quad c_f = \frac{\tau_w}{\frac{1}{2}\rho U^2} \tag{14.69}$$

Combining these two equations permits expressing u_τ as a function of u_e and c_f, both of which are given in the problem statement. Thus,

$$u_\tau = u_e\sqrt{\frac{c_f}{2}} = \left(33\,\frac{\text{m}}{\text{sec}}\right)\sqrt{\frac{0.003}{2}} = 1.278\,\frac{\text{m}}{\text{sec}} \tag{14.70}$$

Therefore, the quantities appearing in the law of the wall are

$$\frac{\bar{u}}{u_\tau} = \frac{15.85 \text{ m/sec}}{1.278 \text{ m/sec}} = 12.40 \tag{14.71}$$

$$\frac{u_\tau y}{\nu} = \frac{(1.278 \text{ m/sec})(0.006 \text{ m})}{1.5 \cdot 10^{-5} \text{ m}^2/\text{sec}} = 511 \tag{14.72}$$

So, substituting these values of u/u_τ and $u_\tau y/\nu$ into the law of the wall as stated in Equation (14.67) gives

$$C = \frac{\bar{u}}{u_\tau} - \frac{1}{\kappa}\ln\frac{u_\tau y}{\nu} = 12.40 - \frac{1}{0.41}\ln(511) = -2.81 \tag{14.73}$$

This value differs substantially from the smooth-wall value of 5.0. Therefore, **the surface is rough**. In order to determine the surface roughness height, k_s, we begin by rewriting the rough-wall law of the wall, Equation (14.68), as follows.

$$\frac{\bar{u}}{u_\tau} = \frac{1}{\kappa}\ln\frac{y}{k_s} + 8.5 = \frac{1}{\kappa}\ln\frac{u_\tau y}{\nu} + \left[8.5 - \frac{1}{\kappa}\ln\frac{u_\tau k_s}{\nu}\right] \tag{14.74}$$

14.9. LAW OF THE WALL II

Comparison of Equations (14.67) and (14.74) shows that

$$C = 8.5 - \frac{1}{\kappa}\ln\frac{u_\tau k_s}{\nu} \implies k_s^+ \equiv \frac{u_\tau k_s}{\nu} = e^{(8.5-C)\kappa} \quad (14.75)$$

Substituting the computed value of $C = -2.81$ from Equation (14.73),

$$k_s^+ = e^{(8.5+2.81)(0.41)} = 103 \quad (14.76)$$

This large a value of k_s^+ indeed corresponds to a completely-rough surface, for which the law of the wall assumes the form quoted in Equation (14.68). Finally, the roughness height is

$$k_s = k_s^+ \frac{\nu}{u_\tau} = 103\frac{1.5 \cdot 10^{-5} \text{ m}^2/\text{sec}}{1.278 \text{ m/sec}} = 1.21 \cdot 10^{-3} \text{ m} \quad (14.77)$$

Therefore, expressed in millimeters, the surface roughness height is

$$k_s = 1.21 \text{ mm} \quad (14.78)$$

14.9 Law of the Wall II

Statement of the Problem: *At a point y = 0.15 inch above the surface, the velocity in a boundary layer is $\overline{u} = 72$ ft/sec. The freestream velocity is $u_e = 108$ ft/sec and the kinematic viscosity is $\nu = 1.63 \cdot 10^{-4}$ ft^2/sec. Determine the skin friction assuming the surface is:*

(a) *Perfectly smooth*

(b) *Rough with dimensionless average sand-grain roughness height of $k_s^+ = 100$.*

Solution: We can determine the skin friction from the given information by using the law of the wall, viz.,

$$\frac{\overline{u}}{u_\tau} = \frac{1}{\kappa}\ln\frac{u_\tau y}{\nu} + C \quad (14.79)$$

where $\overline{u}$ is mean velocity, u_τ is friction velocity, $\kappa = 0.41$ is Kármán's constant, y is distance normal to the surface, ν is kinematic viscosity and C is a constant that depends on the nature of the surface. We can rewrite this equation in terms of u_τ/u_e as follows.

$$\frac{\overline{u}}{u_e}\frac{u_e}{u_\tau} = \frac{1}{\kappa}\ln\left(\frac{u_\tau}{u_e}\frac{u_e y}{\nu}\right) + C \quad (14.80)$$

But, from the given values, we have

$$\frac{\overline{u}}{u_e} = \frac{72 \text{ ft/sec}}{108 \text{ ft/sec}} = \frac{2}{3} \quad \text{and} \quad \frac{u_e y}{\nu} = \frac{(108 \text{ ft/sec})(0.15 \text{ in})/(12 \text{ in/ft})}{1.63 \cdot 10^{-4} \text{ ft}^2/\text{sec}} = 8282 \quad (14.81)$$

Therefore, for the given information, we have

$$\frac{2}{3}\left(\frac{u_e}{u_\tau}\right) = \frac{1}{\kappa}\ln\left(8282\frac{u_\tau}{u_e}\right) + C = \frac{1}{\kappa}\ln\left(\frac{u_\tau}{u_e}\right) + (22 + C) \quad (14.82)$$

When the surface is perfectly smooth, $C = 5.0$. When the dimensionless roughness height, k_s^+, exceeds 50, the surface is completely rough and the law of the wall assumes the following form.
$$\frac{\bar{u}}{u_\tau} = \frac{1}{\kappa} \ell n \frac{y}{k_s} + 8.5 \tag{14.83}$$

Comparison of Equations (14.79) and (14.83) shows that

$$C = 8.5 - \frac{1}{\kappa} \ell n \frac{u_\tau k_s}{\nu} = 8.5 - \frac{1}{\kappa} \ell n k_s^+; \qquad k_s^+ \equiv \frac{u_\tau k_s}{\nu} \tag{14.84}$$

Finally, we can determine the skin friction, c_f, once the ratio of u_τ to u_e is known. That is, noting that the friction velocity and skin friction are given by $u_\tau = \sqrt{\tau_w/\rho}$ and $c_f = \tau_w/(\frac{1}{2}\rho u_e^2)$, we have

$$c_f = 2 \left(\frac{u_\tau}{u_e}\right)^2 \tag{14.85}$$

14.9(a): As noted above, $C = 5.0$ for a perfectly-smooth surface. Thus, after a little rearrangement of terms, Equation (14.82) becomes

$$\frac{2}{3}\frac{u_e}{u_\tau} = \frac{1}{0.41} \ell n \frac{u_\tau}{u_e} + 27 = -\frac{1}{0.41} \ell n \frac{u_e}{u_\tau} + 27 \tag{14.86}$$

This is a transcendental equation for u_e/u_τ, which can be solved using a trial-end-error approach or, more systematically, with Newton's iterations. Either way, it is worthwhile to rewrite Equation (14.86) in terms of $x \equiv u_e/u_\tau$ as follows.

$$f(x) = \frac{2}{3}x + \frac{1}{0.41}\ell nx - 27 = 0 \tag{14.87}$$

Trial and error: The following table lists several values of $f(x)$ as a function of x.

x	$f(x)$
28.00	$-2.060 \cdot 10^{-1}$
28.20	$-5.531 \cdot 10^{-2}$
28.27	$-2.599 \cdot 10^{-3}$
28.30	$1.999 \cdot 10^{-2}$
29.00	$5.462 \cdot 10^{-1}$

Thus, we conclude that
$$\frac{u_e}{u_\tau} \approx 28.27 \tag{14.88}$$

Newton's iterations: First, note that

$$f'(x) = \frac{2}{3} + \frac{1}{0.41x} \tag{14.89}$$

Using Newton's iterations, we expand according to

$$f(x + \Delta x) = f(x) + f'(x)\Delta x + \cdots = 0 \quad \Longrightarrow \quad \Delta x \approx -\frac{f(x)}{f'(x)} \tag{14.90}$$

14.9. LAW OF THE WALL II

So, we make an (especially bad) initial guess of $x = 10$. Hence,

$$\begin{aligned}
f(10) &= \frac{2}{3}10 + \frac{1}{0.41}\ell n\,10 - 27 = -14.7173 \\
f'(10) &= \frac{2}{3} + \frac{1}{0.41 \cdot 10} = 0.91057 \\
\Delta x &= -\frac{-14.7173}{0.91057} = 16.1627
\end{aligned}$$

Thus, as the next guess, we try $x = 10 + 16.1627 = 26.1627$. Then,

$$\begin{aligned}
f(26.1627) &= \frac{2}{3}26.1627 + \frac{1}{0.41}\ell n\,26.1627 - 27 = -1.5964 \\
f'(26.1627) &= \frac{2}{3} + \frac{1}{0.41 \cdot 26.1627} = 0.75989 \\
\Delta x &= -\frac{-1.5964}{0.75989} = 2.1008
\end{aligned}$$

For the third guess, we try $x = 26.1627 + 2.1008 = 28.2635$. Then,

$$\begin{aligned}
f(28.2635) &= \frac{2}{3}28.2635 + \frac{1}{0.41}\ell n\,28.2635 - 27 = -7.49 \cdot 10^{-3} \\
f'(28.2635) &= \frac{2}{3} + \frac{1}{0.41 \cdot 28.2635} = 0.75296 \\
\Delta x &= -\frac{-7.49 \cdot 10^{-3}}{0.75296} = 0.0100
\end{aligned}$$

wherefore $x = 28.2635 + 0.0100 = 28.2735$. Therefore, as determined above by trial and error,

$$x \approx 28.27 \tag{14.91}$$

Finally, recalling Equation (14.85) and the fact that $x = u_e/u_\tau$, the skin friction is

$$c_f = 2\left(\frac{1}{28.27}\right)^2 = 0.00250 \tag{14.92}$$

14.9(b): Recalling Equation (14.84), the constant C for $k_s^+ = 100$ is

$$C = 8.5 - \frac{1}{0.41}\ell n\,100 = -2.73 \tag{14.93}$$

so that the transcendental equation for u_e/u_τ becomes

$$\frac{2}{3}\frac{u_e}{u_\tau} = \frac{1}{0.41}\ell n\frac{u_\tau}{u_e} + 19.27 \tag{14.94}$$

As in Part (a), we define $x \equiv u_e/u_\tau$ and rewrite Equation (14.94) as follows.

$$f(x) = \frac{2}{3}x + \frac{1}{0.41}\ell n\,x - 19.27 = 0 \tag{14.95}$$

Trial and error: The following table lists several values of $f(x)$ as a function of x.

x	$f(x)$
18.00	$-2.203 \cdot 10^{-1}$
18.20	$-6.003 \cdot 10^{-2}$
18.27	$-3.999 \cdot 10^{-3}$
18.30	$2.000 \cdot 10^{-2}$
19.00	$5.782 \cdot 10^{-1}$

Thus, we conclude that

$$\frac{u_e}{u_\tau} \approx 18.27 \tag{14.96}$$

Newton's iterations: This time, we begin with an initial guess of $x = 28$. Hence,

$$f(28) = \frac{2}{3}28 + \frac{1}{0.41}\ell n\, 28 - 19.27 = 7.5240$$

$$f'(28) = \frac{2}{3} + \frac{1}{0.41 \cdot 28} = 0.75377$$

$$\Delta x = -\frac{7.5240}{0.75377} = -9.9818$$

Thus, as the next guess, we try $x = 28 - 9.9818 = 18.0182$. Then,

$$f(18.0182) = \frac{2}{3}18.0182 + \frac{1}{0.41}\ell n\, 18.0182 - 19.27 = -0.20585$$

$$f'(18.0182) = \frac{2}{3} + \frac{1}{0.41 \cdot 18.0182} = 0.80203$$

$$\Delta x = -\frac{-0.20585}{0.80203} = 0.2567$$

For the third guess, we try $x = 18.0182 + 0.2567 = 18.2749$. Then,

$$f(18.2749) = \frac{2}{3}18.2749 + \frac{1}{0.41}\ell n\, 18.2749 - 19.27 = -7.84 \cdot 10^{-5}$$

$$f'(18.2749) = \frac{2}{3} + \frac{1}{0.41 \cdot 18.2749} = 0.80010$$

$$\Delta x = -\frac{-7.74 \cdot 10^{-5}}{0.80010} = 9.80 \cdot 10^{-5} \approx 0.0001$$

wherefore $x = 18.2749 + 0.0001 = 18.2750$. Therefore, as determined above by trial and error,

$$x \approx 18.27 \tag{14.97}$$

Finally, recalling Equation (14.85) and the fact that $x = u_e/u_\tau$, the skin friction is

$$c_f = 2\left(\frac{1}{18.27}\right)^2 = 0.00599 \tag{14.98}$$

This is more than double the smooth-wall value determined in Part (a) [see Equation (14.92)].

14.10 CFD—Truncation Error

Statement of the Problem: *Verify that the following formula for equally-spaced grid points is third-order accurate.* **HINT:** *Be sure to expand in Taylor series about (x_{i+1}, y_j).*

$$\left(\frac{\partial u}{\partial x}\right)_{i+1,j} \approx \frac{11u_{i+1,j} - 18u_{i,j} + 9u_{i-1,j} - 2u_{i-2,j}}{6\Delta x}$$

Solution: In order to verify that the proposed discretization approximation is third-order accurate, we must retain quartic terms in the Taylor-series expansions. Expanding about (x_{i+1}, y_j), we have

$$u_{i+1,j} = u \tag{14.99}$$

$$u_{i,j} = u - u_x \Delta x + \frac{1}{2}u_{xx}(\Delta x)^2 - \frac{1}{6}u_{xxx}(\Delta x)^3 + \frac{1}{24}u_{xxxx}(\Delta x)^4 + \cdots \tag{14.100}$$

$$u_{i-1,j} = u - 2u_x \Delta x + \frac{1}{2}u_{xx}(2\Delta x)^2 - \frac{1}{6}u_{xxx}(2\Delta x)^3 + \frac{1}{24}u_{xxxx}(2\Delta x)^4 + \cdots \tag{14.101}$$

$$u_{i-2,j} = u - 3u_x \Delta x + \frac{1}{2}u_{xx}(3\Delta x)^2 - \frac{1}{6}u_{xxx}(3\Delta x)^3 + \frac{1}{24}u_{xxxx}(3\Delta x)^4 + \cdots \tag{14.102}$$

Hence, combining these terms according to the given discretization approximation,

$$\frac{11u_{i+1,j} - 18u_{i,j} + 9u_{i-1,j} - 2u_{i-2,j}}{6\Delta x} = \frac{11}{6\Delta x}[u]$$
$$- \frac{18}{6\Delta x}\left[u - u_x \Delta x + \frac{1}{2}u_{xx}(\Delta x)^2 - \frac{1}{6}u_{xxx}(\Delta x)^3 + \frac{1}{24}u_{xxxx}(\Delta x)^4 + \cdots\right]$$
$$+ \frac{9}{6\Delta x}\left[u - 2u_x \Delta x + 2u_{xx}(\Delta x)^2 - \frac{4}{3}u_{xxx}(\Delta x)^3 + \frac{2}{3}u_{xxxx}(\Delta x)^4 + \cdots\right]$$
$$- \frac{2}{6\Delta x}\left[u - 3u_x \Delta x + \frac{9}{2}u_{xx}(\Delta x)^2 - \frac{9}{2}u_{xxx}(\Delta x)^3 + \frac{27}{8}u_{xxxx}(\Delta x)^4 + \cdots\right]$$
$$= \left(\frac{11 - 18 + 9 - 2}{6}\right)\frac{u}{\Delta x} + \left(\frac{18 - 18 + 6}{6}\right)u_x + \left(\frac{-9 + 18 - 9}{6}\right)u_{xx}\Delta x$$
$$+ \left(\frac{3 - 12 + 9}{6}\right)u_{xxx}(\Delta x)^2 + \left(\frac{-3/4 + 6 - 27/4}{6}\right)u_{xxxx}(\Delta x)^3 + \cdots \tag{14.103}$$

Therefore, we conclude that

$$\frac{11u_{i+1,j} - 18u_{i,j} + 9u_{i-1,j} - 2u_{i-2,j}}{6\Delta x} = \left(\frac{\partial u}{\partial x}\right)_{i+1,j} - \frac{1}{4}\left(\frac{\partial^4 u}{\partial x^4}\right)_{i+1,j}(\Delta x)^3 + \cdots$$

Since the first neglected term is of third order, we conclude that this formula is third-order accurate.

Chapter 15

Viscous and 2-D Compressible Flow

This chapter covers effects of viscosity, heat transfer and two-dimensionality in compressible flows. The chapter begins with Fanno and Rayleigh flow. Fanno flow is adiabatic flow in a pipe or duct with constant area including viscous effects. In Rayleigh flow we ignore viscous effects while including heat transfer. The chapter also addresses two-dimensional waves, including oblique shocks and the Prandtl-Meyer expansion. An oblique shock forms whenever a supersonic flow turns *into itself* and undergoes compression. Flow properties change abruptly across a shock wave (whose thickness is a few mean free paths) and the entropy increases. This occurs, for example, at the tip of a wedge-shaped body. A Prandtl-Meyer expansion occurs when a supersonic flow turns *away from itself* and undergoes expansion. This occurs on the upper surface of an aileron that is deflected downward. In this case the flow is isentropic. Appendix C includes tables for Fanno flow, Rayleigh flow, shock waves and the Prandtl-Meyer expansion that can be used for gases with a specific-heat ratio $\gamma = 1.4$. Algebraic formulas must be used for other gases.

Section 15.1 Fanno Flow I: The problem in this section focuses on Fanno flow. The solution makes use of the Fanno-flow *tables*.

Section 15.2 Fanno Flow II: Unlike the problem in the previous section, this one involves a gas with $\gamma \neq 1.4$, thus precluding use of the Fanno-flow tables. Rather, the algebraic *equations* must be used.

Section 15.3 Rayleigh Flow I: This is a problem using the Rayleigh-flow *tables* to determine the heat that must be added to a subsonic inlet flow in order to achieve sonic conditions.

Section 15.4 Rayleigh Flow II: This is a Rayleigh-flow problem whose solution uses the Rayleigh-flow *equations* for a flow with a supersonic inlet.

Section 15.5 Oblique Shock Wave: An oblique shock wave problem that uses the normal-shock *tables* to calculate oblique-shock properties.

Section 15.6 Reflection of an Oblique Shock Wave: This problem is similar to that of Section 15.5, but has an added complication, i.e., the shock reflects from a solid boundary.

Section 15.7 Prandtl-Meyer Expansion I: This is the first of two problems dealing with a Prandtl Meyer expansion. This one uses the Prandtl Meyer *tables* to calculate flow properties in an isentropic expansion of a gas with $\gamma = 1.4$.

Section 15.8 Prandtl-Meyer Expansion II: This second Prandtl-Meyer application uses the Prandtl Meyer *equations* to calculate flow properties in an isentropic expansion of a gas with $\gamma \neq 1.4$.

Section 15.9 Compressible Law of the Wall: This problem deals with the compressible law of the wall for turbulent flow over an adiabatic surface. Given measured skin friction, the problem's solution compares the velocity close to the surface with values determined from the incompressible law of the wall.

Section 15.10 CFD—Numerical Dissipation and Dispersion: A computation that determines the amount of numerical dissipation and dispersion for a specified finite-difference scheme.

15.1 Fanno Flow I

Statement of the Problem: *Air flows adiabatically through a pipe of diameter $D = 6$ inches. The flow enters with a pressure, $p_i = 100$ psi, average flow velocity, $\bar{u} = 1650$ ft/sec and temperature, $T_i = -4°$ F. If the pipe length is $L = 45$ inches and the exit pressure is $p_e = 145$ psi, what is the average friction factor, $\bar{f}$, and the exit Mach number, M_e?*

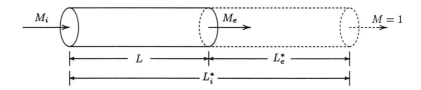

Figure 15.1: *Adiabatic flow in a pipe with friction.*

Solution: We are given the temperature at the inlet, viz., $T_i = -4°$ F $= 455.67°$ R. Hence, the speed of sound is

$$a = \sqrt{\gamma R T} = \sqrt{1.4 \left(1716 \, \frac{\text{ft} \cdot \text{lb}}{\text{slug}°\text{R}}\right)(455.67° \text{R})} = 1046 \, \frac{\text{ft}}{\text{sec}} \quad (15.1)$$

where the specific-heat ratio, γ, and the perfect-gas constant, R, for air are listed in Table A.1. Thus, the inlet Mach number is

$$M_i = \frac{\bar{u}}{a} = \frac{1650 \text{ ft/sec}}{1046 \text{ ft/sec}} = 1.57 \quad (15.2)$$

Because $\gamma = 1.4$, we can use the Fanno-flow tables (see Section C.3). Using linear interpolation, we have

M_i	$\bar{f}L_i^*/D$	p_i/p^*	Source
1.55	0.1543	0.5808	Table
1.57	0.1615	0.5712	Interpolation
1.60	0.1724	0.5568	Table

Therefore, $\bar{f}L_i^*/D = 0.1615$ and $p_i/p^* = 0.5712$. Hence, the sonic pressure is

$$p^* = \frac{p_i}{p_i/p^*} = \frac{100 \text{ psi}}{0.5712} = 175 \text{ psi} \quad (15.3)$$

15.2. FANNO FLOW II

Now, we are given $p_e = 145$ psi, wherefore

$$\frac{p_e}{p^*} = \frac{145 \text{ psi}}{175 \text{ psi}} = 0.8286 \tag{15.4}$$

Again using the Fanno-flow tables, linear interpolation yields

p_e/p^*	M_e	$\overline{f}L_e^*/D$	Source
0.8471	1.15	0.0205	Table
0.8286	1.17	0.0262	Interpolation
0.8044	1.20	0.0336	Table

Therefore, $\overline{f}L_e^*/D = 0.0262$ and $M_e = 1.17$. Finally, as shown in Figure 15.1, the length of the pipe is related to L_i^* and L_e^* by

$$L_i^* = L + L_e^* \implies \frac{\overline{f}L}{D} = \frac{\overline{f}L_i^*}{D} - \frac{\overline{f}L_e^*}{D} \tag{15.5}$$

Using the values computed above,

$$\frac{\overline{f}L}{D} = 0.1615 - 0.0262 = 0.1353 \tag{15.6}$$

So, since $D = 6$ inches and $L = 45$ inches, the average friction factor is

$$\overline{f} = 0.1353 \frac{D}{L} = 0.1353 \left(\frac{6 \text{ in}}{45 \text{ in}} \right) = 0.018 \tag{15.7}$$

Thus, in summary, the average friction factor and exit Mach number are

$$\overline{f} = 0.018 \quad \text{and} \quad M_e = 1.17 \tag{15.8}$$

15.2 Fanno Flow II

Statement of the Problem: Carbon Dioxide flows adiabatically through a smooth rectangular channel of sides 9 cm by 20 cm, and length $L = 10$ m. At the inlet, the pressure is $p = 100$ kPa, the velocity is $\overline{u} = 665$ m/sec and the temperature is $T = 15°$ C. Determine the Mach number, pressure and temperature at the exit.

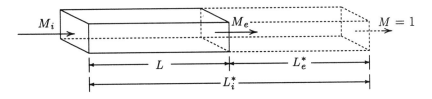

Figure 15.2: *Adiabatic flow in a rectangular duct with friction.*

Solution: Reference to Table A.1 shows that $\gamma = 1.30$ and $R = 189$ J/(kg·K) for CO_2. We are given the temperature at the inlet, viz., $T = 15°$ C $= 288.16$ K. Hence, the speed of sound is

$$a = \sqrt{\gamma RT} = \sqrt{1.30 \left(189 \frac{\text{J}}{\text{kg} \cdot \text{K}} \right) (288.16 \text{ K})} = 266 \frac{\text{m}}{\text{sec}} \tag{15.9}$$

Thus, the inlet Mach number is

$$M_i = \frac{\overline{u}}{a} = \frac{665 \text{ m/sec}}{266 \text{ m/sec}} = 2.5 \tag{15.10}$$

To determine the friction factor, we must first determine the Reynolds number. From Table A.5, the kinematic viscosity of carbon dioxide at 15° C is $\nu = 7.84 \cdot 10^{-6}$ m^2/sec. Because the duct has a non-circular cross section, we must determine its hydraulic diameter before we can use the Moody diagram to determine the friction factor. For this geometry, we have

$$D_h = \frac{4A}{P} = \frac{4(0.09 \text{ m})(0.20 \text{ m})}{2(0.09 \text{ m} + 0.20 \text{ m})} = 0.124 \text{ m} \tag{15.11}$$

So, the Reynolds number based on D_h is

$$Re_{D_h} = \frac{\overline{u} D_h}{\nu} = \frac{(665 \text{ m/sec})(0.124 \text{ m})}{7.84 \cdot 10^{-6} \text{ m}^2/\text{sec}} = 1.05 \cdot 10^7 \tag{15.12}$$

From the Moody diagram, Figure (B.10), the friction factor for incompressible flow through a smooth pipe of diameter D_h at this Reynolds number is

$$f_{inc} = 0.008 \tag{15.13}$$

So, taking account of the Mach number, the average friction factor is

$$\overline{f} \approx \frac{f_{inc}}{\sqrt{1 + \frac{\gamma-1}{2} r M^2}} \tag{15.14}$$

Because the flow is turbulent, we assume a recovery factor of $r = 0.89$, wherefore

$$\overline{f} = \frac{0.008}{\sqrt{1 + 0.15(0.89)(2.5)^2}} = 0.0059 \tag{15.15}$$

Because $\gamma \neq 1.4$, we cannot use the Fanno-flow tables. Rather, we must use the Fanno-flow formulas. First, note that

$$\frac{\overline{f} L_i^*}{D} = \frac{1 - M_i^2}{\gamma M_i^2} + \frac{\gamma+1}{2\gamma} \ell n \left[\frac{(\gamma+1)M_i^2}{2 + (\gamma-1)M_i^2} \right] \tag{15.16}$$

So, with $M_i = 2.5$ and $\gamma = 1.30$,

$$\frac{\overline{f} L_i^*}{D} = \frac{1 - (2.5)^2}{1.30(2.5)^2} + \frac{2.30}{2(1.30)} \ell n \left[\frac{(2.30)(2.5)^2}{2 + 0.30(2.5)^2} \right] = 0.5135 \tag{15.17}$$

Also, the inlet-pressure ratio is

$$\frac{p_i}{p^*} = \frac{1}{M_i} \left[\frac{\gamma+1}{2 + (\gamma-1)M_i^2} \right]^{1/2} = \frac{1}{2.5} \left[\frac{2.30}{2 + 0.30(2.5)^2} \right]^{1/2} = 0.3082 \tag{15.18}$$

and the inlet-temperature ratio is

$$\frac{T_i}{T^*} = \frac{\gamma+1}{2 + (\gamma-1)M_i^2} = \frac{2.30}{2 + 0.30(2.5)^2} = 0.5935 \tag{15.19}$$

15.2. FANNO FLOW II

Now, the lengths from the inlet and outlet to the sonic point are related by

$$L_e^* = L_i^* - L \quad \Longrightarrow \quad \frac{\overline{f}L_e^*}{D_h} = \frac{\overline{f}L_i^*}{D_h} - \frac{\overline{f}L}{D_h} \tag{15.20}$$

So, using $\overline{f}L_i^*/D_h = 0.5135$ and $\overline{f} = 0.0059$ as computed above along with $L = 10$ m and $D_h = 0.124$ m, we have

$$\frac{\overline{f}L_e^*}{D_h} = 0.5135 - \frac{(0.0059)(10 \text{ m})}{0.124 \text{ m}} = 0.0377 \tag{15.21}$$

To determine M_e, we can solve the following equation, which defines M_e implicitly, by trial and error.

$$\frac{\overline{f}L_e^*}{D_h} = \frac{1 - M_e^2}{\gamma M_e^2} + \frac{\gamma + 1}{2\gamma} \ell n \left[\frac{(\gamma + 1)M_e^2}{2 + (\gamma - 1)M_e^2} \right] \tag{15.22}$$

The following table summarizes $\overline{f}L_e^*/D_h$ as a function of M_e.

M_e	$\overline{f}L_e^*/D_h$
1.10	0.0112
1.19	0.0350
1.20	0.0382
1.21	0.0414
1.30	0.0739

Therefore, we conclude that $M_e \approx 1.20$. So, the exit-pressure and exit-temperature ratios are

$$\frac{p_e}{p^*} = \frac{1}{M_e} \left[\frac{\gamma + 1}{2 + (\gamma - 1)M_e^2} \right]^{1/2} = \frac{1}{1.20} \sqrt{\frac{2.30}{2 + 0.30(1.20)^2}} = 0.8104 \tag{15.23}$$

$$\frac{T_e}{T^*} = \frac{\gamma + 1}{2 + (\gamma - 1)M_e^2} = \frac{2.30}{2 + 0.30(1.20)^2} = 0.9457 \tag{15.24}$$

Finally, the values of p and T at the exit are

$$p_e = \frac{p_e/p^*}{p_i/p^*} p_i = \frac{0.8104}{0.3082}(100 \text{ kPa}) = 263 \text{ kPa} \tag{15.25}$$

$$T_e = \frac{T_e/T^*}{T_i/T^*} T_i = \frac{0.9457}{0.5935}(288.16 \text{ K}) = 459.16 \text{ K} = 186° \text{ C} \tag{15.26}$$

Therefore, in summary, the Mach number, pressure and temperature at the exit are

$$M_e = 1.20, \quad p_e = 263 \text{ kPa}, \quad T_e = 186° \text{ C} \tag{15.27}$$

15.3 Rayleigh Flow I

Statement of the Problem: *Air with velocity $\bar{u} = 75$ m/sec enters a pipe of circular cross section at a pressure $p = 200$ kPa and a temperature $T = -10°$ C. Find the heat per unit mass that must be added to achieve sonic conditions at the exit, and compute the corresponding exit pressure.*

Solution: We are given the temperature at the inlet, viz., $T = -10°$ C $= 263.16$ K. Hence, the speed of sound is

$$a = \sqrt{\gamma RT} = \sqrt{1.4 \left(287 \, \frac{\text{J}}{\text{kg} \cdot \text{K}}\right)(263.16 \text{ K})} = 325 \, \frac{\text{m}}{\text{sec}} \quad (15.28)$$

where the specific-heat ratio, γ, and the perfect-gas constant, R, for air are listed in Table A.1. Thus, the inlet Mach number is

$$M_1 = \frac{\bar{u}}{a} = \frac{75 \text{ m/sec}}{325 \text{ m/sec}} = 0.23 \quad (15.29)$$

Because $\gamma = 1.4$, we can use the Rayleigh-flow tables (see Section C.4). Using linear interpolation, we have

M_1	T_{t1}/T_t^*	p_1/p^*	Source
0.20	0.1736	2.2727	Table
0.23	0.2235	2.2332	Interpolation
0.25	0.2568	2.2069	Table

Therefore, $T_{t1}/T_t^* = 0.2235$ and $p_1/p^* = 2.2332$. Also, the inlet total temperature is

$$T_{t1} = T_1 \left[1 + \frac{\gamma - 1}{2} M_1^2\right] = (263.16 \text{ K})\left[1 + 0.2(0.23)^2\right] = 266 \text{ K} \quad (15.30)$$

Thus, the total temperature at the sonic point for this pipe, T_t^*, is

$$T_t^* = \frac{T_{t1}}{T_{t1}/T_t^*} = \frac{266 \text{ K}}{0.2235} = 1190 \text{ K} \quad (15.31)$$

Now, for Rayleigh flow, the total temperature increases because of the heat that is added according to

$$T_t^* = T_{t1} + \frac{q}{c_p} \quad \Longrightarrow \quad q = c_p \left(T_t^* - T_{t1}\right) \quad (15.32)$$

where c_p is the specific-heat coefficient. For air, $c_p = 1004$ J/(kg·K) (see table A.1), wherefore

$$q = \left(1004 \, \frac{\text{J}}{\text{kg} \cdot \text{K}}\right)(1190 \text{ K} - 266 \text{ K}) = 9.28 \cdot 10^5 \, \frac{\text{J}}{\text{kg}} \quad (15.33)$$

Also, the pressure at the exit, for sonic conditions, is

$$p^* = \frac{p_1}{p_1/p^*} = \frac{200 \text{ kPa}}{2.2332} = 89.6 \text{ kPa} \quad (15.34)$$

Thus, in summary, the heat added and the exit pressure are

$$q = 9.28 \cdot 10^5 \, \frac{\text{J}}{\text{kg}} \quad \text{and} \quad p^* = 89.6 \text{ kPa} \quad (15.35)$$

15.4 Rayleigh Flow II

Statement of the Problem: *You have just purchased a blast-gas dispenser, which shoots any loaded gas out of a tube at high velocity. The gas you have chosen to work with is helium. The dispenser includes a heat-control unit that helps regulate the exit temperature. The gas exhausts to the atmosphere for which the pressure is 14.7 psi. The speed of the exit gas is 4500 ft/sec, while the static and total temperature are 70° F and 90° F, respectively. The static temperature of the helium as it enters the heat-control unit is 110° F. You may assume the flow through the heat-control unit can be approximated as Rayleigh flow, and that the inlet flow is supersonic.*

(a) *Find the flow speed at the inlet to the heat-control unit.*

(b) *Compute the heat transfer rate as the gas flows through the heat-control unit and determine if the unit cools or heats the gas.*

Solution: Because the specific-heat ratio of helium is $\gamma = 1.66 \neq 1.40$, we cannot use the Rayleigh-flow tables to solve this problem. Rather, we must use the Rayleigh-flow equations.

15.4(a): Since we are given the inlet static temperature, T_i, we can compute the inlet Mach number, M_i, from the following Rayleigh-Flow relation:

$$\frac{T_i}{T^*} = M_i^2 \left[\frac{\gamma+1}{1+\gamma M_i^2}\right]^2 \tag{15.36}$$

where T^* is the static temperature at the sonic point. However, we must first compute T^* from the given conditions at the exit. Since we are given the temperature at the outlet, viz., $T_e = 70° \text{F} + 459.67° \text{R} = 529.67° \text{R}$, the speed of sound at the exit is

$$a_e = \sqrt{\gamma R T} = \sqrt{1.66 \left(12419 \,\frac{\text{ft} \cdot \text{lb}}{\text{slug} \cdot °\text{R}}\right)(529.67° \text{R})} = 3304 \,\frac{\text{ft}}{\text{sec}} \tag{15.37}$$

where the perfect-gas constant, R, for helium is listed in Table A.1. Thus, the outlet Mach number is

$$M_e = \frac{u_e}{a_e} = \frac{4500 \text{ ft/sec}}{3304 \text{ ft/sec}} = 1.362 \tag{15.38}$$

Using Equation (15.36) with T_e and M_e replacing T_i and M_i, we find

$$\frac{T_e}{T^*} = (1.362)^2 \left[\frac{1.66+1}{1+1.66(1.362)^2}\right]^2 = 0.7887 \tag{15.39}$$

Therefore, the temperature at the sonic point is

$$T^* = \frac{T_e}{T_e/T^*} = \frac{529.67° \text{R}}{0.7887} = 671.57° \text{R} \tag{15.40}$$

wherefore the ratio of $T_i = 110° \text{F} + 459.67° \text{R} = 569.67° \text{R}$ to T^* is

$$\frac{T_i}{T^*} = \frac{569.67° \text{R}}{671.57° \text{R}} = 0.8483 \tag{15.41}$$

Combining Equations (15.36) and (15.41) yields

$$0.8483 = M_i^2 \left[\frac{\gamma+1}{1+\gamma M_i^2}\right]^2 \implies 0.9210 = \frac{(\gamma+1)M_i}{1+\gamma M_i^2} \quad (15.42)$$

Multiplying both sides of this equation by $(1+\gamma M_i^2)$ and substituting $\gamma = 1.66$ leads to the following quadratic equation for M_i.

$$1.5289 M_i^2 - 2.66 M_i + 0.9210 = 0 \implies M_i = 0.477, 1.263 \quad (15.43)$$

Since the inlet flow is known from the problem statement to be supersonic, the inlet Mach number is therefore

$$M_i = 1.263 \quad (15.44)$$

Finally, the inlet velocity is given by

$$\bar{u}_i = M_i a_i = M_i \sqrt{\gamma R T_i} \quad (15.45)$$

Therefore, using the given inlet temperature and the value of M_i from Equation (15.44), the flow speed at the inlet to the heat-control unit is

$$\bar{u}_i = 1.263 \sqrt{1.66 \left(12419 \, \frac{\text{ft} \cdot \text{lb}}{\text{slug} \cdot {}^\circ \text{R}}\right)(569.67^\circ \text{R})} = 4328 \, \frac{\text{ft}}{\text{sec}} \quad (15.46)$$

15.4(b): To determine the heat added, we must compute the change in total temperature. We are given the total temperature at the exit, and we computed the exit Mach number in Part (a). Hence, we can use the Rayleigh-flow total-temperature equation to determine the total temperature at the sonic point, T_t^*, viz.,

$$\frac{T_t}{T_t^*} = \frac{(\gamma+1)M^2}{(1+\gamma M^2)^2}\left[2 + (\gamma-1)M^2\right] \quad (15.47)$$

Then, using the same equation at the inlet, we can substitute the known values of M_i and T_t^* to determine the total temperature at the inlet, T_{t_i}. We then have sufficient information to compute the heat added from the following equation:

$$q = c_p (T_{t_i} - T_{t_e}) \quad (15.48)$$

where c_p is the specific-heat coefficient for constant pressure.

So, to begin, recall that in Part (a), we have shown in Equation (15.38) that the exit Mach number is $M_e = 1.362$. Hence, the total-temperature ratio at the exit is

$$\frac{T_{t_e}}{T_t^*} = \frac{(1.66+1)(1.362)^2}{[1+1.66(1.362)^2]^2}\left[2 + (1.66-1)(1.362)^2\right] = 0.9560 \quad (15.49)$$

Since the total temperature at the exit is $T_{t_e} = 90^\circ \text{F} + 459.67^\circ \text{R} = 549.67^\circ \text{R}$, there follows

$$T_t^* = \frac{T_{t_e}}{T_{t_e}/T_t^*} = \frac{549.67^\circ \text{R}}{0.9560} = 574.97^\circ \text{R} \quad (15.50)$$

Thus, combining Equations (15.44), (15.47) and (15.50), the total temperature at the inlet is

$$T_{t_i} = (574.97^\circ \text{R})\frac{(1.66+1)(1.263)^2}{[1+1.66(1.263)^2]^2}\left[2 + (1.66-1)(1.263)^2\right] = 559.67^\circ \text{R} \quad (15.51)$$

15.5. OBLIQUE SHOCK WAVE

Therefore, since $c_p = 31240$ ft·lb/(slug°R) for helium (see Table A.1), Equation (15.48) tells us that the heat added is

$$q = \left(31240 \, \frac{\text{ft} \cdot \text{lb}}{\text{slug}°\text{R}}\right)(549.67° \, \text{R} - 559.67° \, \text{R})\left(\frac{1}{778} \, \frac{\text{Btu}}{\text{ft} \cdot \text{lb}}\right) = -402 \, \frac{\text{Btu}}{\text{slug}} \quad (15.52)$$

Hence, because q is negative, the heat-control unit removes heat and thus cools the gas.

15.5 Oblique Shock Wave

Statement of the Problem: *A maneuverable vehicle intended for exploration of Mars uses a rudder to guide its motion. When the rudder is deflected, the flow on one side of the rudder can be approximated as a supersonic compression corner as depicted in Figure 15.3. If the vehicle is moving at Mach 5 and the rudder is deflected 40°, determine the shock angle, β, and the Mach number behind the shock, M_2. Do your computations for model tests done in Earth's atmosphere and on the mission to Mars, assuming the Martian atmosphere consists primarily of carbon dioxide.*

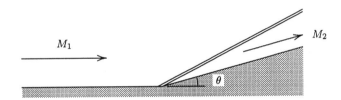

Figure 15.3: *Model for supersonic flow past a deflected rudder.*

Solution: One the one hand, in the case of Earth's atmosphere, the vehicle moves through air, for which the specific-heat ratio is $\gamma = 1.40$. Hence, we can use the information of Appendix C that is presented in tables and figures to aid in obtaining a solution. On the other hand, since the gas in the Martian atmosphere is carbon dioxide for which $\gamma = 1.30$ (see Table A.1), we must work directly with the oblique-shock relations.

Earth: As a first approximation to the shock angle, β, reference to the β-θ-M plot (Figures C.1 and C.2) gives

$$\beta \approx 60° \quad (15.53)$$

Now, we know from the oblique-shock relations that

$$\tan\theta = 2\cot\beta\left[\frac{M_1^2 \sin^2\beta - 1}{2 + M_1^2(\gamma + \cos 2\beta)}\right] \quad (15.54)$$

The following table lists several values of θ as a function of β.

β	θ
60.00°	39.92°
60.26°	40.00°
60.30°	40.01°

Hence, the angle of the shock is
$$\beta = 60.26° \tag{15.55}$$
The normal Mach number upstream of the shock is given by
$$M_{n1} = M_1 \sin\beta = 5\sin(60.26°) = 4.34 \tag{15.56}$$
Then, from the shock tables, we have

M_{n1}	M_{n2}	Source
4.30	0.4277	Table
4.34	0.4268	Interpolation
4.40	0.4255	Table

Therefore, the normal Mach number behind the shock is $M_{n2} = 0.4268$. But, we know that
$$M_{n2} = M_2 \sin(\beta - \theta) \implies M_2 = \frac{M_{n2}}{\sin(\beta - \theta)} \tag{15.57}$$
Thus, the Mach number behind the shock is
$$M_2 = \frac{0.4268}{\sin(60.26° - 40°)} = 1.23 \tag{15.58}$$

Mars: Since the fact that $\gamma \neq 1.40$ precludes the use of the β-θ-M plot (Figures C.1 and C.2), we must deal exclusively with Equation (15.54). Solving by trial and error, we obtain the following.

β	θ
54.2°	39.94°
54.3°	40.00°
54.4°	40.06°

Thus, the shock angle in the Martian atmosphere is
$$\beta = 54.3° \tag{15.59}$$
To determine the Mach number behind the shock, we use the following oblique-shock relation.
$$M_2^2 \sin^2(\beta - \theta) = \frac{1 + \frac{\gamma - 1}{2}M_1^2 \sin^2\beta}{\gamma M_1^2 \sin^2\beta - \frac{\gamma - 1}{2}} \tag{15.60}$$
Thus, using the calculated value of β along with the given values of M_1 and θ, we find
$$M_2^2 \sin^2(54.3° - 40°) = \frac{1 + \frac{1.30-1}{2}(5^2)\sin^2(54.3°)}{1.30(5^2)\sin^2(54.3°) - \frac{1.30-1}{2}} \implies 0.0610 M_2^2 = 0.1632 \tag{15.61}$$
Hence, solving for M_2 gives
$$M_2 = 1.64 \tag{15.62}$$
Summarizing, the shock angles and Mach numbers on Earth and on Mars are as follows.

Planet	β	M_2
Earth	60.3°	1.23
Mars	54.3°	1.64

15.6 Reflection of an Oblique Shock Wave

Statement of the Problem: *A supersonic air stream moving at Mach 3.25 flows past a compression corner, and the shock makes an angle of 30° with the incident stream. The incident shock wave is reflected from an opposite wall which is parallel to the upstream supersonic flow. Calculate the angle of the reflected shock relative to the straight wall and the static pressure behind the reflected shock. The static pressure of the incident flow is $p_1 = 13.2$ psi.*

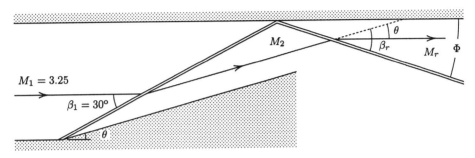

Figure 15.4: *Reflection of an oblique shock wave from a solid boundary.*

Solution: We solve this problem in two steps. First, we calculate wedge angle, θ, the Mach number behind the incident shock, M_2, and the pressure ratio, p_2/p_1. Then, we compute the angle the reflected shock makes with the incident flow direction, β_r, and the pressure ratio, p_r/p_2. The angle Φ follows from the geometry as shown in Figure 15.4, and p_r follows from the computed pressure ratios.

Incident shock: We are given $M_1 = 3.25$ and $\beta_1 = 30°$. Now, we know that

$$\tan\theta = 2\cot\beta_1 \left[\frac{M_1^2 \sin^2\beta_1 - 1}{2 + M_1^2(\gamma + \cos 2\beta_1)} \right] \tag{15.63}$$

Substituting the given values yields

$$\tan\theta = 2\cot 30° \left[\frac{(3.25)^2 \sin^2 30° - 1}{2 + (3.25)^2(1.4 + \cos 60°)} \right] = 0.2575 \tag{15.64}$$

Hence, the wedge angle, θ, is

$$\theta = \tan^{-1}(0.2575) = 14.4° \tag{15.65}$$

The Mach number normal to the incident shock is

$$M_{n1} = M_1 \sin\beta_1 = 3.25 \sin 30° = 1.625 \tag{15.66}$$

Then, from the shock tables, we find

M_{n1}	M_{n2}	p_2/p_1	Source
1.620	0.6625	2.8951	Table
1.625	0.6611	2.9141	Interpolation
1.640	0.6568	2.9712	Table

Finally, we know that

$$M_{n2} = M_2 \sin(\beta_1 - \theta) \implies M_2 = \frac{M_{n2}}{\sin(\beta_1 - \theta)} \qquad (15.67)$$

Hence, the Mach number behind the incident shock and the pressure ratio are

$$M_2 = \frac{0.6611}{\sin(30° - 14.4°)} = 2.46 \quad \text{and} \quad p_2/p_1 = 2.9141 \qquad (15.68)$$

Reflected shock: The reflected shock must turn the flow back by $\theta = 14.4°$ so that the flow downstream of the reflected shock is parallel to the upper wall. Since the incident flow has a Mach number of $M_2 = 2.46$, reference to the β-θ-M plot (Figures C.1 and C.2) gives

$$\beta_r \approx 37° \qquad (15.69)$$

where $\beta_r = \Phi + \theta$ is the angle between the reflected shock and the incident flow direction (see Figure 15.4). Again using Equation (15.63), we can refine the value of β_r. The following table lists several values of θ as a function of β_r.

β_r	θ
36.7°	14.3°
36.8°	14.4°
36.9°	14.5°

Hence, the reflected shock angle relative to the incident flow direction is

$$\beta_r = 36.8° \qquad (15.70)$$

Finally, the angle between the reflected shock and the upper wall is

$$\Phi = \beta_r - \theta = 36.8° - 14.4° = 22.4° \qquad (15.71)$$

So, since the reflected shock is at an angle β_r to the incident flow from Region 2, the Mach number normal to the shock is

$$\tilde{M}_{n1} = M_2 \sin \beta_r = 2.46 \sin 36.8° = 1.474 \qquad (15.72)$$

From the shock tables, we have

$\tilde{M}_{n1}$	p_r/p_2	Source
1.460	2.3202	Table
1.474	2.3682	Interpolation
1.480	2.3888	Table

Hence, the pressure p_r is

$$p_r = p_1 \frac{p_2}{p_1} \frac{p_r}{p_2} = (13.2 \text{ psi})(2.9141)(2.3682) = 91.1 \text{ psi} \qquad (15.73)$$

15.7 Prandtl-Meyer Expansion I

Statement of the Problem: *The Mach number upstream of a Prandtl-Meyer expansion is 2 and the pressure ratio across the wave is $p_2/p_1 = 0.5$. Calculate the angles of the forward and rearward Mach lines of the expansion fan relative to the freestream direction. The gas is hydrogen.*

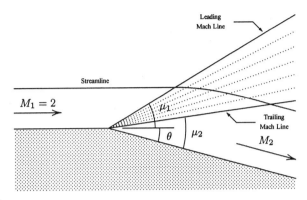

Figure 15.5: *Prandtl-Meyer expansion wave in hydrogen; $p_2/p_1 = 0.5$.*

Solution: Reference to Table A.1 shows that the specific-heat ratio, γ, for hydrogen is 1.40. Therefore, we can use the Prandtl-Meyer tables of Section C.2. From the tables, the Mach angle, μ_1, and the Prandtl-Meyer function, ν_1, upstream of the expansion are

$$M_1 = 2 \implies \mu_1 = 30.00°, \quad \nu_1 = 26.38° \tag{15.74}$$

Since the flow is isentropic, total pressure is constant. Hence, noting that total pressure is given by $p_t = p[1 + \frac{\gamma-1}{2}M^2]^{\gamma/(\gamma-1)}$,

$$\frac{p_2}{p_1} = \left[\frac{1 + \frac{\gamma-1}{2}M_1^2}{1 + \frac{\gamma-1}{2}M_2^2}\right]^{\gamma/(\gamma-1)} \tag{15.75}$$

Solving for M_2, we obtain

$$M_2 = \sqrt{\frac{2}{\gamma-1}\left[\frac{1 + \frac{\gamma-1}{2}M_1^2}{(p_2/p_1)^{(\gamma-1)/\gamma}} - 1\right]} \tag{15.76}$$

For the given values,

$$M_2 = \sqrt{\frac{2}{0.4}\left[\frac{1 + 0.2(2)^2}{(0.5)^{0.4/1.4}} - 1\right]} = 2.44 \tag{15.77}$$

Again appealing to the Prandtl-Meyer tables, linear interpolation gives

M_2	μ_2	ν_2
2.40	24.62°	36.75°
2.44	24.20°	37.71°
2.45	24.09°	37.95°

so that the Mach angle, μ_2, and Prandtl-Meyer function, ν_2, after the expansion are

$$\mu_2 = 24.20° \quad \text{and} \quad \nu_2 = 37.71° \tag{15.78}$$

Thus, by definition of the Prandtl-Meyer function, the flow is turned through an angle θ given by

$$\theta = \nu_2 - \nu_1 = 37.71° - 26.38° = 11.33° \tag{15.79}$$

Finally, the angle of the forward, or leading, Mach line of the expansion wave relative to the freestream direction is the Mach angle, $\mu_1 = 30.0°$. Also, the angle of the rearward, or trailing, Mach line relative to the flow direction is μ_2. Hence, the angle relative to the freestream direction is

$$\mu = \mu_2 - \theta = 24.20° - 11.33° = 12.87° \tag{15.80}$$

Thus, in summary, we have

$$\mu = \begin{cases} 30.0°, & \text{Forward Mach line} \\ 12.9°, & \text{Rearward Mach line} \end{cases} \tag{15.81}$$

15.8 Prandtl-Meyer Expansion II

Statement of the Problem: *The Mach number upstream of a Prandtl-Meyer expansion is 4.5, and the gas is methane. If the flow turns through an angle $\theta = 45°$, what is the Mach number downstream of the expansion? Calculate the angles of the forward and rearward Mach lines of the expansion fan relative to the freestream direction. Make a sketch depicting expansion-wave geometry.*

Solution: Using Table A.1 shows that methane has a specific-heat ratio of $\gamma = 1.31 \neq 1.40$. Therefore, we cannot use the Prandtl-Meyer tables. Rather, we must use the Prandtl-Meyer equations. First, note that the flow expansion angle, θ, is

$$\theta = \nu(M_2) - \nu(M_1) \tag{15.82}$$

where the Prandtl-Meyer function, $\nu(M)$, is

$$\nu(M) = \sqrt{\frac{\gamma+1}{\gamma-1}} \tan^{-1} \sqrt{\frac{\gamma-1}{\gamma+1}(M^2-1)} - \tan^{-1}\sqrt{M^2-1} \tag{15.83}$$

Upstream of the expansion, the Mach number is $M_1 = 4.5$, so that the Prandtl-Meyer function is

$$\nu(M_1) = \sqrt{\frac{1.31+1}{1.31-1}} \tan^{-1} \sqrt{\frac{1.31-1}{1.31+1}(4.5^2-1)} - \tan^{-1}\sqrt{4.5^2-1} = 81.5° \tag{15.84}$$

Thus, the Prandtl-Meyer function downstream of the expansion is

$$\nu(M_2) = \theta + \nu(M_1) = 45° + 81.5° = 126.5° \tag{15.85}$$

Substituting $\gamma = 1.31$, $M = M_2$ and $\nu = 126.5°$ into Equation (15.83), we obtain the following transcendental equation for M_2.

$$2.7298 \tan^{-1} \sqrt{0.1342(M_2^2-1)} - \tan^{-1}\sqrt{M_2^2-1} = 126.5° \tag{15.86}$$

Solving by trial and error, we obtain:

M_2	$\nu(M_2)$
12.0	125.4°
12.5	126.5°
13.0	127.6°

Therefore, the Mach number downstream of the expansion is

$$M_2 = 12.5 \qquad (15.87)$$

The Mach angles are

$$\mu_1 = \sin^{-1}\left(\frac{1}{M_1}\right) = \sin^{-1}\left(\frac{1}{4.5}\right) = 12.8° \qquad (15.88)$$

$$\mu_2 = \sin^{-1}\left(\frac{1}{M_2}\right) = \sin^{-1}\left(\frac{1}{12.5}\right) = 4.6° \qquad (15.89)$$

Figure 15.6 depicts the geometry of the expansion wave.

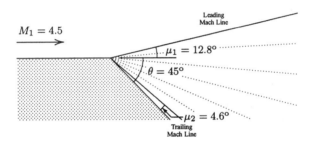

Figure 15.6: *Prandtl-Meyer expansion wave in methane.*

15.9 Compressible Law of the Wall

Statement of the Problem: *Air flows over an adiabatic surface with a freestream Mach number of 4. The boundary layer is turbulent and the measured skin friction is $c_f = 0.001$. Measured velocity values are accurately correlated by the compressible law of the wall. Assuming the turbulent Prandtl number, Pr_t, and the recovery factor, r, are both equal to 0.89, compare the velocity close to the surface, i.e., for $10 \leq u_\tau y/\nu_w \leq 500$, with values determined from the incompressible law of the wall.*

Solution: We know that the compressible law of the wall is

$$\frac{u^*}{u_\tau} = \frac{1}{\kappa}\ell n\left(\frac{u_\tau y}{\nu_w}\right) + C \qquad (15.90)$$

where u^* is the transformed velocity defined by

$$u^* \equiv \frac{u_e}{A}\sin^{-1}\left(A\frac{\tilde{u}}{u_e}\right), \qquad A^2 = \frac{\gamma-1}{2}Pr_t\frac{T_e}{T_w}M_e^2 \qquad (15.91)$$

The values of the quantities κ and C are very nearly the same for both compressible and incompressible flows, viz., 0.41 and 5.0, respectively. Also, u_τ is friction velocity, u_e, T_e and M_e are the freestream velocity, temperature and Mach number, and T_w is the temperature at the surface. The incompressible law of the wall is identical to Equation (15.90) with the mass-averaged velocity, $\tilde{u}$, replacing u^*.

First, note that for an adiabatic wall, the surface temperature is given by

$$\frac{T_w}{T_e} = 1 + \frac{\gamma - 1}{2} r M_e^2 \qquad (15.92)$$

where r is the recovery factor. Since $\gamma = 1.4$ and we are given $r = 0.89$ and $M_e = 4$, the ratio of surface to freestream temperature is

$$\frac{T_w}{T_e} = 1 + \frac{1.4 - 1}{2}(0.89)(4)^2 = 3.848 \qquad (15.93)$$

Combining the second of Equations (15.91) with Equation (15.93) and using the fact that we are given $Pr_t = 0.89$, we have

$$A^2 = \frac{1.4 - 1}{2}(0.89)\left(\frac{1}{3.848}\right)(4)^2 = 0.7401 \quad \Longrightarrow \quad A = 0.86 \qquad (15.94)$$

So, substituting Equation (15.94) into the first of Equations (15.91), the transformed velocity for this boundary layer is

$$u^* \equiv \frac{u_e}{0.86}\sin^{-1}\left(0.86\frac{\tilde{u}}{u_e}\right) \quad \Longrightarrow \quad \frac{u^*}{u_\tau}\frac{u_\tau}{u_e} = \frac{1}{0.86}\sin^{-1}\left(0.86\frac{u_\tau}{u_e}\frac{\tilde{u}}{u_\tau}\right) \qquad (15.95)$$

Next, observe that by definition,

$$c_f = \frac{\tau_w}{\frac{1}{2}\rho_e u_e^2} = \frac{2\rho_w u_\tau^2}{\rho_e u_e^2} \qquad (15.96)$$

Then, since pressure is constant across a boundary layer, the perfect-gas law gives

$$\rho_w T_w = \rho_e T_e \quad \Longrightarrow \quad \frac{\rho_w}{\rho_e} = \frac{T_e}{T_w} \qquad (15.97)$$

Combining Equations (15.96) and (15.97) yields

$$c_f = 2\frac{T_e}{T_w}\left(\frac{u_\tau}{u_e}\right)^2 \quad \Longrightarrow \quad \frac{u_\tau}{u_e} = \sqrt{\frac{T_w}{T_e}\frac{c_f}{2}} \qquad (15.98)$$

Since $T_w/T_e = 3.848$ and we are given $c_f = 0.001$, we find

$$\frac{u_\tau}{u_e} = \sqrt{3.848\frac{0.001}{2}} = 0.04386 \qquad (15.99)$$

Substituting Equation (15.99) into Equation (15.95) yields the following relationship between mass-averaged and transformed velocities.

$$0.04386\frac{u^*}{u_\tau} = \frac{1}{0.86}\sin^{-1}\left[0.86(0.04386)\frac{\tilde{u}}{u_\tau}\right] \quad \Longrightarrow \quad \frac{u^*}{u_\tau} = 26.51\sin^{-1}\left(\frac{\tilde{u}/u_\tau}{26.51}\right) \qquad (15.100)$$

15.10. CFD—NUMERICAL DISSIPATION AND DISPERSION

Thus, taking the sine of both sides of this equation, we arrive at

$$\frac{\tilde{u}}{u_\tau} = 26.51 \sin\left(\frac{u^*/u_\tau}{26.51}\right) \quad (15.101)$$

As indicated in the statement of the problem, measured velocities correlate well with the compressible law of the wall, Equation (15.90). Thus, we can select a value of $y^+ \equiv u_\tau y/\nu_w$, compute the transformed velocity from Equation (15.90) and the mass-averaged velocity from Equation (15.101). The following table includes the computed values.

y^+	u^*/u_τ	$\tilde{u}/u_\tau$	Difference
10	10.6	10.3	2.8%
20	12.3	11.9	3.3%
50	14.5	13.8	4.8%
100	16.2	15.2	6.2%
200	17.9	16.6	7.3%
500	20.2	18.3	9.4%

15.10 CFD—Numerical Dissipation and Dispersion

Statement of the Problem: *Determine the artificial viscosity, ν_a, and numerical-dispersion coefficient, $\mathcal{D}_a$, for Richardson's method applied to the one-dimensional wave equation,*

$$\frac{u_i^{n+1} - u_i^{n-1}}{2\Delta t} + a \frac{u_{i+1}^n - u_{i-1}^n}{2\Delta x} = 0$$

where a is a positive constant.

Solution: Expanding about (i, n), we have

$$u_i^{n\pm 1} = u_i^n \pm u_t \Delta t + \frac{1}{2} u_{tt}(\Delta t)^2 \pm \frac{1}{6} u_{ttt}(\Delta t)^3 + \cdots \quad (15.102)$$

$$u_{i\pm 1}^n = u_i^n \pm u_x \Delta x + \frac{1}{2} u_{xx}(\Delta x)^2 \pm \frac{1}{6} u_{xxx}(\Delta x)^3 + \cdots \quad (15.103)$$

We retain terms up to the third derivatives as the numerical dissipation appears in the *modified equation* as a second derivative and dispersion as a third derivative. The first term in Richardson's method is

$$\frac{u_i^{n+1} - u_i^{n-1}}{2\Delta t} = u_t + \frac{1}{6} u_{ttt}(\Delta t)^2 + \cdots \quad (15.104)$$

while the second term is

$$a \frac{u_{i+1}^n - u_{i-1}^n}{2\Delta x} = au_x + \frac{1}{6} au_{xxx}(\Delta x)^2 + \cdots \quad (15.105)$$

Substituting into the original equation yields

$$u_t + \frac{1}{6} u_{ttt}(\Delta t)^2 + au_x + \frac{1}{6} au_{xxx}(\Delta x)^2 + \cdots = 0 \quad (15.106)$$

or,

$$u_t + au_x = -\frac{(\Delta t)^2}{6} u_{ttt} - \frac{a(\Delta x)^2}{6} u_{xxx} + \cdots \quad (15.107)$$

Now, as can be easily demonstrated by manipulating Equation (15.107),

$$u_{ttt} = -a^3 u_{xxx} + \cdots \tag{15.108}$$

Combining Equations (15.107) and (15.108) gives

$$u_t + au_x = \frac{1}{6}\left[a^3(\Delta t)^2 - a(\Delta x)^2\right] u_{xxx} + \cdots \tag{15.109}$$

To simplify, define the dimensionless parameter, λ, as follows.

$$\lambda = \frac{a\Delta t}{\Delta x} \implies \Delta t = \frac{\lambda}{a}\Delta x \tag{15.110}$$

So, the *modified equation* becomes

$$\begin{aligned}
u_t + au_x &= \frac{1}{6}\left[a^3\left(\frac{\lambda}{a}\right)^2 (\Delta x)^2 - a(\Delta x)^2\right] u_{xxx} + \cdots \\
&= \frac{a(\Delta x)^2}{6}\left(\lambda^2 - 1\right) u_{xxx} + \cdots
\end{aligned} \tag{15.111}$$

Therefore, the modified equation implied by the Richardson method is

$$\frac{\partial u}{\partial t} + a\frac{\partial u}{\partial x} = \nu_a \frac{\partial^2 u}{\partial x^2} + \mathcal{D}_a \frac{\partial^3 u}{\partial x^3} \tag{15.112}$$

where the artificial viscosity, ν_a, and numerical-dispersion coefficient, $\mathcal{D}_a$, are

$$\nu_a = 0, \quad \mathcal{D}_a = \frac{a(\Delta x)^2}{6}\left(\lambda^2 - 1\right) \tag{15.113}$$

Appendix A

Fluid Properties

A.1 Perfect-Gas Properties

The pressure, p, density, ρ, and temperature, T, are related by the perfect-gas equation of state,

$$p = \rho R T \qquad (A.1)$$

where R is the **perfect-gas constant**. Table A.1 includes values of R for common gases. The table also includes the specific heat coefficient, c_p, and the specific-heat ratio, γ.

Table A.1: *Thermodynamic Properties of Common Gases*

Gas	γ	J/(kg·K)		ft·lb/(slug·°R)	
		R	c_p	R	c_p
Air	1.40	287	1,004	1,716	6,003
Carbon dioxide	1.30	189	841	1,130	5,028
Helium	1.66	2,077	5,225	12,419	31,240
Hydrogen	1.40	4,124	14,180	24,677	84,783
Methane	1.31	518	2,208	3,098	13,783
Nitrogen	1.40	297	1,039	1,776	6,212
Oxygen	1.40	260	910	1,555	5,440

A.2 Pressure

The standard value of the pressure in the atmosphere is

$$p = \begin{cases} 1 & \text{atm} & \text{(atmospheres)} \\ 14.7 & \text{psi} & \text{(pounds per square inch)} \\ 101 & \text{kPa} & \text{(kiloPascals)} \\ 760 & \text{mmHg} & \text{(millimeters of mercury)} \\ 2116.8 & \text{psf} & \text{(pounds per square foot)} \end{cases} \qquad (A.2)$$

where one Pascal = one Newton per square meter. Pressure is often quoted in terms of **absolute** (psia) and **gage** (psig) values, with the latter relative to the atmospheric pressure. Thus, $p = 14.9$ psi is an absolute pressure of 14.9 psia and a gage pressure of 0.2 psig.

A.3 Density

Values of the density of air and water at atmospheric pressure and a temperature of 20° C (68° F) are as follows.

$$\rho = \begin{cases} 1.20 \text{ kg/m}^3 \ (.00234 \text{ slug/ft}^3), & \text{Air} \\ 998 \text{ kg/m}^3 \ (1.94 \text{ slug/ft}^3), & \text{Water} \end{cases} \quad (A.3)$$

Table A.2 includes densities of common liquids, while Table A.3 includes the variation of density with temperature for water.

Table A.2: *Densities of Common Liquids*

Liquid	T (°C)	ρ (kg/m³)	T (°F)	ρ (slug/ft³)
Ethyl alcohol	20	789	68	1.53
Carbon tetrachloride	20	1,590	68	3.09
Glycerin	20	1,260	68	2.44
Kerosene	20	814	68	1.58
Mercury	20	13,550	68	26.30
Oil: SAE 10W	38	870	100	1.69
Oil: SAE 10W-30	38	880	100	1.71
Oil: SAE 30	38	880	100	1.71
Seawater	16	1,030	60	1.99

Table A.3: *Density of Water as a Function of Temperature*

T (°C)	ρ (kg/m³)	T (°F)	ρ (slug/ft³)
0	1,000	32	1.94
10	1,000	50	1.94
20	998	68	1.94
30	996	86	1.93
40	992	104	1.93
50	988	122	1.92
60	983	140	1.91
70	978	158	1.90
80	972	176	1.89
90	965	194	1.87
100	958	212	1.86

A.4 Compressibility and Speed of Sound

Table A.4 lists compressibility, τ, and the speed of sound, a, for several common fluids. In general, for a perfect gas, the speed of sound is

$$a = \sqrt{\gamma R T} \quad (A.4)$$

A.5. VISCOSITY

Table A.4: *Compressibility and Sound Speed of Common Fluids for a Pressure of 1 atm*

Fluid	T (°C)	τ (m²/N)	a (m/sec)	T (°F)	τ (ft²/lb)	a (ft/sec)
Air	15.6	$1.00 \cdot 10^{-5}$	341	60.0	$4.79 \cdot 10^{-4}$	1119
Ether	15.0	$1.63 \cdot 10^{-9}$	1032	59.0	$7.82 \cdot 10^{-8}$	3386
Ethyl alcohol	20.0	$9.43 \cdot 10^{-10}$	1213	68.0	$4.52 \cdot 10^{-8}$	3980
Glycerin	15.6	$2.21 \cdot 10^{-10}$	1860	60.0	$1.06 \cdot 10^{-8}$	6102
Helium	15.6	$1.00 \cdot 10^{-5}$	998	60.0	$4.79 \cdot 10^{-4}$	3274
Mercury	15.6	$0.35 \cdot 10^{-10}$	1450	60.0	$0.17 \cdot 10^{-8}$	4757
Water	15.6	$4.65 \cdot 10^{-10}$	1481	60.0	$2.23 \cdot 10^{-8}$	4859

A.5 Viscosity

Table A.5 includes kinematic viscosities of common liquids and gases, while Table A.6 includes the variation of ν with temperature for water. For air and water at atmospheric pressure and 20° C (68° F), the kinematic viscosity is

$$\nu = \begin{cases} 1.51 \cdot 10^{-5} \text{ m}^2/\text{sec} & (1.62 \cdot 10^{-4} \text{ ft}^2/\text{sec}), \quad \text{Air} \\ 1.00 \cdot 10^{-6} \text{ m}^2/\text{sec} & (1.08 \cdot 10^{-5} \text{ ft}^2/\text{sec}), \quad \text{Water} \end{cases} \tag{A.5}$$

Table A.5: *Kinematic Viscosities of Common Liquids and Gases*

Liquid	T (°C)	$10^6 \nu$ (m²/sec)	T (°F)	$10^5 \nu$ (ft²/sec)
Ethyl alcohol	20	1.51	68	1.62
Carbon tetrachloride	20	0.60	68	0.65
Gasoline	16	0.46	60	0.49
Glycerin	20	1190.00	68	1280.00
Kerosene	20	2.37	68	2.55
Mercury	20	0.12	68	0.13
Oil: SAE 10W	38	41.00	100	44.00
Oil: SAE 10W-30	38	76.00	100	82.00
Oil: SAE 30	38	110.00	100	118.00
Air	15	14.60	59	15.71
Carbon dioxide	15	7.84	59	8.44
Helium	15	114.00	59	123.00
Hydrogen	15	101.00	59	109.00
Methane	15	15.90	59	17.10
Nitrogen	15	14.50	59	15.60
Oxygen	15	15.00	59	16.10

The viscosity of air is well approximated by **Sutherland's Law**, which is an empirical equation that is quite accurate for a wide range of temperatures. The formula is

$$\mu = \frac{AT^{3/2}}{T + S} \tag{A.6}$$

where A and S are empirical constants. Note that T is the absolute temperature, and is thus given either in Kelvins (K) or degrees Rankine (° R). The values of A and S are

$$A = 1.46 \cdot 10^{-6} \frac{\text{kg}}{\text{m} \cdot \text{sec} \cdot \text{K}^{1/2}}, \qquad S = 110.3 \text{ K} \tag{A.7}$$

$$A = 2.27 \cdot 10^{-8} \frac{\text{slug}}{\text{ft} \cdot \text{sec} \cdot (^\circ\text{R})^{1/2}}, \quad S = 198.6\ ^\circ\text{R} \tag{A.8}$$

Table A.6: *Kinematic Viscosity of Water as a Function of Temperature*

$T\ (^\circ\text{C})$	$10^6 \nu\ (\text{m}^2/\text{sec})$	$T\ (^\circ\text{F})$	$10^5 \nu\ (\text{ft}^2/\text{sec})$
0	1.79	32	1.93
10	1.31	50	1.41
20	1.00	68	1.08
30	0.80	86	0.86
40	0.66	104	0.71
50	0.55	122	0.59
60	0.47	140	0.51
70	0.41	158	0.44
80	0.36	176	0.39
90	0.33	194	0.36
100	0.29	212	0.31

A.6 Surface Tension

In general, surface tension, σ, depends upon the surface material as well as the nature of the two fluids involved, and it is a function of pressure and temperature. It is also considerably affected by dirt on the surface. Table A.7 lists surface tension for several common liquids.

Table A.7: *Surface Tension of Common Fluids*

Fluid$_1$	Fluid$_2$	$T\ (^\circ\text{C})$	σ (N/m)	$T\ (^\circ\text{F})$	σ (lb/ft)
Water	Air	0	.076	32	.0052
Water	Air	20	.073	68	.0050
Water	Air	100	.059	212	.0040
Mercury	Air	20	.466	68	.0319
Mercury	Water	20	.375	68	.0257
Carbon tetrachloride	Air	20	.027	68	.0018
Carbon tetrachloride	Water	20	.045	68	.0030
Ethyl alcohol	Air	20	.023	68	.0016
Gasoline	Air	16	.022	60	.0015
Glycerin	Air	20	.063	68	.0043
SAE 10 Oil	Air	16	.036	60	.0025
SAE 30 Oil	Air	16	.035	60	.0024
Seawater	Air	16	.073	60	.0050

Appendix B

Hydraulics Properties

B.1 Fluid Statics

The center of pressure of a planar object whose centroid lies a distance $\bar{z}$ below the surface is

$$z_{cp} = \bar{z} + \frac{I}{\bar{z}A} \tag{B.1}$$

Figure B.1 gives the area, A, moment of inertia, I, and centroid location for several geometric shapes.

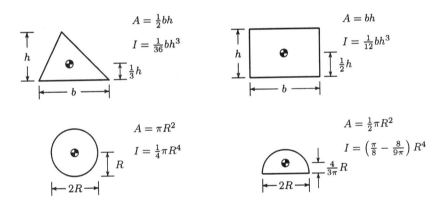

Figure B.1: *Areas, moments of inertia and centroids for common geometries; in all cases, the moment of inertia is relative to a horizontal axis passing through the centroid.*

B.2 Open-Channel Flow

The *Chézy-Manning equation* for open-channel flow is

$$\bar{u} = \frac{\chi}{n} S_o^{1/2} R_h^{2/3}, \qquad \chi = 1.49 \text{ ft}^{1/3}/\text{sec} = 1.00 \text{ m}^{1/3}/\text{sec} \qquad (B.2)$$

where $\bar{u}$ is average flow speed, n is the *Manning roughness coefficient*, S_o is bottom slope and R_h is the hydraulic radius. The hydraulic radius is defined by

$$R_h \equiv \frac{A}{P} \qquad (B.3)$$

where A is cross-sectional area and P is the perimeter of the wetted, solid surface. Figure B.2 shows hydraulic radii for simple geometries.

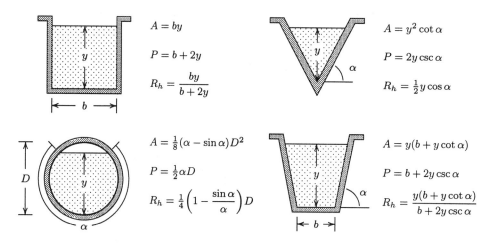

Figure B.2: *Areas, perimeters and hydraulic radii for common geometries.*

B.2. OPEN-CHANNEL FLOW

Table B.1 lists the Manning roughness coefficient, n, for several types of surfaces. The quantity k_s in the table is the effective surface-roughness height.

Table B.1: *Experimental Values of the Manning Roughness Coefficient*

Type of Channel	n	k_s (ft)	k_s (mm)
Artificially lined channels:			
Glass	.010 ± .002	0.0011	0.3
Brass	.011 ± .002	0.0019	0.6
Smooth steel	.012 ± .002	0.0032	1.0
Cast iron	.013 ± .003	0.0051	1.6
Brickwork	.015 ± .002	0.0120	3.7
Asphalt	.016 ± .003	0.0180	5.4
Corrugated metal	.022 ± .005	0.1200	37.0
Rubble masonry	.025 ± .005	0.2600	80.0
Excavated earth channels:			
Clean	.022 ± .004	0.12	37
Gravelly	.025 ± .005	0.26	80
Weedy	.030 ± .005	0.80	240
Stony, large cobbles	.035 ± .010	1.50	500
Natural channels:			
Clean and straight	.030 ± .005	0.80	240
Major rivers	.035 ± .010	1.50	500
Sluggish, deep pools	.040 ± .010	3.00	900
Floodplains:			
Pasture, farmland	.035 ± .010	1.5	500
Light brush	.050 ± .020	6.0	2000
Heavy brush	.075 ± .025	15.0	5000

B.3 Pipe Flow

The following figures and tables can be used to determine minor losses for flow in a pipe. In most cases, the dimensionless *head-loss coefficient*, K, is given where the head loss, h_L is

$$h_L = K \frac{\bar{u}^2}{2g} \tag{B.4}$$

Inlet. For flow at an inlet with radius of curvature $\mathcal{R}$, the head-loss coefficient is as shown in Figures B.3 and B.4.

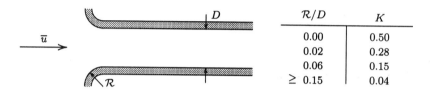

$\mathcal{R}/D$	K
0.00	0.50
0.02	0.28
0.06	0.15
$\geq$ 0.15	0.04

Figure B.3: *Head loss coefficient at an inlet.*

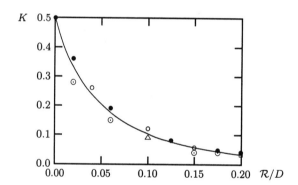

Figure B.4: *Loss coefficient at an inlet: ○, ●, ⊙, △, □ measurements; —— approximate curve fit to measurements.*

Contraction. For flow at a contraction of angle θ, the head-loss coefficient is as shown in Figure B.5. Table B.2 includes additional loss-coefficient data valid for both round and rectangular ducts as a function of A_2/A_1. The areas A_1 and A_2 denote cross-sectional area upstream and downstream of the contraction, respectively.

D_2/D_1	$K_{\theta=60°}$	$K_{\theta=180°}$
0.20	0.08	0.49
0.40	0.07	0.42
0.60	0.06	0.32
0.80	0.05	0.18

Figure B.5: *Head loss coefficient at a contraction.*

B.3. PIPE FLOW

Table B.2: *Loss Coefficients for Gradual Contractions*

A_2/A_1	\multicolumn{7}{c}{Included angle, θ}
	10°	15°-40°	50°-60°	90°	120°	150°	180°
0.10	0.05	0.05	0.08	0.19	0.29	0.37	0.43
0.25	0.05	0.04	0.07	0.17	0.27	0.35	0.41
0.50	0.05	0.05	0.06	0.12	0.18	0.24	0.26

For a sudden contraction, i.e., for $\theta = 180°$, the loss coefficient can be approximated by

$$K \approx \frac{1}{2}\left[1 - \left(\frac{D_2}{D_1}\right)^2\right] \quad \text{(sudden contraction)} \tag{B.5}$$

Expansion. For flow at a expansion of angle θ, the head-loss coefficient is as shown in Figure B.6. For a sudden expansion, i.e., for $\theta = 180°$, the loss coefficient can be approximated by

$$K \approx \left[1 - \left(\frac{D_1}{D_2}\right)^2\right]^2 \quad \text{(sudden expansion)} \tag{B.6}$$

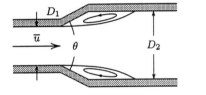

D_1/D_2	$K_{\theta=10°}$	$K_{\theta=180°}$
0.20	0.13	0.92
0.40	0.11	0.72
0.60	0.06	0.42
0.80	0.03	0.16

Figure B.6: *Head loss coefficient at an expansion.*

90° Smooth Bend. For flow at a 90° smooth bend with radius of curvature $\mathcal{R}$, the head-loss coefficient is as shown in Figure B.7.

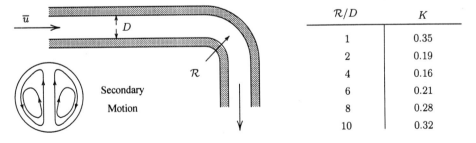

$\mathcal{R}/D$	K
1	0.35
2	0.19
4	0.16
6	0.21
8	0.28
10	0.32

Figure B.7: *Head loss coefficient at a 90° smooth bend.*

Figure B.8 can be used to determine the head loss for 90° bends, including effects of surface roughness. The figure shows the increment in loss coefficient, ΔK, that must be added to the contribution due to bend length, so that

$$K = \frac{\pi}{2}f\frac{\mathcal{R}}{D} + \Delta K \tag{B.7}$$

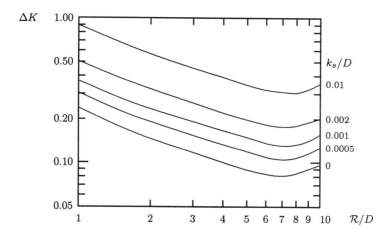

Figure B.8: *Loss coefficient for a 90° bend.*

Other Minor Losses. Figure B.9 provides an assortment of data for loss-coefficient, K, and equivalent-length, L_e. The head loss is given in terms of equivalent length as follows.

$$h_L = \frac{\bar{u}^2}{2g} f \frac{L_e}{D} \tag{B.8}$$

where f is the friction factor. Figure B.10 is the Moody diagram, from which the friction factor can be determined as a function of Reynolds number, Re_D, and roughness height to diameter ratio, k_s/D.

Major Losses. For straight pipes of length L and diameter D, the major loss is

$$h_L = \frac{\bar{u}^2}{2g} f \frac{L}{D} \tag{B.9}$$

where $\bar{u}$ is the average pipe velocity. The Moody diagram, Figure B.10, presents friction factor as a function of Reynolds number with surface roughness, k_s/D, as a parameter. The inset in the figure lists values of k_s characteristic of several materials. Alternatively, f can be determined from the Colebrook formula, viz.,

$$\frac{1}{\sqrt{f}} = -2 \log_{10} \left(\frac{k_s/D}{3.7} + \frac{2.51}{Re_D \sqrt{f}} \right) \tag{B.10}$$

B.3. PIPE FLOW

Reentrant Inlet	ℓ/D	$K_{t/D=0}$	$K_{t/D=0.02}$	
	0.0	0.50	0.50	
	0.1	0.83	0.60	
	0.2	0.92	0.66	
	0.3	0.97	0.69	
	0.4	1.00	0.71	
Beveled Inlet	L/D	$K_{\theta=10°}$	$K_{\theta=30°}$	$K_{\theta=50°}$
	0.00	0.50	0.50	0.50
	0.05	0.39	0.30	0.35
	0.10	0.32	0.19	0.28
	0.15	0.27	0.14	0.24
	0.20	0.24	0.12	0.23

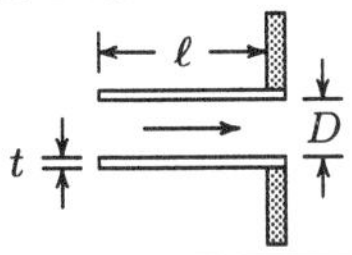

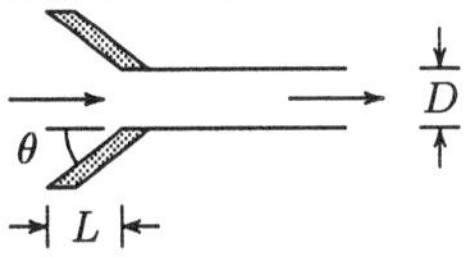

90° Miter Bend

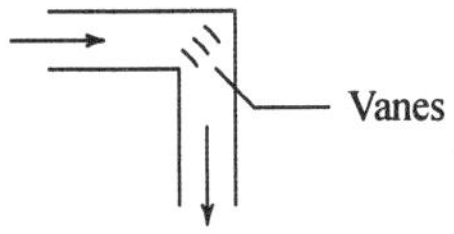

	K
Without vanes	1.1
With vanes	0.2

General Miter Bend

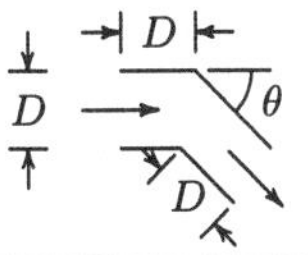

θ	L_e/D
0°	2
30°	8
60°	24
90°	58

Elbow

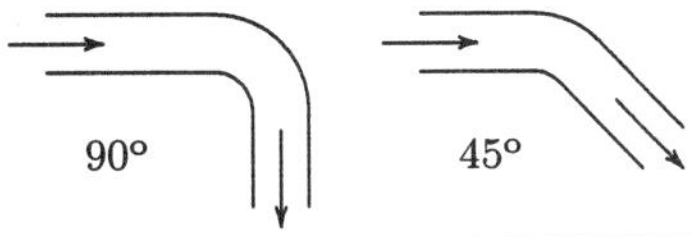

Regular:	K	Long radius:	K
90° – flanged	0.3	90° – flanged	0.2
90° – threaded	1.5	90° – threaded	0.7
45° – threaded	0.4	45° – flanged	0.2

Return Bend

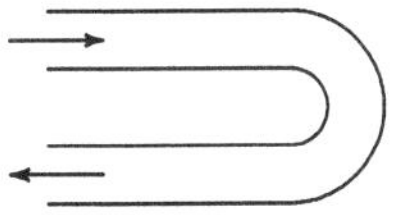

	K
Flanged	0.2
Threaded	1.5

Tee

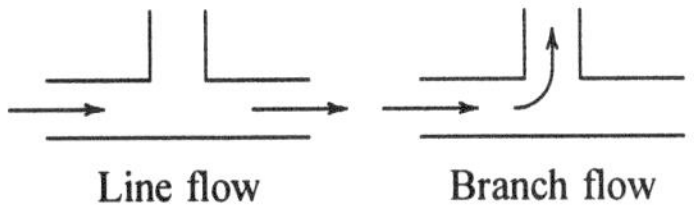

Line flow:	K	Branch flow:	K
Flanged	0.2	Flanged	1.0
Threaded	0.9	Threaded	2.0

Threaded Union

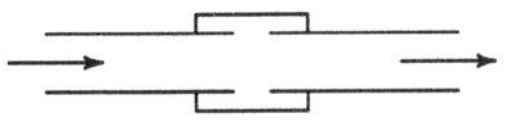

$K = 0.08$

Assorted Valves

	K		K		K
Globe – fully open	10	Gate – fully open	0.15 – 0.20	Ball – fully open	0.05
Angle – fully open	2 – 5	Gate – $\frac{1}{4}$ closed	0.26	Ball – $\frac{1}{3}$ closed	5.5
Swing check – forward flow	2	Gate – $\frac{1}{2}$ closed	2.1 – 5.6	Ball – $\frac{2}{3}$ closed	210
Swing check – backward flow	∞	Gate – $\frac{3}{4}$ closed	17		

Figure B.9: *Loss coefficient data for various pipe-system elements.*

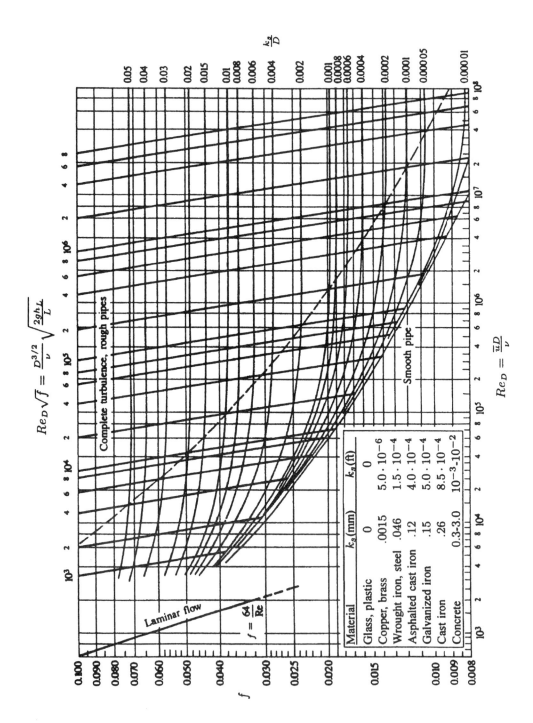

Figure B.10: *The Moody diagram [From Moody (1944)—used with permission of the ASME]*.

Appendix C

Compressible-Flow Properties

This appendix includes several tables that are valid for gases with a specific-heat ratio, $\gamma = 1.4$.

C.1 Isentropic Flow and Normal-Shock Relations

Notation used in the following tables is as follows.

Isentropic flow:

M	=	local Mach number
p/p_t	=	ratio of static to total pressure
ρ/ρ_t	=	ratio of static to total density
T/T_t	=	ratio of static to total temperature
A/A^*	=	ratio of local area to reference sonic-flow area

Normal shock waves:

M_1	=	Mach number ahead of the shock wave
M_2	=	Mach number behind the shock wave
p_2/p_1	=	pressure ratio across the shock wave
T_2/T_1	=	temperature ratio across the shock wave
p_{t2}/p_{t1}	=	total pressure ratio across the shock wave

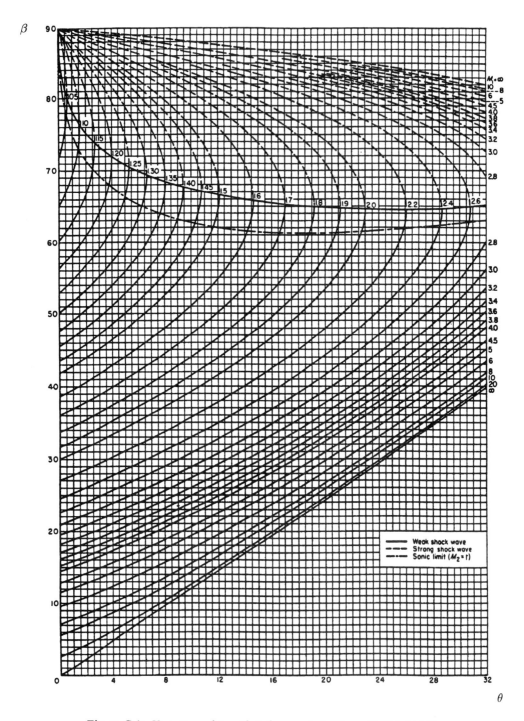

Figure C.1: *Variation of β with θ for $\gamma = 1.4$ [From NACA-1135].*

C.1. ISENTROPIC FLOW AND NORMAL-SHOCK RELATIONS

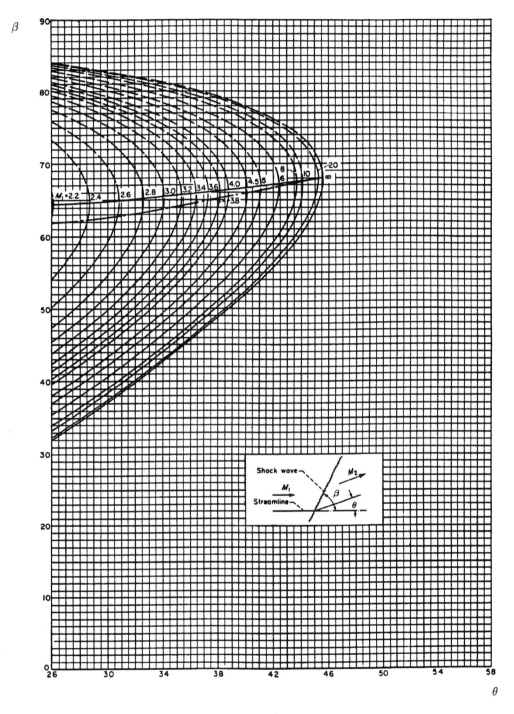

Figure C.2: *Variation of β with θ for $\gamma = 1.4$ [From NACA-1135].*

Subsonic Flow of a Perfect Gas with $\gamma = 1.4$

M	p/p_t	ρ/ρ_t	T/T_t	A/A^*	M	p/p_t	ρ/ρ_t	T/T_t	A/A^*
0.00	1.0000	1.0000	1.0000	∞	0.50	0.8430	0.8852	0.9524	1.3398
0.02	0.9997	0.9998	0.9999	28.9421	0.52	0.8317	0.8766	0.9487	1.3034
0.04	0.9989	0.9992	0.9997	14.4815	0.54	0.8201	0.8679	0.9449	1.2703
0.06	0.9975	0.9982	0.9993	9.6659	0.56	0.8082	0.8589	0.9410	1.2403
0.08	0.9955	0.9968	0.9987	7.2616	0.58	0.7962	0.8498	0.9370	1.2130
0.10	0.9930	0.9950	0.9980	5.8218	0.60	0.7840	0.8405	0.9328	1.1882
0.12	0.9900	0.9928	0.9971	4.8643	0.62	0.7716	0.8310	0.9286	1.1656
0.14	0.9864	0.9903	0.9961	4.1824	0.64	0.7591	0.8213	0.9243	1.1451
0.16	0.9823	0.9873	0.9949	3.6727	0.66	0.7465	0.8115	0.9199	1.1265
0.18	0.9776	0.9840	0.9936	3.2779	0.68	0.7338	0.8016	0.9153	1.1097
0.20	0.9725	0.9803	0.9921	2.9635	0.70	0.7209	0.7916	0.9107	1.0944
0.22	0.9668	0.9762	0.9904	2.7076	0.72	0.7080	0.7814	0.9061	1.0806
0.24	0.9607	0.9718	0.9886	2.4956	0.74	0.6951	0.7712	0.9013	1.0681
0.26	0.9541	0.9670	0.9867	2.3173	0.76	0.6821	0.7609	0.8964	1.0570
0.28	0.9470	0.9619	0.9846	2.1656	0.78	0.6691	0.7505	0.8915	1.0471
0.30	0.9395	0.9564	0.9823	2.0351	0.80	0.6560	0.7400	0.8865	1.0382
0.32	0.9315	0.9506	0.9799	1.9219	0.82	0.6430	0.7295	0.8815	1.0305
0.34	0.9231	0.9445	0.9774	1.8229	0.84	0.6300	0.7189	0.8763	1.0237
0.36	0.9143	0.9380	0.9747	1.7358	0.86	0.6170	0.7083	0.8711	1.0179
0.38	0.9052	0.9313	0.9719	1.6587	0.88	0.6041	0.6977	0.8659	1.0129
0.40	0.8956	0.9243	0.9690	1.5901	0.90	0.5913	0.6870	0.8606	1.0089
0.42	0.8857	0.9170	0.9659	1.5289	0.92	0.5785	0.6764	0.8552	1.0056
0.44	0.8755	0.9094	0.9627	1.4740	0.94	0.5658	0.6658	0.8498	1.0031
0.46	0.8650	0.9016	0.9594	1.4246	0.96	0.5532	0.6551	0.8444	1.0014
0.48	0.8541	0.8935	0.9559	1.3801	0.98	0.5407	0.6445	0.8389	1.0003
0.50	0.8430	0.8852	0.9524	1.3398	1.00	0.5283	0.6339	0.8333	1.0000

C.1. ISENTROPIC FLOW AND NORMAL-SHOCK RELATIONS

Supersonic Flow of a Perfect Gas with $\gamma = 1.4$

	Isentropic Flow				Normal Shock Wave				
M_1	p/p_t	ρ/ρ_t	T/T_t	A/A^*	M_2	p_2/p_1	ρ_2/ρ_1	T_2/T_1	p_{t2}/p_{t1}
1.00	0.5283	0.6339	0.8333	1.0000	1.0000	1.0000	1.0000	1.0000	1.0000
1.02	0.5160	0.6234	0.8278	1.0003	0.9805	1.0471	1.0334	1.0132	1.0000
1.04	0.5039	0.6129	0.8222	1.0013	0.9620	1.0952	1.0671	1.0263	0.9999
1.06	0.4919	0.6024	0.8165	1.0029	0.9444	1.1442	1.1009	1.0393	0.9998
1.08	0.4800	0.5920	0.8108	1.0051	0.9277	1.1941	1.1349	1.0522	0.9994
1.10	0.4684	0.5817	0.8052	1.0079	0.9118	1.2450	1.1691	1.0649	0.9989
1.12	0.4568	0.5714	0.7994	1.0113	0.8966	1.2968	1.2034	1.0776	0.9982
1.14	0.4455	0.5612	0.7937	1.0153	0.8820	1.3495	1.2378	1.0903	0.9973
1.16	0.4343	0.5511	0.7879	1.0198	0.8682	1.4032	1.2723	1.1029	0.9961
1.18	0.4232	0.5411	0.7822	1.0248	0.8549	1.4578	1.3069	1.1154	0.9946
1.20	0.4124	0.5311	0.7764	1.0304	0.8422	1.5133	1.3416	1.1280	0.9928
1.22	0.4017	0.5213	0.7706	1.0366	0.8300	1.5698	1.3764	1.1405	0.9907
1.24	0.3912	0.5115	0.7648	1.0432	0.8183	1.6272	1.4112	1.1531	0.9884
1.26	0.3809	0.5019	0.7590	1.0504	0.8071	1.6855	1.4460	1.1657	0.9857
1.28	0.3708	0.4923	0.7532	1.0581	0.7963	1.7448	1.4808	1.1783	0.9827
1.30	0.3609	0.4829	0.7474	1.0663	0.7860	1.8050	1.5157	1.1909	0.9794
1.32	0.3512	0.4736	0.7416	1.0750	0.7760	1.8661	1.5505	1.2035	0.9758
1.34	0.3417	0.4644	0.7358	1.0842	0.7664	1.9282	1.5854	1.2162	0.9718
1.36	0.3323	0.4553	0.7300	1.0940	0.7572	1.9912	1.6202	1.2290	0.9676
1.38	0.3232	0.4463	0.7242	1.1042	0.7483	2.0551	1.6549	1.2418	0.9630
1.40	0.3142	0.4374	0.7184	1.1149	0.7397	2.1200	1.6897	1.2547	0.9582
1.42	0.3055	0.4287	0.7126	1.1262	0.7314	2.1858	1.7243	1.2676	0.9531
1.44	0.2969	0.4201	0.7069	1.1379	0.7235	2.2525	1.7589	1.2807	0.9476
1.46	0.2886	0.4116	0.7011	1.1501	0.7157	2.3202	1.7934	1.2938	0.9420
1.48	0.2804	0.4032	0.6954	1.1629	0.7083	2.3888	1.8278	1.3069	0.9360
1.50	0.2724	0.3950	0.6897	1.1762	0.7011	2.4583	1.8621	1.3202	0.9298
1.52	0.2646	0.3869	0.6840	1.1899	0.6941	2.5288	1.8963	1.3336	0.9233
1.54	0.2570	0.3789	0.6783	1.2042	0.6874	2.6002	1.9303	1.3470	0.9166
1.56	0.2496	0.3710	0.6726	1.2190	0.6809	2.6725	1.9643	1.3606	0.9097
1.58	0.2423	0.3633	0.6670	1.2344	0.6746	2.7458	1.9981	1.3742	0.9026
1.60	0.2353	0.3557	0.6614	1.2502	0.6684	2.8200	2.0317	1.3880	0.8952
1.62	0.2284	0.3483	0.6558	1.2666	0.6625	2.8951	2.0653	1.4018	0.8877
1.64	0.2217	0.3409	0.6502	1.2836	0.6568	2.9712	2.0986	1.4158	0.8799
1.66	0.2151	0.3337	0.6447	1.3010	0.6512	3.0482	2.1318	1.4299	0.8720
1.68	0.2088	0.3266	0.6392	1.3190	0.6458	3.1261	2.1649	1.4440	0.8639
1.70	0.2026	0.3197	0.6337	1.3376	0.6405	3.2050	2.1977	1.4583	0.8557
1.72	0.1966	0.3129	0.6283	1.3567	0.6355	3.2848	2.2304	1.4727	0.8474
1.74	0.1907	0.3062	0.6229	1.3764	0.6305	3.3655	2.2629	1.4873	0.8389
1.76	0.1850	0.2996	0.6175	1.3967	0.6257	3.4472	2.2952	1.5019	0.8302
1.78	0.1794	0.2931	0.6121	1.4175	0.6210	3.5298	2.3273	1.5167	0.8215
1.80	0.1740	0.2868	0.6068	1.4390	0.6165	3.6133	2.3592	1.5316	0.8127

Supersonic Flow of a Perfect Gas with $\gamma = 1.4$ (continued)

	Isentropic Flow				Normal Shock Wave				
M_1	p/p_t	ρ/ρ_t	T/T_t	A/A^*	M_2	p_2/p_1	ρ_2/ρ_1	T_2/T_1	p_{t2}/p_{t1}
1.80	0.1740	0.2868	0.6068	1.4390	0.6165	3.6133	2.3592	1.5316	0.8127
1.82	0.1688	0.2806	0.6015	1.4610	0.6121	3.6978	2.3909	1.5466	0.8038
1.84	0.1637	0.2745	0.5963	1.4836	0.6078	3.7832	2.4224	1.5617	0.7948
1.86	0.1587	0.2686	0.5910	1.5069	0.6036	3.8695	2.4537	1.5770	0.7857
1.88	0.1539	0.2627	0.5859	1.5308	0.5996	3.9568	2.4848	1.5924	0.7765
1.90	0.1492	0.2570	0.5807	1.5553	0.5956	4.0450	2.5157	1.6079	0.7674
1.92	0.1447	0.2514	0.5756	1.5804	0.5918	4.1341	2.5463	1.6236	0.7581
1.94	0.1403	0.2459	0.5705	1.6062	0.5880	4.2242	2.5767	1.6394	0.7488
1.96	0.1360	0.2405	0.5655	1.6326	0.5844	4.3152	2.6069	1.6553	0.7395
1.98	0.1318	0.2352	0.5605	1.6597	0.5808	4.4071	2.6369	1.6713	0.7302
2.00	0.1278	0.2300	0.5556	1.6875	0.5774	4.5000	2.6667	1.6875	0.7209
2.02	0.1239	0.2250	0.5506	1.7160	0.5740	4.5938	2.6962	1.7038	0.7115
2.04	0.1201	0.2200	0.5458	1.7451	0.5707	4.6885	2.7255	1.7203	0.7022
2.06	0.1164	0.2152	0.5409	1.7750	0.5675	4.7842	2.7545	1.7369	0.6928
2.08	0.1128	0.2104	0.5361	1.8056	0.5643	4.8808	2.7833	1.7536	0.6835
2.10	0.1094	0.2058	0.5313	1.8369	0.5613	4.9783	2.8119	1.7705	0.6742
2.12	0.1060	0.2013	0.5266	1.8690	0.5583	5.0768	2.8402	1.7875	0.6649
2.14	0.1027	0.1968	0.5219	1.9018	0.5554	5.1762	2.8683	1.8046	0.6557
2.16	0.9956^{-1}	0.1925	0.5173	1.9354	0.5525	5.2765	2.8962	1.8219	0.6464
2.18	0.9649^{-1}	0.1882	0.5127	1.9698	0.5498	5.3778	2.9238	1.8393	0.6373
2.20	0.9352^{-1}	0.1841	0.5081	2.0050	0.5471	5.4800	2.9512	1.8569	0.6281
2.22	0.9064^{-1}	0.1800	0.5036	2.0409	0.5444	5.5831	2.9784	1.8746	0.6191
2.24	0.8785^{-1}	0.1760	0.4991	2.0777	0.5418	5.6872	3.0053	1.8924	0.6100
2.26	0.8514^{-1}	0.1721	0.4947	2.1153	0.5393	5.7922	3.0319	1.9104	0.6011
2.28	0.8251^{-1}	0.1683	0.4903	2.1538	0.5368	5.8981	3.0584	1.9285	0.5921
2.30	0.7997^{-1}	0.1646	0.4859	2.1931	0.5344	6.0050	3.0845	1.9468	0.5833
2.32	0.7751^{-1}	0.1609	0.4816	2.2333	0.5321	6.1128	3.1105	1.9652	0.5745
2.34	0.7512^{-1}	0.1574	0.4773	2.2744	0.5297	6.2215	3.1362	1.9838	0.5658
2.36	0.7281^{-1}	0.1539	0.4731	2.3164	0.5275	6.3312	3.1617	2.0025	0.5572
2.38	0.7057^{-1}	0.1505	0.4688	2.3593	0.5253	6.4418	3.1869	2.0213	0.5486
2.40	0.6840^{-1}	0.1472	0.4647	2.4031	0.5231	6.5533	3.2119	2.0403	0.5401
2.42	0.6630^{-1}	0.1439	0.4606	2.4479	0.5210	6.6658	3.2367	2.0595	0.5317
2.44	0.6426^{-1}	0.1408	0.4565	2.4936	0.5189	6.7792	3.2612	2.0788	0.5234
2.46	0.6229^{-1}	0.1377	0.4524	2.5403	0.5169	6.8935	3.2855	2.0982	0.5152
2.48	0.6038^{-1}	0.1346	0.4484	2.5880	0.5149	7.0088	3.3095	2.1178	0.5071
2.50	0.5853^{-1}	0.1317	0.4444	2.6367	0.5130	7.1250	3.3333	2.1375	0.4990
2.52	0.5674^{-1}	0.1288	0.4405	2.6865	0.5111	7.2421	3.3569	2.1574	0.4911
2.54	0.5500^{-1}	0.1260	0.4366	2.7372	0.5092	7.3602	3.3803	2.1774	0.4832
2.56	0.5332^{-1}	0.1232	0.4328	2.7891	0.5074	7.4792	3.4034	2.1976	0.4754
2.58	0.5169^{-1}	0.1205	0.4289	2.8420	0.5056	7.5991	3.4263	2.2179	0.4677
2.60	0.5012^{-1}	0.1179	0.4252	2.8960	0.5039	7.7200	3.4490	2.2383	0.4601

n^{-p} is shorthand for $n \cdot 10^{-p}$

Supersonic Flow of a Perfect Gas with $\gamma = 1.4$ (continued)

	Isentropic Flow				Normal Shock Wave				
M_1	p/p_t	ρ/ρ_t	T/T_t	A/A^*	M_2	p_2/p_1	ρ_2/ρ_1	T_2/T_1	p_{t2}/p_{t1}
2.60	0.5012^{-1}	0.1179	0.4252	2.8960	0.5039	7.720	3.449	2.238	0.4601
2.62	0.4859^{-1}	0.1153	0.4214	2.9511	0.5022	7.842	3.471	2.259	0.4526
2.64	0.4711^{-1}	0.1128	0.4177	3.0073	0.5005	7.965	3.494	2.280	0.4452
2.66	0.4568^{-1}	0.1103	0.4141	3.0647	0.4988	8.088	3.516	2.301	0.4379
2.68	0.4429^{-1}	0.1079	0.4104	3.1233	0.4972	8.213	3.537	2.322	0.4307
2.70	0.4295^{-1}	0.1056	0.4068	3.1830	0.4956	8.338	3.559	2.343	0.4236
2.72	0.4165^{-1}	0.1033	0.4033	3.2440	0.4941	8.465	3.580	2.364	0.4166
2.74	0.4039^{-1}	0.1010	0.3998	3.3061	0.4926	8.592	3.601	2.386	0.4097
2.76	0.3917^{-1}	0.9885^{-1}	0.3963	3.3695	0.4911	8.721	3.622	2.407	0.4028
2.78	0.3799^{-1}	0.9671^{-1}	0.3928	3.4342	0.4896	8.850	3.643	2.429	0.3961
2.80	0.3685^{-1}	0.9463^{-1}	0.3894	3.5001	0.4882	8.980	3.664	2.451	0.3895
2.82	0.3574^{-1}	0.9259^{-1}	0.3860	3.5674	0.4868	9.111	3.684	2.473	0.3829
2.84	0.3467^{-1}	0.9059^{-1}	0.3827	3.6359	0.4854	9.243	3.704	2.496	0.3765
2.86	0.3363^{-1}	0.8865^{-1}	0.3794	3.7058	0.4840	9.376	3.724	2.518	0.3701
2.88	0.3263^{-1}	0.8675^{-1}	0.3761	3.7771	0.4827	9.510	3.743	2.540	0.3639
2.90	0.3165^{-1}	0.8489^{-1}	0.3729	3.8498	0.4814	9.645	3.763	2.563	0.3577
2.92	0.3071^{-1}	0.8307^{-1}	0.3696	3.9238	0.4801	9.781	3.782	2.586	0.3517
2.94	0.2980^{-1}	0.8130^{-1}	0.3665	3.9993	0.4788	9.918	3.801	2.609	0.3457
2.96	0.2891^{-1}	0.7957^{-1}	0.3633	4.0763	0.4776	10.055	3.820	2.632	0.3398
2.98	0.2805^{-1}	0.7788^{-1}	0.3602	4.1547	0.4764	10.194	3.839	2.656	0.3340
3.00	0.2722^{-1}	0.7623^{-1}	0.3571	4.2346	0.4752	10.333	3.857	2.679	0.3283
3.02	0.2642^{-1}	0.7461^{-1}	0.3541	4.3160	0.4740	10.474	3.875	2.703	0.3227
3.04	0.2564^{-1}	0.7303^{-1}	0.3511	4.3989	0.4729	10.615	3.893	2.726	0.3172
3.06	0.2489^{-1}	0.7149^{-1}	0.3481	4.4835	0.4717	10.758	3.911	2.750	0.3118
3.08	0.2416^{-1}	0.6999^{-1}	0.3452	4.5696	0.4706	10.901	3.929	2.774	0.3065
3.10	0.2345^{-1}	0.6852^{-1}	0.3422	4.6573	0.4695	11.045	3.947	2.799	0.3012
3.12	0.2276^{-1}	0.6708^{-1}	0.3393	4.7467	0.4685	11.190	3.964	2.823	0.2960
3.14	0.2210^{-1}	0.6568^{-1}	0.3365	4.8377	0.4674	11.336	3.981	2.848	0.2910
3.16	0.2146^{-1}	0.6430^{-1}	0.3337	4.9304	0.4664	11.483	3.998	2.872	0.2860
3.18	0.2083^{-1}	0.6296^{-1}	0.3309	5.0248	0.4654	11.631	4.015	2.897	0.2811
3.20	0.2023^{-1}	0.6165^{-1}	0.3281	5.1210	0.4643	11.780	4.031	2.922	0.2762
3.22	0.1964^{-1}	0.6037^{-1}	0.3253	5.2189	0.4634	11.930	4.048	2.947	0.2715
3.24	0.1908^{-1}	0.5912^{-1}	0.3226	5.3186	0.4624	12.081	4.064	2.972	0.2668
3.26	0.1853^{-1}	0.5790^{-1}	0.3199	5.4201	0.4614	12.232	4.080	2.998	0.2622
3.28	0.1799^{-1}	0.5671^{-1}	0.3173	5.5234	0.4605	12.385	4.096	3.023	0.2577
3.30	0.1748^{-1}	0.5554^{-1}	0.3147	5.6286	0.4596	12.538	4.112	3.049	0.2533
3.32	0.1698^{-1}	0.5440^{-1}	0.3121	5.7358	0.4587	12.693	4.128	3.075	0.2489
3.34	0.1649^{-1}	0.5329^{-1}	0.3095	5.8448	0.4578	12.848	4.143	3.101	0.2446
3.36	0.1602^{-1}	0.5220^{-1}	0.3069	5.9558	0.4569	13.005	4.158	3.127	0.2404
3.38	0.1557^{-1}	0.5113^{-1}	0.3044	6.0687	0.4560	13.162	4.173	3.154	0.2363
3.40	0.1512^{-1}	0.5009^{-1}	0.3019	6.1837	0.4552	13.320	4.188	3.180	0.2322

n^{-p} is shorthand for $n \cdot 10^{-p}$

Supersonic Flow of a Perfect Gas with $\gamma = 1.4$ (continued)

	Isentropic Flow				Normal Shock Wave				
M_1	p/p_t	ρ/ρ_t	T/T_t	A/A^*	M_2	p_2/p_1	ρ_2/ρ_1	T_2/T_1	p_{t2}/p_{t1}
3.4	0.1512^{-1}	0.5009^{-1}	0.3019	6.18	0.4552	13.32	4.188	3.180	0.2322
3.5	0.1311^{-1}	0.4523^{-1}	0.2899	6.79	0.4512	14.13	4.261	3.315	0.2129
3.6	0.1138^{-1}	0.4089^{-1}	0.2784	7.45	0.4474	14.95	4.330	3.454	0.1953
3.7	0.9903^{-2}	0.3702^{-1}	0.2675	8.17	0.4439	15.81	4.395	3.596	0.1792
3.8	0.8629^{-2}	0.3355^{-1}	0.2572	8.95	0.4407	16.68	4.457	3.743	0.1645
3.9	0.7532^{-2}	0.3044^{-1}	0.2474	9.80	0.4377	17.58	4.516	3.893	0.1510
4.0	0.6586^{-2}	0.2766^{-1}	0.2381	10.72	0.4350	18.50	4.571	4.047	0.1388
4.1	0.5769^{-2}	0.2516^{-1}	0.2293	11.71	0.4324	19.44	4.624	4.205	0.1276
4.2	0.5062^{-2}	0.2292^{-1}	0.2208	12.79	0.4299	20.41	4.675	4.367	0.1173
4.3	0.4449^{-2}	0.2090^{-1}	0.2129	13.95	0.4277	21.41	4.723	4.532	0.1080
4.4	0.3918^{-2}	0.1909^{-1}	0.2053	15.21	0.4255	22.42	4.768	4.702	0.9948^{-1}
4.5	0.3455^{-2}	0.1745^{-1}	0.1980	16.56	0.4236	23.46	4.812	4.875	0.9170^{-1}
4.6	0.3053^{-2}	0.1597^{-1}	0.1911	18.02	0.4217	24.52	4.853	5.052	0.8459^{-1}
4.7	0.2701^{-2}	0.1464^{-1}	0.1846	19.58	0.4199	25.60	4.893	5.233	0.7809^{-1}
4.8	0.2394^{-2}	0.1343^{-1}	0.1783	21.26	0.4183	26.71	4.930	5.418	0.7214^{-1}
4.9	0.2126^{-2}	0.1233^{-1}	0.1724	23.07	0.4167	27.84	4.966	5.607	0.6670^{-1}
5.0	0.1890^{-2}	0.1134^{-1}	0.1667	25.00	0.4152	29.00	5.000	5.800	0.6172^{-1}
5.1	0.1683^{-2}	0.1044^{-1}	0.1612	27.07	0.4138	30.18	5.033	5.997	0.5715^{-1}
5.2	0.1501^{-2}	0.9620^{-2}	0.1561	29.28	0.4125	31.38	5.064	6.197	0.5297^{-1}
5.3	0.1341^{-2}	0.8875^{-2}	0.1511	31.65	0.4113	32.60	5.093	6.401	0.4913^{-1}
5.4	0.1200^{-2}	0.8197^{-2}	0.1464	34.17	0.4101	33.85	5.122	6.610	0.4560^{-1}
5.5	0.1075^{-2}	0.7578^{-2}	0.1418	36.87	0.4090	35.13	5.149	6.822	0.4236^{-1}
5.6	0.9643^{-3}	0.7012^{-2}	0.1375	39.74	0.4079	36.42	5.175	7.038	0.3938^{-1}
5.7	0.8663^{-3}	0.6496^{-2}	0.1334	42.80	0.4069	37.74	5.200	7.258	0.3664^{-1}
5.8	0.7794^{-3}	0.6023^{-2}	0.1294	46.05	0.4059	39.08	5.224	7.481	0.3412^{-1}
5.9	0.7021^{-3}	0.5590^{-2}	0.1256	49.51	0.4050	40.44	5.246	7.709	0.3179^{-1}
6.0	0.6334^{-3}	0.5194^{-2}	0.1220	53.18	0.4042	41.83	5.268	7.941	0.2965^{-1}
6.1	0.5721^{-3}	0.4829^{-2}	0.1185	57.08	0.4033	43.24	5.289	8.176	0.2767^{-1}
6.2	0.5173^{-3}	0.4495^{-2}	0.1151	61.21	0.4025	44.68	5.309	8.415	0.2584^{-1}
6.3	0.4684^{-3}	0.4187^{-2}	0.1119	65.59	0.4018	46.14	5.329	8.658	0.2416^{-1}
6.4	0.4247^{-3}	0.3904^{-2}	0.1088	70.23	0.4011	47.62	5.347	8.905	0.2259^{-1}
6.5	0.3855^{-3}	0.3643^{-2}	0.1058	75.13	0.4004	49.13	5.365	9.156	0.2115^{-1}
6.6	0.3503^{-3}	0.3402^{-2}	0.1030	80.32	0.3997	50.65	5.382	9.411	0.1981^{-1}
6.7	0.3187^{-3}	0.3180^{-2}	0.1002	85.80	0.3991	52.20	5.399	9.670	0.1857^{-1}
6.8	0.2902^{-3}	0.2974^{-2}	0.0976	91.59	0.3985	53.78	5.415	9.933	0.1741^{-1}
6.9	0.2646^{-3}	0.2785^{-2}	0.0950	97.70	0.3979	55.38	5.430	10.199	0.1634^{-1}
7.0	0.2416^{-3}	0.2609^{-2}	0.0926	104.14	0.3974	57.00	5.444	10.469	0.1535^{-1}
7.1	0.2207^{-3}	0.2446^{-2}	0.0902	110.93	0.3968	58.64	5.459	10.744	0.1443^{-1}
7.2	0.2019^{-3}	0.2295^{-2}	0.0880	118.08	0.3963	60.31	5.472	11.022	0.1357^{-1}
7.3	0.1848^{-3}	0.2155^{-2}	0.0858	125.60	0.3958	62.00	5.485	11.304	0.1277^{-1}
7.4	0.1694^{-3}	0.2025^{-2}	0.0837	133.52	0.3954	63.72	5.498	11.590	0.1202^{-1}
7.5	0.1554^{-3}	0.1904^{-2}	0.0816	141.84	0.3949	65.46	5.510	11.879	0.1133^{-1}

n^{-p} is shorthand for $n \cdot 10^{-p}$

Supersonic Flow of a Perfect Gas with $\gamma = 1.4$ (continued)

	Isentropic Flow				Normal Shock Wave				
M_1	p/p_t	ρ/ρ_t	T/T_t	A/A^*	M_2	p_2/p_1	ρ_2/ρ_1	T_2/T_1	p_{t2}/p_{t1}
7.5	0.1554^{-3}	0.1904^{-2}	0.8163^{-1}	141.8	0.3949	65.46	5.510	11.88	0.1133^{-1}
8.0	0.1024^{-3}	0.1414^{-2}	0.7246^{-1}	190.1	0.3929	74.50	5.565	13.39	0.8488^{-2}
8.5	0.6898^{-4}	0.1066^{-2}	0.6472^{-1}	251.1	0.3912	84.13	5.612	14.99	0.6449^{-2}
9.0	0.4739^{-4}	0.8150^{-3}	0.5814^{-1}	327.2	0.3898	94.33	5.651	16.69	0.4964^{-2}
9.5	0.3314^{-4}	0.6313^{-3}	0.5249^{-1}	421.1	0.3886	105.13	5.685	18.49	0.3866^{-2}
10.0	0.2356^{-4}	0.4948^{-3}	0.4762^{-1}	535.9	0.3876	116.50	5.714	20.39	0.3045^{-2}

n^{-p} is shorthand for $n \cdot 10^{-p}$

C.2 Prandtl-Meyer Function and Mach Angle

Prandtl-Meyer Function, $\nu(M)$, and Mach angle, $\mu(M)$, for $\gamma = 1.4$

M	$\mu(M)$	$\nu(M)$	M	$\mu(M)$	$\nu(M)$	M	$\mu(M)$	$\nu(M)$
1.00	90.00	0.00	2.00	30.00	26.38	3.00	19.47	49.76
1.05	72.25	0.49	2.05	29.20	27.75	3.05	19.14	50.71
1.10	65.38	1.34	2.10	28.44	29.10	3.10	18.82	51.65
1.15	60.41	2.38	2.15	27.72	30.43	3.15	18.51	52.57
1.20	56.44	3.56	2.20	27.04	31.73	3.20	18.21	53.47
1.25	53.13	4.83	2.25	26.39	33.02	3.25	17.92	54.35
1.30	50.28	6.17	2.30	25.77	34.28	3.30	17.64	55.22
1.35	47.79	7.56	2.35	25.18	35.53	3.35	17.37	56.07
1.40	45.58	8.99	2.40	24.62	36.75	3.40	17.10	56.91
1.45	43.60	10.44	2.45	24.09	37.95	3.45	16.85	57.73
1.50	41.81	11.91	2.50	23.58	39.12	3.50	16.60	58.53
1.55	40.18	13.38	2.55	23.09	40.28	3.55	16.36	59.32
1.60	38.68	14.86	2.60	22.62	41.41	3.60	16.13	60.09
1.65	37.31	16.34	2.65	22.17	42.53	3.65	15.90	60.85
1.70	36.03	17.81	2.70	21.74	43.62	3.70	15.68	61.60
1.75	34.85	19.27	2.75	21.32	44.69	3.75	15.47	62.33
1.80	33.75	20.73	2.80	20.92	45.75	3.80	15.26	63.04
1.85	32.72	22.16	2.85	20.54	46.78	3.85	15.05	63.75
1.90	31.76	23.59	2.90	20.17	47.79	3.90	14.86	64.44
1.95	30.85	24.99	2.95	19.81	48.78	3.95	14.66	65.12
2.00	30.00	26.38	3.00	19.47	49.76	4.00	14.48	65.78

Values for $\mu(M)$ and $\nu(M)$ are in degrees

Prandtl-Meyer Function, $\nu(M)$, and Mach angle, $\mu(M)$, for $\gamma = 1.4$ (continued)

M	$\mu(M)$	$\nu(M)$	M	$\mu(M)$	$\nu(M)$	M	$\mu(M)$	$\nu(M)$
4.00	14.48	65.78	6.00	9.59	84.96	8.00	7.18	95.62
4.05	14.29	66.44	6.05	9.51	85.30	8.05	7.14	95.83
4.10	14.12	67.08	6.10	9.44	85.63	8.10	7.09	96.03
4.15	13.94	67.71	6.15	9.36	85.97	8.15	7.05	96.23
4.20	13.77	68.33	6.20	9.28	86.29	8.20	7.00	96.43
4.25	13.61	68.94	6.25	9.21	86.62	8.25	6.96	96.63
4.30	13.45	69.54	6.30	9.13	86.94	8.30	6.92	96.82
4.35	13.29	70.13	6.35	9.06	87.25	8.35	6.88	97.01
4.40	13.14	70.71	6.40	8.99	87.56	8.40	6.84	97.20
4.45	12.99	71.27	6.45	8.92	87.87	8.45	6.80	97.39
4.50	12.84	71.83	6.50	8.85	88.17	8.50	6.76	97.57
4.55	12.70	72.38	6.55	8.78	88.47	8.55	6.72	97.76
4.60	12.56	72.92	6.60	8.71	88.76	8.60	6.68	97.94
4.65	12.42	73.45	6.65	8.65	89.05	8.65	6.64	98.12
4.70	12.28	73.97	6.70	8.58	89.33	8.70	6.60	98.29
4.75	12.15	74.48	6.75	8.52	89.62	8.75	6.56	98.47
4.80	12.02	74.99	6.80	8.46	89.89	8.80	6.52	98.64
4.85	11.90	75.48	6.85	8.39	90.17	8.85	6.49	98.81
4.90	11.78	75.97	6.90	8.33	90.44	8.90	6.45	98.98
4.95	11.66	76.45	6.95	8.27	90.71	8.95	6.42	99.15
5.00	11.54	76.92	7.00	8.21	90.97	9.00	6.38	99.32
5.05	11.42	77.38	7.05	8.15	91.23	9.05	6.34	99.48
5.10	11.31	77.84	7.10	8.10	91.49	9.10	6.31	99.65
5.15	11.20	78.29	7.15	8.04	91.75	9.15	6.27	99.81
5.20	11.09	78.73	7.20	7.98	92.00	9.20	6.24	99.97
5.25	10.98	79.17	7.25	7.93	92.24	9.25	6.21	100.12
5.30	10.88	79.60	7.30	7.87	92.49	9.30	6.17	100.28
5.35	10.77	80.02	7.35	7.82	92.73	9.35	6.14	100.44
5.40	10.67	80.43	7.40	7.77	92.97	9.40	6.11	100.59
5.45	10.57	80.84	7.45	7.71	93.21	9.45	6.07	100.74
5.50	10.48	81.24	7.50	7.66	93.44	9.50	6.04	100.89
5.55	10.38	81.64	7.55	7.61	93.67	9.55	6.01	101.04
5.60	10.29	82.03	7.60	7.56	93.90	9.60	5.98	101.19
5.65	10.19	82.42	7.65	7.51	94.12	9.65	5.95	101.33
5.70	10.10	82.80	7.70	7.46	94.34	9.70	5.92	101.48
5.75	10.02	83.17	7.75	7.41	94.56	9.75	5.89	101.62
5.80	9.93	83.54	7.80	7.37	94.78	9.80	5.86	101.76
5.85	9.84	83.90	7.85	7.32	95.00	9.85	5.83	101.90
5.90	9.76	84.26	7.90	7.27	95.21	9.90	5.80	102.04
5.95	9.68	84.61	7.95	7.23	95.42	9.95	5.77	102.18
6.00	9.59	84.96	8.00	7.18	95.62	10.00	5.74	102.32

Values for $\mu(M)$ and $\nu(M)$ are in degrees

C.3 Fanno Flow

Notation used in the Fanno-flow tables of this section is as follows.

M	=	local Mach number
fL^*/D	=	dimensionless length to sonic point
p/p^*	=	ratio of static to sonic pressure
ρ/ρ^*	=	ratio of static to sonic density
T/T^*	=	ratio of static to sonic temperature
p_t/p_t^*	=	ratio of local total to sonic total pressure

Fanno Flow for $\gamma = 1.4$

M	fL^*/D	p/p^*	ρ/ρ^*	T/T^*	p_t/p_t^*
0.00	∞	∞	∞	1.2000	∞
0.05	280.0203	21.9034	18.2620	1.1994	11.5914
0.10	66.9216	10.9435	9.1378	1.1976	5.8218
0.15	27.9320	7.2866	6.0995	1.1946	3.9103
0.20	14.5333	5.4554	4.5826	1.1905	2.9635
0.25	8.4834	4.3546	3.6742	1.1852	2.4027
0.30	5.2993	3.6191	3.0702	1.1788	2.0351
0.35	3.4525	3.0922	2.6400	1.1713	1.7780
0.40	2.3085	2.6958	2.3184	1.1628	1.5901
0.45	1.5664	2.3865	2.0693	1.1533	1.4487
0.50	1.0691	2.1381	1.8708	1.1429	1.3398
0.55	0.7281	1.9341	1.7092	1.1315	1.2549
0.60	0.4908	1.7634	1.5753	1.1194	1.1882
0.65	0.3246	1.6183	1.4626	1.1065	1.1356
0.70	0.2081	1.4935	1.3665	1.0929	1.0944
0.75	0.1273	1.3848	1.2838	1.0787	1.0624
0.80	0.0723	1.2893	1.2119	1.0638	1.0382
0.85	0.0363	1.2047	1.1489	1.0485	1.0207
0.90	0.0145	1.1291	1.0934	1.0327	1.0089
0.95	0.0033	1.0613	1.0440	1.0165	1.0021
1.00	0.0000	1.0000	1.0000	1.0000	1.0000

Fanno Flow for $\gamma = 1.4$ (continued)

M	fL^*/D	p/p^*	ρ/ρ^*	T/T^*	p_t/p_t^*
1.00	0.0000	1.0000	1.0000	1.0000	1.0000
1.05	0.0027	0.9443	0.9605	0.9832	1.0020
1.10	0.0099	0.8936	0.9249	0.9662	1.0079
1.15	0.0205	0.8471	0.8926	0.9490	1.0175
1.20	0.0336	0.8044	0.8633	0.9317	1.0304
1.25	0.0486	0.7649	0.8367	0.9143	1.0468
1.30	0.0648	0.7285	0.8123	0.8969	1.0663
1.35	0.0820	0.6947	0.7899	0.8794	1.0890
1.40	0.0997	0.6632	0.7693	0.8621	1.1149
1.45	0.1178	0.6339	0.7503	0.8448	1.1440
1.50	0.1361	0.6065	0.7328	0.8276	1.1762
1.55	0.1543	0.5808	0.7166	0.8105	1.2116
1.60	0.1724	0.5568	0.7016	0.7937	1.2502
1.65	0.1902	0.5342	0.6876	0.7770	1.2922
1.70	0.2078	0.5130	0.6745	0.7605	1.3376
1.75	0.2250	0.4929	0.6624	0.7442	1.3865
1.80	0.2419	0.4741	0.6511	0.7282	1.4390
1.85	0.2583	0.4562	0.6404	0.7124	1.4952
1.90	0.2743	0.4394	0.6305	0.6969	1.5553
1.95	0.2899	0.4234	0.6211	0.6816	1.6193
2.00	0.3050	0.4082	0.6124	0.6667	1.6875
2.05	0.3197	0.3939	0.6041	0.6520	1.7600
2.10	0.3339	0.3802	0.5963	0.6376	1.8369
2.15	0.3476	0.3673	0.5890	0.6235	1.9185
2.20	0.3609	0.3549	0.5821	0.6098	2.0050
2.25	0.3738	0.3432	0.5756	0.5963	2.0964
2.30	0.3862	0.3320	0.5694	0.5831	2.1931
2.35	0.3983	0.3213	0.5635	0.5702	2.2953
2.40	0.4099	0.3111	0.5580	0.5576	2.4031
2.45	0.4211	0.3014	0.5527	0.5453	2.5168
2.50	0.4320	0.2921	0.5477	0.5333	2.6367
2.55	0.4425	0.2832	0.5430	0.5216	2.7630
2.60	0.4526	0.2747	0.5385	0.5102	2.8960
2.65	0.4624	0.2666	0.5342	0.4991	3.0359
2.70	0.4718	0.2588	0.5301	0.4882	3.1830
2.75	0.4809	0.2513	0.5262	0.4776	3.3377
2.80	0.4898	0.2441	0.5225	0.4673	3.5001
2.85	0.4983	0.2373	0.5189	0.4572	3.6707
2.90	0.5065	0.2307	0.5155	0.4474	3.8498
2.95	0.5145	0.2243	0.5123	0.4379	4.0376
3.00	0.5222	0.2182	0.5092	0.4286	4.2346

Fanno Flow for $\gamma = 1.4$ (continued)

M	fL^*/D	p/p^*	ρ/ρ^*	T/T^*	p_t/p_t^*
3.00	0.5222	0.2182	0.5092	0.4286	4.2346
3.05	0.5296	0.2124	0.5062	0.4195	4.4410
3.10	0.5368	0.2067	0.5034	0.4107	4.6573
3.15	0.5437	0.2013	0.5007	0.4021	4.8838
3.20	0.5504	0.1961	0.4980	0.3937	5.1210
3.25	0.5569	0.1911	0.4955	0.3855	5.3691
3.30	0.5632	0.1862	0.4931	0.3776	5.6286
3.35	0.5693	0.1815	0.4908	0.3699	5.9000
3.40	0.5752	0.1770	0.4886	0.3623	6.1837
3.45	0.5809	0.1727	0.4865	0.3550	6.4801
3.50	0.5864	0.1685	0.4845	0.3478	6.7896
3.55	0.5918	0.1645	0.4825	0.3409	7.1128
3.60	0.5970	0.1606	0.4806	0.3341	7.4501
3.65	0.6020	0.1568	0.4788	0.3275	7.8020
3.70	0.6068	0.1531	0.4770	0.3210	8.1691
3.75	0.6115	0.1496	0.4753	0.3148	8.5517
3.80	0.6161	0.1462	0.4737	0.3086	8.9506
3.85	0.6206	0.1429	0.4721	0.3027	9.3661
3.90	0.6248	0.1397	0.4706	0.2969	9.7990
3.95	0.6290	0.1366	0.4691	0.2912	10.2496
4.00	0.6331	0.1336	0.4677	0.2857	10.7188
4.05	0.6370	0.1307	0.4663	0.2803	11.2069
4.10	0.6408	0.1279	0.4650	0.2751	11.7147
4.15	0.6445	0.1252	0.4637	0.2700	12.2427
4.20	0.6481	0.1226	0.4625	0.2650	12.7916
4.25	0.6516	0.1200	0.4613	0.2602	13.3622
4.30	0.6550	0.1175	0.4601	0.2554	13.9549
4.35	0.6583	0.1151	0.4590	0.2508	14.5706
4.40	0.6615	0.1128	0.4579	0.2463	15.2099
4.45	0.6646	0.1105	0.4569	0.2419	15.8735
4.50	0.6676	0.1083	0.4559	0.2376	16.5622
4.55	0.6706	0.1062	0.4549	0.2334	17.2767
4.60	0.6734	0.1041	0.4539	0.2294	18.0178
4.65	0.6762	0.1021	0.4530	0.2254	18.7862
4.70	0.6790	0.1001	0.4521	0.2215	19.5828
4.75	0.6816	0.0982	0.4512	0.2177	20.4084
4.80	0.6842	0.0964	0.4504	0.2140	21.2637
4.85	0.6867	0.0946	0.4495	0.2104	22.1497
4.90	0.6891	0.0928	0.4487	0.2068	23.0671
4.95	0.6915	0.0911	0.4480	0.2034	24.0169
5.00	0.6938	0.0894	0.4472	0.2000	25.0000

C.4 Rayleigh Flow

Notation used in the Rayleigh-flow tables of this section is as follows.

M	=	local Mach number
T_t/T_t^*	=	ratio of local total to sonic total temperature
p/p^*	=	ratio of static to sonic pressure
ρ/ρ^*	=	ratio of static to sonic density
T/T^*	=	ratio of static to sonic temperature
p_t/p_t^*	=	ratio of local total to sonic total pressure

Rayleigh Flow for $\gamma = 1.4$

M	T_t/T_t^*	p/p^*	ρ/ρ^*	T/T^*	p_t/p_t^*
0.00	0.0000	2.4000	∞	0.0000	1.2679
0.05	0.0119	2.3916	167.2500	0.0143	1.2657
0.10	0.0468	2.3669	42.2500	0.0560	1.2591
0.15	0.1020	2.3267	19.1019	0.1218	1.2486
0.20	0.1736	2.2727	11.0000	0.2066	1.2346
0.25	0.2568	2.2069	7.2500	0.3044	1.2177
0.30	0.3469	2.1314	5.2130	0.4089	1.1985
0.35	0.4389	2.0487	3.9847	0.5141	1.1779
0.40	0.5290	1.9608	3.1875	0.6151	1.1566
0.45	0.6139	1.8699	2.6409	0.7080	1.1351
0.50	0.6914	1.7778	2.2500	0.7901	1.1141
0.55	0.7599	1.6860	1.9607	0.8599	1.0940
0.60	0.8189	1.5957	1.7407	0.9167	1.0753
0.65	0.8683	1.5080	1.5695	0.9608	1.0582
0.70	0.9085	1.4235	1.4337	0.9929	1.0431
0.75	0.9401	1.3427	1.3241	1.0140	1.0301
0.80	0.9639	1.2658	1.2344	1.0255	1.0193
0.85	0.9810	1.1931	1.1600	1.0285	1.0109
0.90	0.9921	1.1246	1.0977	1.0245	1.0049
0.95	0.9981	1.0603	1.0450	1.0146	1.0012
1.00	1.0000	1.0000	1.0000	1.0000	1.0000

Rayleigh Flow for $\gamma = 1.4$ (continued)

M	T_t/T_t^*	p/p^*	ρ/ρ^*	T/T^*	p_t/p_t^*
1.00	1.0000	1.0000	1.0000	1.0000	1.0000
1.05	0.9984	0.9436	0.9613	0.9816	1.0012
1.10	0.9939	0.8909	0.9277	0.9603	1.0049
1.15	0.9872	0.8417	0.8984	0.9369	1.0109
1.20	0.9787	0.7958	0.8727	0.9118	1.0194
1.25	0.9689	0.7529	0.8500	0.8858	1.0303
1.30	0.9580	0.7130	0.8299	0.8592	1.0437
1.35	0.9464	0.6758	0.8120	0.8323	1.0594
1.40	0.9343	0.6410	0.7959	0.8054	1.0777
1.45	0.9218	0.6086	0.7815	0.7787	1.0983
1.50	0.9093	0.5783	0.7685	0.7525	1.1215
1.55	0.8967	0.5500	0.7568	0.7268	1.1473
1.60	0.8842	0.5236	0.7461	0.7017	1.1756
1.65	0.8718	0.4988	0.7364	0.6774	1.2066
1.70	0.8597	0.4756	0.7275	0.6538	1.2402
1.75	0.8478	0.4539	0.7194	0.6310	1.2767
1.80	0.8363	0.4335	0.7119	0.6089	1.3159
1.85	0.8250	0.4144	0.7051	0.5877	1.3581
1.90	0.8141	0.3964	0.6988	0.5673	1.4033
1.95	0.8036	0.3795	0.6929	0.5477	1.4516
2.00	0.7934	0.3636	0.6875	0.5289	1.5031
2.05	0.7835	0.3487	0.6825	0.5109	1.5579
2.10	0.7741	0.3345	0.6778	0.4936	1.6162
2.15	0.7649	0.3212	0.6735	0.4770	1.6780
2.20	0.7561	0.3086	0.6694	0.4611	1.7434
2.25	0.7477	0.2968	0.6656	0.4458	1.8128
2.30	0.7395	0.2855	0.6621	0.4312	1.8860
2.35	0.7317	0.2749	0.6588	0.4172	1.9634
2.40	0.7242	0.2648	0.6557	0.4038	2.0451
2.45	0.7170	0.2552	0.6527	0.3910	2.1311
2.50	0.7101	0.2462	0.6500	0.3787	2.2218
2.55	0.7034	0.2375	0.6474	0.3669	2.3173
2.60	0.6970	0.2294	0.6450	0.3556	2.4177
2.65	0.6908	0.2216	0.6427	0.3448	2.5233
2.70	0.6849	0.2142	0.6405	0.3344	2.6343
2.75	0.6793	0.2071	0.6384	0.3244	2.7508
2.80	0.6738	0.2004	0.6365	0.3149	2.8731
2.85	0.6685	0.1940	0.6346	0.3057	3.0014
2.90	0.6635	0.1879	0.6329	0.2969	3.1359
2.95	0.6586	0.1820	0.6312	0.2884	3.2768
3.00	0.6540	0.1765	0.6296	0.2803	3.4245

Rayleigh Flow for $\gamma = 1.4$ (continued)

M	T_t/T_t^*	p/p^*	ρ/ρ^*	T/T^*	p_t/p_t^*
3.00	0.6540	0.1765	0.6296	0.2803	3.4245
3.05	0.6495	0.1711	0.6281	0.2725	3.5790
3.10	0.6452	0.1660	0.6267	0.2650	3.7408
3.15	0.6410	0.1612	0.6253	0.2577	3.9101
3.20	0.6370	0.1565	0.6240	0.2508	4.0871
3.25	0.6331	0.1520	0.6228	0.2441	4.2721
3.30	0.6294	0.1477	0.6216	0.2377	4.4655
3.35	0.6258	0.1436	0.6205	0.2315	4.6674
3.40	0.6224	0.1397	0.6194	0.2255	4.8783
3.45	0.6190	0.1359	0.6183	0.2197	5.0984
3.50	0.6158	0.1322	0.6173	0.2142	5.3280
3.55	0.6127	0.1287	0.6164	0.2088	5.5676
3.60	0.6097	0.1254	0.6155	0.2037	5.8173
3.65	0.6068	0.1221	0.6146	0.1987	6.0776
3.70	0.6040	0.1190	0.6138	0.1939	6.3488
3.75	0.6013	0.1160	0.6130	0.1893	6.6314
3.80	0.5987	0.1131	0.6122	0.1848	6.9256
3.85	0.5962	0.1103	0.6114	0.1805	7.2318
3.90	0.5937	0.1077	0.6107	0.1763	7.5505
3.95	0.5914	0.1051	0.6100	0.1722	7.8820
4.00	0.5891	0.1026	0.6094	0.1683	8.2268
4.05	0.5869	0.1002	0.6087	0.1645	8.5853
4.10	0.5847	0.0978	0.6081	0.1609	8.9579
4.15	0.5827	0.0956	0.6075	0.1573	9.3451
4.20	0.5807	0.0934	0.6070	0.1539	9.7473
4.25	0.5787	0.0913	0.6064	0.1506	10.1649
4.30	0.5768	0.0893	0.6059	0.1473	10.5985
4.35	0.5750	0.0873	0.6054	0.1442	11.0486
4.40	0.5732	0.0854	0.6049	0.1412	11.5155
4.45	0.5715	0.0836	0.6044	0.1383	11.9999
4.50	0.5698	0.0818	0.6039	0.1354	12.5023
4.55	0.5682	0.0800	0.6035	0.1326	13.0231
4.60	0.5666	0.0784	0.6030	0.1300	13.5629
4.65	0.5651	0.0767	0.6026	0.1274	14.1223
4.70	0.5636	0.0752	0.6022	0.1248	14.7017
4.75	0.5622	0.0736	0.6018	0.1224	15.3019
4.80	0.5608	0.0722	0.6014	0.1200	15.9234
4.85	0.5594	0.0707	0.6010	0.1177	16.5667
4.90	0.5581	0.0693	0.6007	0.1154	17.2325
4.95	0.5568	0.0680	0.6003	0.1132	17.9213
5.00	0.5556	0.0667	0.6000	0.1111	18.6339

Appendix D

Drag Data

Figure D.1 shows the drag coefficient, C_D, as a function of Reynolds number, Re_D. Letting F_D denote the drag force, C_D is defined as

$$C_D \equiv \frac{F_D}{\frac{1}{2}\rho U^2 A}, \qquad A = \frac{\pi}{4}D^2 \qquad (D.1)$$

where ρ, U and D denote fluid density, freestream velocity and cylinder/sphere diameter.

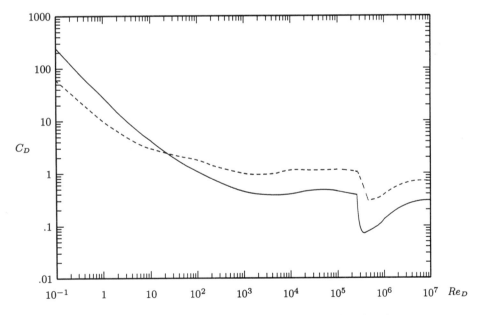

Figure D.1: *Drag on spheres (——) and cylinders (- - -).*

Shape			C_D data
Cube	$A = s^2$	Orientation: (a) Unrotated, (b) Rotated 45°	C_D: 1.08, 0.81
Rectangular plate	$A = bh$	b/h: 1, 5, 10, 20, ∞	C_D: 1.18, 1.20, 1.30, 1.50, 2.00
Hemisphere	$A = \frac{\pi}{4}d^2$	Flat side facing: (a) Upstream, (b) Downstream	C_D: 1.42, 0.38
Ellipsoid	$A = \frac{\pi}{4}d^2$	ℓ/d: 0.75, 1, 2, 4, 8	C_D: 0.20, 0.20, 0.13, 0.10, 0.08
Cone	$A = \frac{\pi}{4}d^2$	θ: 10°, 30°, 60°, 90°	C_D: 0.30, 0.55, 0.80, 1.15
Flat-Faced Cylinder	$A = \frac{\pi}{4}d^2$	ℓ/d: 0.5, 1, 2, 4, 8	C_D: 1.15, 0.90, 0.85, 0.87, 0.99
Thin Disk	$A = \frac{\pi}{4}d^2$		$C_D = 1.10 - 1.17$
Parachute	$A = \frac{\pi}{4}d^2$		$C_D = 1.20 - 1.40$
Waving Flag	$A = \ell h$	ℓ/h: 1, 2, 3	C_D: 0.07, 0.12, 0.15

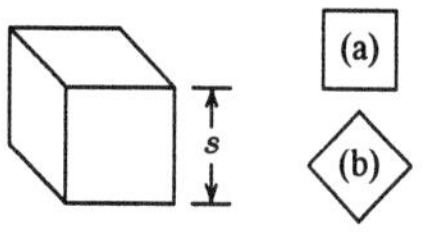

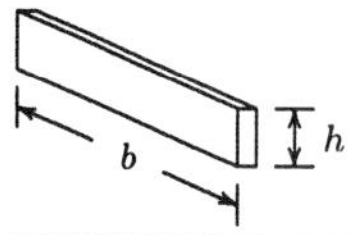

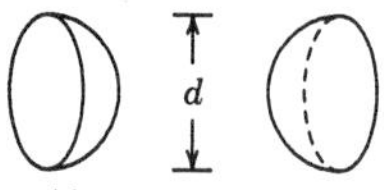

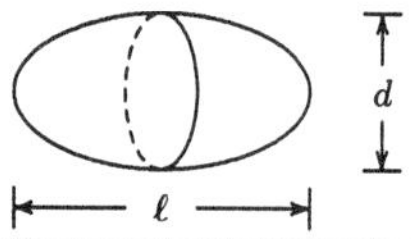

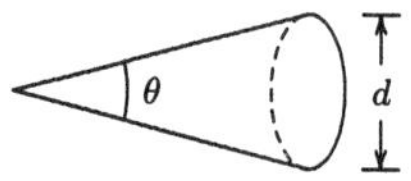

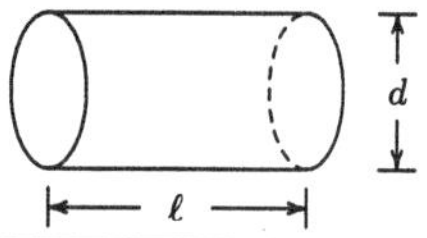

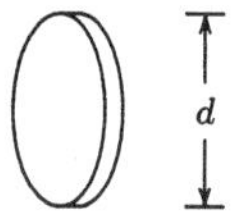

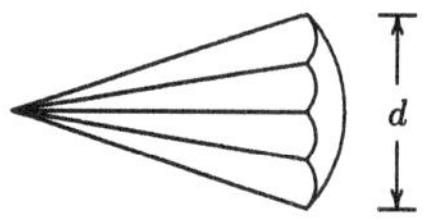

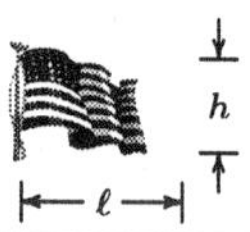

Figure D.2: *Drag-coefficient data for high-Reynolds-number flow past assorted three-dimensional objects.*

Appendix E

Conversion Factors

Mass	1 lbm	=	0.454 kg	Velocity	1 mph	=	1.467 ft/sec
	1 slug	=	14.594 kg		1 mph	=	0.447 m/sec
	1 oz	=	28.35 g		1 knot	=	1.688 ft/sec
	1 kg	=	2.205 lbm		1 knot	=	0.514 m/sec
Length	1 ft	=	0.3048 m	Work	1 Btu	=	778 ft·lb
	1 in	=	25.4 mm		1 ft·lb	=	1.3558 J
	1 mi	=	5280 ft		1 Btu	=	1055 J
Temperature	°C	=	$\frac{5}{9}$(°F−32°)	Power	1 hp	=	550 ft·lb/sec
	K	=	°C + 273.16		1 hp	=	2545 Btu/hr
	°R	=	$\frac{9}{5}$K		1 kW	=	1.341 hp
	°R	=	°F + 459.67°		1 J/sec	=	1 W
Force	1 lb	=	4.448 N	Volume	1 gal	=	0.0037854 m^3
	1 kN	=	224.8 lb		1 gal	=	231 in^3
	1 ton	=	2000 lb		1 gal	=	0.134 ft^3
	1 ton	=	8.897 kN		1 L	=	0.001 m^3
Pressure	1 psi	=	6.895 kPa	Rotation	1 rpm	=	0.1047 sec^{-1}
	1 psf	=	47.88 Pa		1 Hz	=	2π sec^{-1}
	1 atm	=	101 kPa		1 cps		2π sec^{-1}
	1 atm	=	2116.8 psf		1 rev		2π radians
	1 atm	=	760 mmHg				
Gravity	g	=	32.174 ft/sec^2				
	g	=	9.807 m/sec^2				

Other Publications by DCW Industries

Clichés of Liberalism, D. C. Wilcox (1999): This is an often humorous and always insightful discussion and analysis of the trite phrases that pass for political discourse in today's America. The book provides rational arguments refuting the ideas of the leftist radicals who have hijacked the term liberal, which— in classical literature—describes the individualistic, free-enterprise oriented philosophy of men such as John Locke, Thomas Jefferson and Adam Smith. The book consists of 10 essays focusing on political philosophy, economics and individual liberty.

The Low-Down on Entropy and Interpretive Thermodynamics, S. J. Kline (1999): If you feel you do not understand entropy or the Second Law of Thermodynamics as well as you would like to, this is a book you may want to read. The book begins at the origins of thermodynamics in 1824 and comes forward to 1997. You can begin from your level and read as far as you want. The book contains several important new results that many experts and students of thermodynamics will find useful.

Turbulence Modeling for CFD, Second Edition, D. C. Wilcox (1998): A first- or second-year graduate text on modern methods for formulating and analyzing engineering models of turbulence. Presents a comprehensive discussion of algebraic, one-equation, two-equation and stress-transport models. Emphasizes an integrated balance of similarity solutions, perturbation methods and numerical integration schemes to test and formulate rational models. Includes a brief introduction to DNS, LES and Chaos theory. Accompanied by a floppy disk with a variety of useful programs, including an industrial-strength boundary-layer program with a menu driven input-data preparation utility.

Basic Fluid Mechanics, D. C. Wilcox (1997): This book is appropriate for a two-term, junior or senior level undergraduate series of courses, or as an introductory text for graduate students with minimal prior knowledge of fluid mechanics. The first part of the book provides sufficient material for an introductory course, focusing primarily on the control-volume approach. With a combination of dimensional analysis and the control-volume method, the text discusses pipe flow, open-channel flow, elements of turbomachine theory and one-dimensional compressible flows. The balance of the text can be presented in a subsequent course, focusing on the differential equations of fluid mechanics. Topics covered include a rigorous development of the Navier-Stokes equation, potential-flow, exact Navier-Stokes solutions, boundary layers, simple viscous compressible flows, centered expansions and oblique shocks. The book is accompanied by a floppy disk with a variety of useful programs.

Perturbation Methods in the Computer Age, D. C. Wilcox (1995): Advanced undergraduate or first-year graduate text on asymptotic and perturbation methods. Discusses asymptotic expansion of integrals, including Laplace's method, stationary phase and steepest descent. Introduces the general principles of singular perturbation theory, including examples for both ODE's and PDE's. Covers multiple-scale analysis, including the method of averaging and the WKB method. Shows, through a collection of practical examples, how useful asymptotics can be when used in conjunction with computational methods.

Visit our World Wide Web Home Page for complete details about our books, software products and special sales that we conduct from time to time.

DCW Industries, Inc.
5354 Palm Drive, La Cañada, California 91011-1655 USA
Telephone: 818/790-3844   FAX: 818/952-1272
E-Mail: dcwilcox@ix.netcom.com   WWW: http://www.dcwindustries.com